Higher Education's Leadership in Climate Action and Sustainability

Elsa M. Núñez
Eastern Connecticut State University, USA

Patricia Szczys
Eastern Connecticut State University, USA

IGI Global
Scientific Publishing
Publishing Tomorrow's Research Today

Vice President of Editorial Melissa Wagner
Director of Acquisitions Mikaela Felty
Director of Book Development Jocelynn Hessler
Production Manager Mike Brehm
Cover Design Jose Rosado

Published in the United States of America by
IGI Global Scientific Publishing
701 East Chocolate Avenue
Hershey, PA, 17033, USA
Tel: 717-533-8845
Fax: 717-533-7115
Website: https://www.igi-global.com E-mail: cust@igi-global.com

Library of Congress Cataloging-in-Publication Data

Names: Nunez, Elsa editor | Szczys, Patricia, 1975- editor
Title: Higher education's leadership in climate action and sustainability / edited by: Elsa Nunez, Patricia Szczys.
Description: Hershey, PA : IGI Global Scientific Publishing, [2025] | Includes bibliographical references and index. | Summary: "This book aims to chronicle the ways in which higher education was ahead of government and private business in tackling climate change and will inspire the next generation of leaders to take decisive action to fully realize the power of higher education in forging a just transition to a sustainable future"-- Provided by publisher.
Identifiers: LCCN 2025011130 (print) | LCCN 2025011131 (ebook) | ISBN 9798369354544 hardback | ISBN 9798369354551 paperback | ISBN 9798369354568 ebook
Subjects: LCSH: Educational leadership | Education, Higher--Aims and objectives | Climate justice
Classification: LCC LB2806 .H494 2025 (print) | LCC LB2806 (ebook) | DDC 304.2/80711--dc23/eng/20250820
LC record available at https://lccn.loc.gov/2025011130
LC ebook record available at https://lccn.loc.gov/2025011131

British Cataloguing in Publication Data
A Cataloguing in Publication record for this book is available from the British Library.

Table of Contents

Detailed Table of Contents

Higher education plays a critical role in society, from its earliest beginnings formally educating clergy to now serving many different functions in communities around the world. In this chapter we look at how the missions of higher education have evolved through time, and how those missions can be applied to address the climate challenges we are currently facing. We also overview a number of important concepts that connect institutions within the sector together to drive greater impact in addressing climate change. By aligning the core missions and functions of the higher education sector with climate action, senior leaders and decision-makers can better understand how to mobilize all of their institutional strengths to address this important issue.

Multiple actors within America's colleges and universities wish to do something constructive about the world's climate change crisis, both in response to the advocacy of their institutional constituents and out of their own desires to find solutions. Universities are well-positioned to help the world find solutions to climate change. This chapter offers advice to university leaders, faculty, and students, based on the author's experiences in academia and public policy practice, about what they can do to contribute to solving the critical, complex, and multifaceted problem of climate change.

Critics of activism on university campuses argue that academic institutions should stick to their core missions of teaching and research. While there is some merit to this argument and to the related view that teachers should avoid using their academic authority to impose their political views on their students, universities should not shy away from engaging in research and teaching on subjects with policy implications or from basing their own policies on the best available science. Nor should they refrain from supporting extracurricular activities that enrich their students' education with real-world engagement. The complexity and civilizational implications of the climate crisis make it a uniquely rich subject for interdisciplinary study that fosters students' intellectual growth, as well as an appropriate focus of extracurricular activities that complement students' work in the classroom. The chapter proceeds from these theoretical considerations to concrete examples of university programs that effectively support student involvement in climate-related activities.

Colleges and universities in the United States have a unique capability of convening important discussions about difficult topics. Nowhere else in modern society can people discuss challenging issues in an open forum in a civil manner. This capability is based in both the history of higher education, and in the rights and traditions of free speech in the United States. It is supported by promotion of thought and creativity being a major function of higher education.

Incorporating sustainability across the curriculum is a growing component of college and university sustainability commitments. Education for Sustainability aligns with institutional commitments to advance solutions to the world's most pressing challenges. Most sustainability curricular change initiatives take a "bottom up" approach, engaging interested faculty through small learning communities. While effective, this approach is slow and many instructors choose not to engage. This chapter draws on the experience of a technologically focused university to suggest strategies for "scaling up" sustainability across the curriculum. The following strategies supported significant expansion of sustainability course integration: laying the groundwork through a major initiative to enhance undergraduate learning; leveraging strategic planning to win high-level commitment and resources; connecting sustainability education with professional development in teaching and learning; empowering faculty to co-design initiatives; and supporting wellbeing through on-going learning communities.

This essay explores the intersection of two robustly related fields--Africana Studies and the Environmental Humanities. Drawing on expertise in Black feminist literary studies, I argue that the current climate and higher education requires that we develop new pedagogical approaches to engage students about the twin impacts of racial and environmental injustice. Throughout the essay I refer to scholars of Africana Studies, Environmental Humanities, Black Feminism, Caribbean Studies and eco-feminism who have made notable interventions in the field and whose work helps us to understand the overlapping concerns in these interrelated areas of study. I argue that taking into consideration how students' understanding of the manifold connections between racial and environmental justice bolsters their commitment to both, demands that higher education account for and address their perspectives by taking up new pedagogical approaches that dissolve the boundaries between theory and praxis, analysis and experience, and even humans and nature. Of particular interest is how engaged pedagogy and Black feminist pedagogy offer an array of tools for teaching about environmental justice through an intersectional lens in the 21st century.

This chapter examines the intersection of liberal arts education and sustainability initiatives at the University of Minnesota Morris (UMN Morris), a public liberal arts college. Through student voices, institutional history, and concrete examples, it demonstrates how the college has become a leader in sustainability while maintaining its core liberal arts mission. The article argues that the college's rural location, historical context (including the campus site as a Native American boarding school), and commitment to student engagement have created an environment where sustainability serves as both an educational framework and a lived practice. By examining specific initiatives and partnerships, including community collaborations and student-led projects, the piece illustrates how UMN Morris has effectively combined the traditional liberal arts emphasis on broad, interdisciplinary thinking with practical applications in sustainability, creating a model for environmental leadership in higher education.

In the Times Higher Education Impact Ranking 2023, The American College of Greece (ACG) was rated the 1st most sustainable private higher education institution in Greece and the 4th among the 18 public and private universities/institutions rated in the country. 2,031 universities from 121 countries/regions applied in the Impact Ranking 2023 and were assessed and ranked for their contributions to the United Nations' 17 Sustainable Development Goals (SDGs). Among the 2,031 institutions globally, ACG scored in the top 7% for 'Responsible Consumption and Production', top 12% for 'Climate Action' and top 17% for 'Partnerships for the Goals'. In Greece, ACG attained the top scores for all three of these SDGs.

In spring 2024, U.S. News & World Report published an article identifying 20 beautiful college and university campuses. Included were familiar institutional names — Duke, Harvard, Stanford. Also on the list was Aurora University, a private comprehensive college in the far western suburbs of Chicago, where this author served as president for over two decades. According to the magazine author, the beauty of a campus may sway college choice by making a positive first impression. (Claybourn, 2024) Indeed, this likely was true for Aurora. Prospective students and their families remarked frequently on the pleasing nature of the campus environment in post-visit evaluations. Particularly gratifying were the favorable comparisons made to the facilities of other institutions.

Sustainability has been at the heart of UC Merced's mission since its founding. From the outset, the campus was designed to embody a vision that balances environmental, social, and economic impacts. One of the most widely embraced definitions of sustainability comes from the Brundtland Commission's report (1987), Our Common Future, which describes it as: "Meeting the needs of the present without compromising the ability of future generations to meet their own needs." This philosophy aligns perfectly with UC Merced's approach—building a campus that not only serves today's community but also preserves resources and opportunities for those to come. UC Merced's commitment to sustainable development is evident across all aspects of campus life, making it a living example of how higher education can drive meaningful change for a more sustainable future.

This chapter reviews recent environmental initiatives across Connecticut's 12 community college campuses and explores future opportunities as the newly formed Connecticut State Community College, established in July 2023 through the merger of the 12 campuses. It examines a range of sustainability efforts, historically rooted in environmental stewardship. These include academic programs, campus-wide green initiatives, hands-on learning, and community partnerships. Even prior to the merger, campuses collaborated through the Connecticut State Colleges and Universities System Office to implement statewide sustainability projects, achieving cost savings and operational efficiencies. More recently, the college has expanded its focus to include the United Nations' 17 Sustainable Development Goals, which encompass environmental, social, and economic dimensions of sustainability. In this context, the colleges have begun addressing issues like poverty, public health, employment, and education. As with all institutions, CTState faces several challenges that hinder progress including limited funding, institutional resistance, low student engagement, and stakeholder awareness. The chapter discusses each of these barriers and potential strategies to overcome them. As a new institution, Connecticut State Community College must adopt a systematic, statewide approach to sustainability leveraging important opportunities to scale energy efficiency initiatives, share academic programs across campuses, and foster stronger ties with state universities.

As a result of strategic leadership from all levels of campus, the university's geographic advantage, and visionary philanthropy, Virginia Wesleyan University has grown into a higher education leader in environmental sustainability. The university fosters community engagement and inspires responsible environmental action at every level of the institution through targeted fundraising from individuals and grant-making organizations. Its commitment to environmental education and stewardship extends beyond the classroom by integrating green practices into academic programs and campus initiatives from Virginia Beach, Virginia, to Tokyo, Japan.

Preface

INTRODUCTION

This book, conceived at the end of the long university presidency of Dr. Elsa Núñez, emerges at a moment of profound national and global challenge and opportunity. The scientific consensus is clear: the human-driven rapid warming of our atmosphere, land, and oceans requires transformative action to mitigate and adapt to the conditions wrought by the changing climate. We are living in a climate-changed world where the frequency and intensity of fire, flood, drought, and biodiversity loss impact the lives of thousands, millions, or billions every day. These changes impact not only our natural environment, but they are also stressing our social institutions and national and global economies.

> *Sustainability is meeting the needs of the present without compromising the ability of future generations to meet their own needs. Building a more sustainable society requires the integration of environmental concerns alongside plans for economic development. ~ 1987 United Nations Brundtland Commission*

With the adoption of the United Nations Sustainable Development Goals (SDGs) in 2015, leaders across the globe acknowledged the interconnected "wicked problems". We cannot separate the issues of equity, health, poverty, and education, from employment, infrastructure, and economies, or from pollution, biodiversity, and climate. The SDGs offer a dynamic and scalable framework for action.

Yet even with extensive scientific data and this powerful framework for development, progress remains limited not by what we *know* but by the political and institutional will to *act*. In the 21st century, we need everyone – across sectors – working toward an equitable and sustainable future, and this is where institutions of higher education of all stripes are uniquely positioned to lead. Colleges and universities are stewards of knowledge, public trust, social responsibility, and serve as hubs for innovation driving regional and national economies. These institutions are crucibles for leadership development, critical inquiry, research, and community engagement – functions that are indispensable in our collective response to climate change and inequity.

The "wicked problems" of sustainability are problems that defy simple solutions, those that cannot be learned fully in a single academic department or be achieved by a single organization or agency. These problems are multi-faceted and interdisciplinary – perfectly positioned for students to grapple with complexity and work within diverse teams to address our most pressing issues. This learning is transformative and the research impactful. College campuses and the communities in which they are embedded serve as living laboratories for understanding deeply the multifaceted nature of sustainability.

The purpose of this book is to give voice to leaders – presidents, faculty, and directors – who are shaping commitments to sustainability in higher education across the United States and beyond. This is not a traditional research anthology. Rather, it is a collection of reflections, case studies, and institutional narratives told in the contributors' own words. It captures lived experiences, lessons learned, milestones achieved, recognition to valuable collaborations, and shares aspirations for the future. And it offers perspectives from a diversity of institutions both public and private, 2- and 4-year degree-granting, and small liberal arts colleges to large research universities and community college. These voices were drawn from the American Midwest and Northeast to Greece and the American West. They to also represent gender, ethnic, and age diversity.

This book invites the reader into a dynamic conversation in higher education and provides a window into the motivations, strategies, and values that guided the authors' work. Readers will learn how institutions have aligned sustainability goals with institutional mission, weaving them together into the very fabric of program or institutional identity. You will find chapters that present ideas and values that drive community partnerships for societal impact and stories of transformations that were sparked by philanthropy, industry partnerships, investments in sustainable campus operations, and the important integration of sustainability across the curriculum.

In the end, this book is both a reflection and a call to action. Universities as non-profit-driven institutions have the advantage of taking the long view, holding the common public good in sharp focus. As the impacts of climate change accumulate and worsen, the imperative for higher education to lead has never been clearer. The next generation of business leaders, healthcare and social workers, policy makers, storytellers and artist and citizens must be prepared with the knowledge and skills to adapt to the changes already underway, and to build something new - a sustainable civilization. The hope is that these words will guide, challenge, and empower those who read them and together we stride forward toward the equitable and sustainable systems we need.

ORGANIZATION OF THE BOOK

This volume begins with a compelling introduction in Chapter 1, *Missions of Higher Education*, which positions the climate crisis as the defining environmental challenge of our time and representing a powerful opportunity for higher education to lead. The chapter underscores how the diversity, complementarity, and alignment of various institutional missions can catalyze meaningful climate action in higher education. Building on this foundation, Chapter 2, *Advice for Universities Confronting the Climate Crisis*, offers a strategic roadmap for university leaders, faculty, and students eager to take action. Drawing on deep experience in both academia and public policy, the author delivers practical guidance for navigating the complex demands of institutional responsibility and public expectation.

In *Chapter 3: Climate Action and the Ethics of Student Engagement*, the tension between institutional neutrality and climate action is explored through an analysis of the role for climate commitments like the American College & University Presidents' Climate Commitment. The chapter considers Stanley Fish's provocative stance on academic responsibility, stimulating a deeper conversation on how universities can ethically integrate climate consciousness without compromising intellectual rigor. This discussion is complemented by *Chapter 4: Advancing Climate Health through Campus Civil Discourse and Course Integration*, which highlights the unique potential of the college campus as a site for civil, informed discourse. It argues for a holistic model of sustainability education that fosters critical thought, collaboration, and personal responsibility.

Chapter 5: Scaling up Faculty Engagement in Education for Sustainability Across the Curriculum addresses the imperative to mainstream sustainability in higher education pedagogy. The chapter identifies five robust strategies that have enabled one institution to move beyond isolated efforts to promote broad, faculty-led integration of sustainability across disciplines. By leveraging strategic planning and embedding sustainability into teaching development, institutions can increase their impact.

Turning to cultural and disciplinary dimensions, *Chapter 6: Africana Studies and the Environmental Humanities* offers an intimate account of teaching Caribbean literature through an environmental lens. This chapter situates environmental inquiry within Africana Studies, using art, literature, and student engagement to explore the relationship between identity, place, and ecological awareness. The personal narrative underscores the value of interdisciplinary and culturally responsive approaches to environmental education.

In *Chapter 7: Practicality Plus Creativity: The Public Liberal Arts College*, we examine a case study demonstrating the alignment of liberal arts values with sustainability practice. Weaving together historical context, student engagement and leadership, and institutional partnerships within the community, this chapter illustrates how a rural public liberal arts institution has become a national model for integrating sustainability into its academic mission, campus culture, and community impact.

Chapter 8: Leading to Green offers a presidential perspective on institutional sustainability, chronicling a quarter-century of environmental commitments. This chapter reflects on how the role of leadership, engagement of faculty and staff, and a focus on regional impact positioned the university to engage in energy conservation, advance green skills and technology, and integration sustainability into the curriculum. Similarity, *Chapter 9: University Climate Planning: Building Community Partnerships to Drive Action*, presents a powerful model for aligning climate action with justice and equity. We learn how the Climate Justice Action Plan (CJAP) was co-developed with the grassroots organizations, exemplifying how authentic partnerships with frontline communities can advance both environmental and social goals. The chapter demonstrates how the university's sustainability office is not merely a campus resource but an active participant in creating regenerative, inclusive systems.

Chapter 10: Transforming Education for a Sustainable Future: Climate Action in Higher Education, highlights how The American College of Greece uses a systemic and data-driven approach to embedding sustainability within its strategic vision, operational practices, and academic mission. ACG's performance in key Sustainable Development Goals affirms its commitment to educating future leaders equipped to address global environmental challenges.

Chapter 11: Aurora University: Campus Ecosystems in Action, demonstrates interconnected strategies used to advance campus development. The author argues that environmental aesthetics serve as a recruitment asset, but also enhance student enjoyment, promote a sense of place, and encourage the integration of deeper ecological ethos and sustainability commitments. Aurora's story underscores the often-overlooked value of green space and built environments in advancing ecological stewardship.

The University of California Merced was conceived as a sustainable campus, guided by the principles outlined in the Brundtland Commission's landmark definition of sustainability. In *Chapter 12: Sustainability at UC Merced*, authors share how leaders have integrated environmental, economic, and social considerations into every aspect of planning and operations. They have built an identity around dedication to sustainable growth with a focus on preserving resources for future generations.

Chapter 13: Sustainability at Connecticut's Community Colleges: Protecting the Environment Starts Locally, explores how community colleges integrate environmental values into workforce development, credit and non-credit programs, and transfer pathways. The focus on community-based sustainability affirms that meaningful environmental progress begins at the grassroots level, and community colleges are essential partners in advancing this work.

CONCLUSION

This volume concludes with Chapter 14, *Geographic Advantage, Visionary Philanthropy, and Strategic Growth: Virginia Wesleyan University's Rise as a Sustainability Leader in Higher Education*. This chapter is a powerful illustration of what can be achieved when university leadership advances environmental stewardship through targeted academic programs and purpose-built infrastructure. By forging strong partnerships with community organizations and visionary philanthropists, the university placed sustainability at the heart of its mission and recognized its coastal setting not only as a unique vantage point for observing climate change, but also as a compelling call to action.

Patricia Szczys
Eastern Connecticut State University, USA

Chapter 1
Missions of Higher Education

Tim Carter
Second Nature, USA

ABSTRACT

Higher education plays a critical role in society, from its earliest beginnings formally educating clergy to now serving many different functions in communities around the world. In this chapter we look at how the missions of higher education have evolved through time, and how those missions can be applied to address the climate challenges we are currently facing. We also overview a number of important concepts that connect institutions within the sector together to drive greater impact in addressing climate change. By aligning the core missions and functions of the higher education sector with climate action, senior leaders and decision-makers can better understand how to mobilize all of their institutional strengths to address this important issue.

INTRODUCTION

Climate change, as an issue that spans geographies, disciplines, and sectors, is one of the most important wicked environmental problems facing society today. Climate change also provides an important opportunity for higher education to fully embrace all its strengths in service of this grand challenge. These strengths are directly connected to higher education's missions, which have grown and evolved over time. In this chapter, we identify some of the important historical moments where higher education has leveraged its various missions to take a leading role in addressing climate change. We also introduce the relationships between diversity,

DOI: 10.4018/979-8-3693-5454-4.ch001

complementarity and compatibility as key concepts for how higher education can continue to apply these missions to address current and future climate challenges.

The modern university began in late medieval times with an original mission to educate and train theologians and the clergy, eventually expanding into law and medicine (Arbo and Benneworth 2007). This education and training feature, or *knowledge transfer*, formed the first mission of higher education. In the early 19th century, Wilhelm von Humboldt formally introduced a second mission to higher education in founding the University of Berlin and the so-called "Humboldtian reforms" anchored in the concepts of academic freedom. These ideas foundationally shaped the modern research university as a place where the independent pursuit of knowledge takes place (Bongaerts 2022). Research, or *knowledge creation*, was rapidly and widely adopted as the second mission of higher education.

While these two traditional missions of knowledge transfer and knowledge creation served society well, in the 20th century there was increasing interest in the higher education sector directly meeting societal needs outside of teaching and basic research, and thus a "third mission" emerged (Compagnucci and Spigarelli 2020). As Trencher et al (2014) describes, "the concept of a third mission is somewhat ambiguous. Yet in principle, it refers to diverse activities not covered by the first mission (education) and second mission (research) such as technology transfer, continuing education and social engagement in the form of public access to lectures and cultural assets, voluntary work and consultancy…From this perspective, the term 'social contribution' is a useful synonym for describing the core notion of the third mission."

Much of higher education's third mission has concentrated on different types of organizational infrastructure (e.g. incubators, centers, labs, etc) at institutions which encourage technology transfer. This is only part of the third mission, however. A more expansive view includes a move towards the "entrepreneurial university", a phrase that scholars have used for institutions embracing third mission principles (Etzkowitz 1983). Entrepreneurialism illuminates "the distinctive position of the university as an institution capable of leveraging its education, research, knowledge exchange and community engagement activities to create and promote entrepreneurial thinking and actions both internally and externally." (Abreu and Grinevich 2024). In this sense, the whole of the university is being mobilized around a range of socially relevant activities, some of which have business and market-based applications, while others may simply be for the public good.

In recent years, a fourth mission for higher education has also been developed focusing on "co-creation for sustainability" (Trencher et al 2014). The idea with the fourth mission is that the public service orientation of higher education institutions necessitates a posture of responsiveness to community needs and collaboration with communities. This contrasts with the traditional approaches where the locus of power and origin of actions take place within the higher education institution that then is shared (and in its worst sense imposed) on external stakeholders and communities. A truly bi-directional relationship with the community allows for flows of expertise, resources, and other assets to balance and equalize the campus-community partnerships that form and projects that are developed, ultimately producing more durable and impactful results.

Whether described as one of the four missions or framed as entrepreneurialism, the application of these approaches is largely determined by the social problem being addressed. Theoretically, there is no limit to the number of issues that could be tackled but given that some problems may be more difficult to solve than others, many topics often fall into the category of "wicked problems" (Rittel and Weber 1973) that are, by their nature, complex. Solutions to wicked problems are best created when institutions bring a range of approaches to bear on the problem at hand. The entrepreneurial institution, anchored in multiple missions, is well-positioned to address these types of dilemmas. The complexities in solving problems caused by climate change provides the higher education sector with ample opportunities to apply these theories in the real world and in the present moment.

Application of the Missions: A Brief History of Higher Education's Engagement with Climate Change

When considering how higher education has applied its missions to address climate change, the path is not a linear one according to when these missions were created. Research, or second mission activity, is the earliest application of higher education's missions to climate action rather than education and training, the first mission. Higher education's research has been directly connected with climate change since the 19th century with physical science research describing fundamental processes occurring in the environment such as atmospheric chemistry, geology, and oceanography, have a much longer history in higher education than those targeting climate change specifically (e.g. Priestley 1774). The study often referenced as the first to directly consider the science of a changing climate was from the Swedish faculty member Svante Arrhenius in 1896 when he published "On the Influence of Carbonic Acid in the Air upon the Temperature of the Ground" (Arrhenius 1896). In this paper, he hypothesized that a doubling of atmospheric carbon dioxide (CO2) would create a greenhouse effect on earth, and this is often cited as the first climate change academic paper (Adamo et al 2021). It was a female researcher, however, Eunice Foote, who first published about climate change forty years earlier with her work titled, "Circumstances Affecting the Heat of the Sun's Rays." (1856). Using gas-filled glass cylinders, Foote demonstrated connections between carbon dioxide and rapid temperature rise within those cylinders, a phenomenon which is now referred to as the "greenhouse effect" (Foote 1856).

Since those early studies, researchers in many disciplines have continued to apply research methods to all aspects of climate change, climate action, and climate solutions. One of the more recent seminal research projects began in 1958 by Charles Keeling, which began to measure the concentrations of CO2 at a monitoring station in Hawaii (Keeling 1958). This work produced the famous "Keeling curve" which is the longest continually measured dataset of CO2, quantifying the increases in CO2 concentrations over time and supporting the global climate models (GCMs) that are now used to make future climate projections (https://keelingcurve.ucsd.edu/).

First mission activities that explicitly train students through formal education in climate change (curriculum, majors, etc) have their origins in the environmental science and environmental studies programs that began to emerge in the 1960s. The University of California, Santa Barbara has one of the first environmental studies programs in the country, which was catalyzed by a catastrophic oil spill in Santa Barbara that took place the previous year (https://es.ucsb.edu/program-history). Arizona State University started a School of Sustainability in 2006 with the first undergraduate class of students beginning in 2008. This was the first school of its kind in the United States, and anchored ASU as one of the leading institutions to create formal majors in sustainability with a significant amounts of coursework for these majors in fields related to climate change (https://schoolofsustainability.asu.edu/). Many graduate student programs dedicated to climate change were also started around the turn of the 21st Century, such as the University of Washington's Program on Climate Change that began in 2001 as a graduate certificate and now has spread widely across many aspects of the academic enterprise (https://pcc.uw.edu/).

The third and fourth mission applications to climate change in higher education is the most recent development in the sector. These missions picked up steam in the 2000s with the emergence of "grand challenge" programs at colleges and universities. These programs often center around some specific climate goal (transition to 100% renewable energy) or general theme (climate resilience). Many of these grand challenges are anchored in specific locations and connected with a city or state's goals. The University of California, Los Angeles (UCLA) launched the Sustainable LA Grand Challenge in 2013 to align with the city of Los Angeles's sustainability plan in areas of energy, water, and ecosystem health. UCLA mobilized hundreds of research faculty, supporting projects across the city aligned with meeting goals of 100% renewable energy, 100% locally sourced water, and enhanced ecosystem health. In the Midwest, Indiana University launched a university wide, climate solutions-focused institute to accelerate sustainability and resilience work throughout the state. Since its launch, the Environmental Resilience Institute has assisted 49 communities with climate action and preparedness, initiated 16 community climate action plans, and connected businesses, nonprofits, and local governments to 100,000 hours of student service (https://eri.iu.edu/).

In recent years we have begun to see climate change programs more fully embrace, and explicitly articulate, all four missions within large-scale reorganizations of the institutions. One of the best examples of this is the Climate School at Columbia, which was launched in 2020. Columbia had a long history of applied research and education with The Earth Institute at Columbia University that began in 1995, and a masters program in climate and society that began in 2004. This laid a strong foundation for Columbia's leadership to create the Climate School which is designed to, "educate future climate leaders, support groundbreaking research, and foster essential climate solutions from the community to the planetary scale" (https://www.climate.columbia.edu/about). High-level positions dedicated to climate action have become popular as well with vice-president or vice-provost level positions at University of Michigan, Georgetown University, Duke University, Yale University, and others being created.

Not all higher education's climate actions have taken place within individual institutions. The most significant and lasting inter-institutional program on climate change for the sector began in 2006 with the launch of the American College and University Presidents' Climate Commitment (ACUPCC). The ACUPCC was designed as a program where presidents and chancellors of colleges and universities committed their institutions to achieve carbon neutrality in their campus operations, give students an opportunity to engage in climate action, and increase the amount of climate research happening on their campuses. There were 12 presidents who led this effort, and it quickly expanded in 2007 - 2009 to include hundreds of institutions across the United States with many different institution types represented. The program launch was originally supported by three non-profit organizations. In 2008, Second Nature became the sole organization supporting the effort and has continued to lead the effort since that time. In 2015, the program was expanded to include climate resilience and was rebranded as the Presidents' Climate Leadership Commitments (Commitments) and the Climate Leadership Network. This program remains in place today and is the largest and longest-standing voluntary carbon neutrality program in any sector in the world.

Building off the Missions: The Role of Diversity, Complementarity, and Compatibility to Advance Climate Action

The missions of higher education are foundational for understanding how and why higher education is serving a distinct role in helping society create climate solutions. Without mission-alignment, it is highly unlikely that leaders within institutions, or leaders across institution types, will make a sustained case for climate action as a top priority. Mission alignment is a necessary condition, but focusing on only the four missions does not sufficiently capture the ways that the higher education sector as a whole could be better mobilized towards climate action. In this final section, we build off the discussion about the sector's missions by adding the three related concepts for climate action in higher education: diversity, complementarity, and compatibility. We apply these concepts between institutions in the sector, as well as between the higher education sector and other sectors in society.

The higher education sector, particularly in the United States where there are over 4,000 institutions, contains a wide range of institution types. There are large public research universities, small private liberal arts colleges, Historically Black Colleges and Universities (HBCUs) that span all different classifications, regional comprehensive universities, and community colleges among many others. All these institutional types also span geographies and therefore contain within them a range of regional cultural contexts, further distinguishing institutions from each other. Institutions also are located along an urban-to-rural gradient within regions which increases the diversity even more.

Some may view this diversity as a challenging obstacle for scaling climate action - if all the institution types are unique snowflakes, then no one will ever work together! This is where the concept of complementarity comes in, however. Complementarity for climate action views institutional diversity as a *strength* that *allows* for collaboration, not a barrier to it. Madia et al (2024) describe five areas of institutional complementarity related to climate (research, student experiences, community connections, workforce development, and public-private partnerships) and use a series of vignettes to illustrate how the differences between institution types can be utilized as a strength. Under research, for example, they tell the following story:

"In the Pacific Northwest, a pioneering research collaboration has emerged among Evergreen Research University, Cascade Liberal Arts College, Pacific Technical College, and Northwest Regional State University to forge a comprehensive strategy against climate change. Evergreen Research University historically had been viewed as the leader with a basic climate and scientific research groundwork for climate solutions, but hadn't focused on how this new knowledge was being translated in the real world. Cascade Liberal Arts College enriched this foundation by partnering with Evergreen to weave climate change research into its interdisciplinary curriculum, emphasizing policy and societal engagement with students from across disciplinary majors. Pacific Technical Institute adapted its vocational training to spotlight the research underlying renewable energy technologies, transforming Evergreen's research into practical, hands-on solutions that could be deployed in the region. Northwest Regional University, using the place-based ethos of regional comprehensive universities with a strong focus on community collaboration and sustainability, crafted a specialized program in climate change adaptation and mitigation. This initiative combined academic rigor with real-world applicability, preparing students to spearhead effective climate response strategies in their hometowns across the northwest. Together, these institutions built off the research foundation to create a dynamic network, combining education and practical implementation to drive regional climate action." (2024)

Complementarity can be a powerful tool when strong relationships and trust between institutions already exists, and this often means that institutions of the same type tend to work with each other. A second key principle for this type of complementarity to be activated is that potential collaborating institutions must see themselves as *compatible* (Williams 2007). In the case of the higher education sector, the perceive lack of compatibility can be a real barrier to institutional partnerships. Institutions maintain "peer and aspirant" lists of all the institutions they consider similar and institutions that they hope to be someday. Research universities rarely, if ever, see a community college as a peer or aspirant. For the purposes of climate solutions, however, these comparisons would benefit from a refresh. Instead of asking, "who am I similar too and is doing the same thing as me but a little bit better?", the question could be, "who is different than me and can complement my ongoing activity?" This could potentially unlock scaled solutions within the sector that haven't yet been realized and provide an important opportunity for expanding the impact of the sector by encouraging a wider range of institution types to envision their own participation in collaborative projects.

Take the example of offshore wind (OSW) energy in New England. Many of the major research universities in the region have active research in some field either directly connected with wind energy or the wind energy industry. In 2024, a number of these institutions were funded to form the Academic Center for Reliability and Resilience of Offshore Wind (ARROW). ARROW is led by the University of Massachusetts, Amherst and includes seven other higher education institutions, nearly all of which are large, research institutions. ARROW has expressed the following objectives:

- Establish a broadly comprehensive and multi-pronged OSW education and research program designed to prepare the next generation of US OSW professionals and establish global leadership for the US in OSW education,
- Empower the first cohorts of this next-generation US OSW workforce to seed a reliable and resilient, domestically educated, OSW professional workforce.
- Deliver innovations to accelerate achievement of US goals for OSW levelized cost of energy (LCOE) and deployment and ensure that the OSW system delivers energy in a reliable and resilient manner.
- Engage with stakeholder communities to ensure that the energy transition is just, equitable and fair, and that all stakeholders have a voice in the process.

In the same region, Bristol Community College (BCC), located in New Bedford, MA, is the founder and supporter the National Offshore Wind Institute (NOWI) which was launched in 2023 as "specifically designed as a one-stop-shop, state-of-the-art training facility, to offer all required trainings and certifications for a maritime and offshore wind industry on site, year-round" (nowi.org). The connection with workforce training and credentialing has defined community and technical colleges in many locations with BCC's industry connection with some of the largest wind manufacturers a central feature of NOWI.

There are likely plenty of reasons why ARROW and NOWI are running parallel efforts for OSW. Funding sources, ability to apply, manage, and access large federal and state research grants, organizational leadership, strategic priorities at a given organization, or simply the individual knowledge of who to partner with, all create barriers to partnerships that are to be expected. Additionally, this isn't an indictment on what seem to be two great initiatives. What it does point to is the opportunity for greater alignment within the higher education sector to use the complementarity between institution types, given the diversity of types that exist, to effectively and efficiently develop climate solutions. Community colleges have long-standing relationships with industry and workforce training programs, but rarely are part of federally funded research proposals because they generally don't have the organizational infrastructure to support research. Research collaborations are largely inquiry-driven efforts by faculty members who have long-standing professional relationships with another research faculty.

CONCLUSIONS

The missions of higher education have defined the sector's activities since the first institutions were started nearly 1000 years ago. As climate change continues to transform our society, often impacting the most vulnerable members of communities first and worst, these missions can continue to evolve as the higher education sector's strengths are better understood and aligned with needs in the world. The diversity of institution types, particularly in the United States, presents an incredible opportunity for the sector to lead in its teaching, research, social impact, and co-creation, while maximizing institutional strengths to drive climate solutions forwards. Breaking down the black boxes of what all institution types do, identifying how different institution types can complement one another in their climate actions, and dispelling perceptions about a lack of compatibility because of these differences, are key next steps in moving beyond the missions into greater impact.

REFERENCES

Abreu, M., & Grinevich, V. (2024). The entrepreneurial university: Strategies, processes, and competing goals. *The Journal of Technology Transfer*, *49*(6), 1991–2034. DOI: 10.1007/s10961-024-10085-7

Adamo, N., Al-Ansari, N., & Sissakian, V. (2021). Review of climate change impacts on human environment: Past, present and future projections. *Engineering*, *13*(11), 605–630. DOI: 10.4236/eng.2021.1311044

Arbo, P., & Benneworth, P. (2007). Understanding the Regional Contribution of Higher Education Institutions: A Literature Review. OECD Education Working Papers, No. 9. OECD Publishing, Paris.

Arrhenius S. (1896). On the influence of carbonic acid in the air upon the temperature of the ground. Philosophical Magazine and Journal of Science Series 5: 237-276.

Bongaerts, J. C. (2022). The Humboldtian Model of Higher Education and its Significance for the European University on Responsible Consumption and Production. *BHM. Berg- und Hüttenmännische Monatshefte*, *167*(10), 500–507. DOI: 10.1007/s00501-022-01280-w

Compagnucci, L., & Spigarelli, F. (2020). The Third Mission of the university: A systematic literature review on potentials and constraints. *Technological Forecasting and Social Change*, *161*, 120284. DOI: 10.1016/j.techfore.2020.120284

Etzkowitz, H. (1983). Entrepreneurial scientists and entrepreneurial universities in American academic science. *Minerva*, *21*(2-3), 198–233. DOI: 10.1007/BF01097964 PMID: 11611566

Foote, E. (1856). Circumstances Affecting the Heat of the Sun's Rays. *The American Journal of Science and Arts*, *22*, 382–383.

Keeling, C. (1958). The concentration and isotopic abundances of atmospheric carbon dioxide in rural areas. *Geochimica et Cosmochimica Acta, 13*(4), 322–334. DOI: 10.1016/0016-7037(58)90033-4

Madia, M. T. Carter, S Thomas. 2024. Higher education complementarity for climate action. Second Nature: Cambridge, MA. www.secondnature.org

Priestly, J. (1774). *Experiments and Observations on Different Kinds of Air*. Printed for J. Johnson.

Rittel, H., & Webber, M. (1973). Dilemmas in general theory of planning. *Policy Sciences*, *4*(2), 155–169. DOI: 10.1007/BF01405730

Rubens, A., Spigarelli, F., Cavicchi, A., & Rinaldi, C. (2017). Universities' third mission and the entrepreneurial university and the challenges they bring to higher education institutions. *Journal of Enterprising Communities: People and Places in the Global Economy*, *11*(3), 354–372. DOI: 10.1108/JEC-01-2017-0006

Trencher, G., Yarime, M., McCormick, K., Doll, C., & Kraines, S. (2014). Beyond the Third Mission: Exploring the Emerging University Function of Co-creation for Sustainability. *Science & Public Policy*, *41*(2), 151–179. DOI: 10.1093/scipol/sct044

Williams, J. (2017). Collaboration, alliance, and merger among higher education institutions. OECD Education Working Papers, No. 160, OECD Publishing, Paris. DOI: 10.1787/19939019

Chapter 2
Advice for Universities Confronting the Climate Crisis

James R. Gomes
Clark University, USA

ABSTRACT

Multiple actors within America's colleges and universities wish to do something constructive about the world's climate change crisis, both in response to the advocacy of their institutional constituents and out of their own desires to find solutions. Universities are well-positioned to help the world find solutions to climate change. This chapter offers advice to university leaders, faculty, and students, based on the author's experiences in academia and public policy practice, about what they can do to contribute to solving the critical, complex, and multifaceted problem of climate change.

INTRODUCTION

What are universities doing about the world's climate crisis? The answer may appear obvious: research it, deepen and refine our understanding of it, teach students about it and train them for careers in climate-related fields, and disseminate the knowledge they create both to aid in the discovery of new knowledge and to help affected people and places cope with it. This is what universities do. Still, there is the question of how, specifically, universities can be most effective.

There is a growing literature on the subject of what universities can or should do about climate change. Much of this literature consists of suggestions for how universities can reduce their own climate impacts. Piece is different in that it offers

DOI: 10.4018/979-8-3693-5454-4.ch002

recommendations for a wide range of steps universities can take to advance climate solutions. The recommendations are based on personal observations and experiences in academic, non-profit, and governmental institutions in the United States over the last five decades[1].

These decades began with little public awareness that human activities might, over time, contribute to the warming of Earth's atmosphere, continued through a period when a broadening sense emerged of anthropogenic climate change as a present reality, and have culminated in today's widespread recognition that the effects of climate change are already upon us and an increasing concern about a future where these effects will accelerate.

Fortunately, universities have important roles to play in shaping that future.

A VERY BRIEF HISTORY OF THE CLIMATE CRISIS

The current crisis has been more than two centuries in the making. As steam engines powered by coal and later internal combustion engines fueled by oil came into broad use during the Industrial Revolution, the carbon dioxide released from burning fossil fuels began to accumulate in the Earth's atmosphere, along with other gases such as methane, producing a "greenhouse effect," holding in heat that would otherwise escape into space.

The 20th Century would witness several developments that contributed to the rapid accumulation of greenhouse gases in Earth's atmosphere. As the century began, global population stood at about 1.6 billion; by 2000 it had nearly quadrupled to approximately 6.1 billion, (United Nations Department of Economic and Social Affairs, Population Division, 2011). Automobiles went from being novel and hazardous contraptions to nearly ubiquitous vehicles that dominate local and medium-range human transport, along with the trucks that ship all manner of goods from place to place. Air travel, which contributes significantly to greenhouse gas buildup due to the heavy fuel consumption needed to power jet engines, obviously did not exist in 1900. Food production and transport burgeoned to meet the growth in population, leading to increased carbon emissions, as well as increased methane emissions from livestock, (Thompson, 2021).

What would come to be known as "global warming" caused by the buildup of gases emitted by these activities had been predicted by scientific researchers long before its effects began to be felt by humans. As early as the mid-19th century, experiments conducted separately by Eunice Foote in the United States and John Tyndall in England were already suggesting that increasing the concentration of carbon dioxide in the atmosphere would result in a warmer planet, (Kurland et al., 2023). By the beginning of the 20th century, the Swedish Physicist Svante Arrhenius, who would win the Nobel Prize in Chemistry in 1903, calculated that doubling the atmosphere's concentration of CO2 would raise Earth's surface temperature by about 4 degrees C., a prediction that, disturbingly, matches up well with measurements over the last century, (noaa.gov).

In 1958, the Mauna Loa Observatory, operated by the National Oceanic and Atmospheric Administration (NOAA), began to measure CO2 levels in Earth's atmosphere. Its first reading, taken in March of that year, was 315 ppm, (noaa.gov, n.d.). Ice core samples reveal that CO2 levels in Earth's atmosphere had already risen slowly from approximately 280 ppm in the mid-1700s to roughly 300 ppm in the early 1900s, (Lindsey, 2024).

Broad public awareness of and concern about Earth's environment is a relatively recent phenomenon. Most scholars would likely date the dawning of the modern environmental movement – as distinct from the earlier conservation movement – to the publication of Rachel Carson's *Silent Spring* in the early 1960s. Carson's book focused attention on the unintended consequences of pesticide use, particularly DDT, on birds and other living systems, (Carson, 1962).

April 22, 1970, was the first Earth Day. It comprised a nationwide series of seminars, teach-ins, and demonstrations, largely held at universities. (I attended one such event while still a high school student.) (richardhowe.com). Earth Day signaled the emergence of the environment as a significant public issue. While climate change had not yet been recognized as a threat, the growing awareness and demand for action would result in the United States enacting a variety of statutes over the next decade intended to ameliorate other unintended consequences of modern, industrialized societies on the nation's air, water, land, and health, Clean Air Amendments of 1970).

In the 1980s climate change began to penetrate public consciousness as an issue of concern. In 1987, the United Nations' World Commission on Environment and Development issued a report citing "the serious probability of climate change generated by the 'greenhouse effect' of gases emitted to the atmosphere," (The World Commission on Environment and Development, 1987). A year later, NASA Scientist James Hansen testified before the Senate Energy and Natural Resources Committee, stating that "Global warming has reached a level such that we can ascribe with a high degree of confidence a cause and effect relationship between the greenhouse effect and observed warming." Dr. Hansen's testimony was front page news in the *New York Times*, (nytimes.com). At the time of Hansen's testimony, Mauna Loa's measurement of CO2 level stood at 353 ppm.

Within a few more years, the United Nations would begin hosting annual global climate Conferences of the Parties ("COPs"), scholarly and popular authors published an increasing number of books and articles on the subject, perhaps most notably former Vice President Al Gore's book (and film) *An Inconvenient Truth,* (Gore,

2006), and public officials in the U.S. and overseas would introduce numerous proposals for new laws aimed at reducing global warming or adapting to its effects.

All of these efforts were well-intended, but the net effect of all the conferences, books, and legislation would prove minimal in the face of the steady growth in greenhouse gas emissions, (statista.com). Furthermore, the problem of a warming Earth was no longer just a possibility or a future concern: it had arrived. Record setting heat waves, "thousand year" floods, crop failures, deadly wildfires, and climate refugees abandoning areas that would no longer support human habitation became regular occurrences.

While activities such as cutting down forests and raising livestock also contribute to the problem, the burning of fossil fuels for transport, heating, cooling and lighting, and manufacturing, is by far the largest driver of increasing CO2 buildup in the atmosphere and the resulting rise in the planet's temperature. Absent the deployment at scale of technologies that are unproven and uncertain, such as carbon capture and storage or geoengineering, reducing and ultimately stopping human-caused climate change will require decarbonizing the world economy.

While it is not possible to predict precisely the severity of the consequences that would accompany each unit of increase in atmospheric carbon concentration or temperature, the United Nations Intergovernmental Panel on Climate Change's (IPCC) *Climate Change 2023 Synthesis Report* is unequivocal in stating that, "Every increment of global warming will intensify multiple and concurrent hazards," (IPCC, 2023).

As I write this chapter in the fall of 2024, the world's population is approaching 8.2 billion, and Earth's atmospheric CO2 concentration stands at 424. The Earth's surface temperature in the period 2001-20 was approximately 1 degree C. higher than in the period 1850-1900, and had increased faster since the first Earth Day in 1970 than in any fifty year period over at least the last 2000 years, (IPCC, 2023).

A PROBLEM MADE FOR UNIVERSITIES

The world has an increasingly urgent need for solutions to climate change. The various policy tools that were deployed with considerable success during the last quarter of the 20th century to reduce the harms and risks from "first generation" pollution problems are not well suited to the problem of climate change.

Universities can and should be a significant part of solving the problem of climate change. They have a number of strengths and resources that would appear to match up well with this civilizational challenge, superior in most respects to those of governments or businesses or advocacy groups.

It is of course impossible to predict just what forms the eventual solutions to climate change will take. But we can have a fairly high degree of confidence that the blend of characteristics typical of universities will provide substantial value in bringing these solutions about.

First and foremost, the problem of climate change is complex. It is a worldwide problem, and the activities that drive it – chiefly energy production and consumption, transport, construction, agriculture, heating/cooling/lighting – are universal and interconnected. Universities are in the business of helping people understand and make sense out of complexity.

The solutions to climate change will almost surely be multidisciplinary. Take a moment to consider the following list, presented alphabetically:

agronomists, architects, biologists, business analysts, chemists, computer scientists, economists, engineers, geographers, historians, lawyers, marine scientists, materials scientists, mathematicians, operations researchers, philosophers, physicians, policy analysts, political scientists, psychologists, soil scientists, urban planners

You can probably think of others, but the point should be clear: this problem is broad and deep enough to call upon the widest spectrum of human knowledge, technique, and wisdom in devising solutions. Universities are home to all of these experts and more.

Universities can take a long-term perspective. There will be no quick fix for climate change, no vaccine or panacea. Universities have long time horizons. While all have their own financial pressures and requirements, universities are not subject to quarterly earnings reports or biennial or quadrennial elections.

Solving climate change will require new knowledge. It should be obvious by now, nearly four decades since James Hansen's testimony before the U.S. Senate announcing that climate change was, in fact, occurring, that humanity still does not possess the requisite knowledge to solve the problem. This is true at the scientific, technical, economic, and political levels. What is so often said at the end of journal articles is clearly applicable in this case: further study is needed. And even while confidence in institutions has been in decline for some time, universities continue to rank higher in the public's estimation than most, (Brenan, 2004).

Universities are also places where considerable time and attention is devoted to considering questions of human values. Climate is assuredly an issue of economic, geographic, and intergenerational justice just as much as it is an issue of engineering and atmospheric chemistry. Humanity is unlikely to adopt the institutional arrangements necessary to solve climate change unless care has been taken to address these aspects of the problem.

Perhaps most important, students care deeply about the issue of climate. Today's students began learning about the climate problem as children. Many of them are deeply worried about how it will affect their lives, livelihoods, and the world they will live in. They expect their university to prepare them for the climate challenges they and their children will face, as well as to focus the university's resources on finding solutions.

Lastly, governments, industry, and philanthropies are all interested in this issue and will provide funding to universities and their faculty members to research and teach about climate change and work toward solutions.

Again, we cannot predict how, eventually, humanity will solve the climate change problem. But there are enough examples of other high stakes issues where universities have played a major role, including disease prevention, weapons development, agriculture, and space exploration, to expect they will be heavily involved in solving this one.

HOW UNIVERSITIES CAN MAKE A DIFFERENCE

"The world is full of people who are anxious to function in an advisory capacity."
--Charlie Brown

Mostly by happenstance, I have accumulated a range of perspectives about universities and climate change. I spent about a quarter century in colleges and universities as a student, teacher, research institute director, and advisor to senior administrators on climate change. I worked for a similar span of years as a lawyer, political strategist and communicator, state environmental official, and environmental policy advocate. My formal training was in political science, public policy, and law, but my career has been primarily as a practitioner, not an academic[2].

Universities are typically quite decentralized institutions made up of distinct groupings of people, each with a variety of roles. Rather than offering advice to universities as such, I have presented my suggestions to each key grouping or audience in turn: university leadership, academic administration, faculty, and students.

For University Presidents and Boards of Trustees

By all means, adopt a formal university policy on the world's climate crisis. Something like this: you recognize climate change as a severe and present threat to humanity and all life on this planet and are committed to doing all you can, consistent with your roles and resources, to help the world overcome it.

This recommendation may appear to sail against the tide of recent experience. Many universities have pulled back from issuing policy statements on issues that extend beyond the gates of their campuses. In some well-publicized cases, university presidents have lost their jobs because they or their institution became embroiled in an intense public controversy.

Nevertheless, climate change should be an exception to the disengagement trend. *Literally no one* is in favor of large swaths of the Earth becoming uninhabitable, or untold numbers of species being wiped out due to their inability to adapt to changed circumstances, or severe and destructive weather events becoming increasingly common.

Your university's policy should be stated at a fairly high level, which is not to say it should be vague. But you need not and should not take a detailed stand on *how* the world eventually gets control of its thermostat, whether it be by some combination of expanded use of renewable energy, carbon taxes, battery storage, geoengineering, or what have you. The complex interplay of politics, economics, innovation, and technology will sort all of this out over time.

But you should be unequivocal. Taking this stand is not a detour from your university's mission: the preservation of the biosphere and the protection of human lives and communities are preconditions to the free pursuit and dissemination of knowledge that are at the core of your mission.

What about your university's own carbon footprint, chiefly consisting of the release of carbon dioxide into the atmosphere resulting from the university's heating, use of electricity, construction, travel, and everything else? By all means, your university's climate commitment should include reducing its own greenhouse gas emissions, typically described as Scope 1 (direct GHG emissions that occur from sources that are controlled or owned by an organization), Scope 2 (indirect GHG emissions associated with the purchase of electricity, steam, heat, or cooling) and Scope 3 (the result of activities from assets not owned or controlled by the reporting organization, but that the organization indirectly impacts)[3].

But your university's most significant impacts in helping the world solve its climate problem should *not* be the reductions you achieve in the university's own emissions. Although this may be the factor that the student body focuses on more closely than any other in evaluating the university's climate performance, most of your attention should be focused elsewhere.

No one university's emissions contribute more than a flyspeck to the world's climate problem. The sobering fact is that if every university in the world could somehow become carbon neutral tomorrow, that change would have no measurable impact on the world's climate.

There is a lesson here for how we should think about climate change generally. This is not a problem that will be solved "if we all just do our part." The world needs both innovations that make emissions reductions more feasible and less expensive *and* collective action in the form of governmental mandates and nudges that push all of us to behave differently[4]. Universities have an important role to play in meeting both of these needs. The scientists, engineers, architects, political scientists, lawyers, economists, business leaders and many other professionals that universities both train and employ will all be needed in the years ahead to solve this complex global problem.

Once you have adopted your climate policy, don't put it in a drawer: speak out. Use your individual and institutional voices regularly to make it clear to the entire university community that the commitment is real. In particular, encourage (i.e., *nag*) all the various schools, departments, and programs throughout your university to do what *they* can do to fight climate change; I offer some ideas about this below.

However:

You don't need to divest from fossil fuel stocks in the university's investment portfolio. Even the largest university endowments hold only a tiny sliver of any fossil fuel company's outstanding shares. If you sell your shares, they will be quickly bought up by other investors at the current market price. This buying and selling of stock will have zero effect on world demand for the underlying products that contribute to climate change – oil, gas, coal, or whatever. And even if a concerted campaign of multiple divestments did somehow harm domestic energy firms, any falloff in production would quickly be taken up by offshore oil and gas companies, including those controlled by autocratic foreign governments that are heavily reliant on fossil fuel revenues.

For a university that aspires to do its part to slow and ultimately stop climate change, divesting can become a "feel good" substitute for effective action. It risks being an exercise in virtue signaling, perhaps effective at mollifying some of your students, faculty, and alums, but little more.

Much better than divesting would be to earmark any and all profits from fossil fuel holdings in your portfolio to fight climate change. The trustees can adopt a policy to devote any such profits to some combination of climate-related purposes, e.g., scholarships for students studying climate-related subjects, adaptation and relief efforts for countries most impacted by climate change, research clearly designed to aid in mitigating greenhouse gas emissions, and so on. Your fossil fuel profits are fairly simple to calculate – mutual fund companies do this on a daily basis to determine their customers' gains and losses, and your finance people would probably only need to do so annually.

Furthermore, you can afford to take this step. You may be surprised what a small slice of your university's investment returns come from oil and gas stocks. The energy sector comprises less than 5% of the capitalization of the S&P 500[5]. Furthermore, many universities keep sizable portions of their endowments in bonds, real estate, cash, and other classes of investments, the returns from which would remain available to meet other spending priorities.

By dedicating any profits derived from fossil fuel investments to climate action, you will reclaim the moral high ground from any critics asserting that the university is profiting from climate change. This act of financial jiu jitsu would turn the very profits of fossil fuel companies toward a future where those companies will no longer be contributing to global warming.

To sum up, actions that your university can take have the power to affect the Earth's future climate. Stand up proudly for the future, and aim high.

For Academic Leadership

Provosts, deans, and department heads have several means available to give effect to the climate commitments made by the president and the trustees.

Fundamentally, you should come together to commit to making your university a knowledge and action hub on issues of climate change. Knowing you've succeeded at this goal may be akin to Justice Potter Stewart's test for pornography in *Jacobellis v. Ohio*: "I know it when I see it." (1964). But maybe Stewart was on to something. Institutions, very much including universities, do not develop reputations for strength in particular areas by accident. Rather, they make a conscious decision to become excellent, communicate this expectation broadly and consistently to all internal audiences, and carefully manage performance and accomplishments. When this happens, you *do* know it when you see it. Work to make your university a place and a community where research, teaching, discussion, and activism around climate is so prevalent that anyone spending time on campus will detect its pulse.

The academic administration should offer its full support to faculty members for research related to energy and climate, both "pure" basic research and use-inspired research. Basic research often yields knowledge or insights that were not top of mind for the researchers but that later become the basis for highly beneficial applications, (Bush, 1945). But this is not the only way that practical progress informed by inquiry occurs. In fact, when humans have felt the need to make major advances quickly, be it for the military, satellites and space travel, vaccines, or what have you, a more intentional approach to research, motivated by practical use, has often been employed. What Donald Stokes calls "use-inspired basic research" can produce insights that may be used to improve climate-relevant outcomes, such as lower greenhouse gas emissions, improved energy efficiency, climate adaptation measures, and more effective climate communication strategies, (Stokes, 1997).

Strongly consider establishing a graduation requirement for understanding climate change for undergraduates. The underlying notion is that an educated person in the 21st century must not be ignorant of the fundamental changes going on in our planet's atmosphere, what is causing them, and what effects they are already having and will have in the future.

This requirement need not be met by passing a course, but if it is to be a course it should be a serious and specific one. The goal of preparing the citizens of the future for the challenges that await them will not be satisfied by simply letting students pick one of dozens of existing courses taught across the university, from *Introduction to Botany* to *Novels of the End Times*. One could imagine a new, team-taught course for First or Second Years that combines relevant scientific, sociological, economic, political, and philosophic material. Alternatively, students could be given a curated climate reading list upon admission and be responsible for completing it at some point before graduation.

Yes, this can be a slippery slope – what about a required course or curriculum on racism or computer technology or [insert your idea here]? But you will send a powerful signal by placing the planet's climate future in the foreground of your university's concerns.

However:

Don't create a Climate Change undergraduate major. The college years are a time for students to build out a broad intellectual toolkit, even though many of them don't know that yet. While such a major would be attractive to some students and prospective students, it would be a disservice to them to offer it. The thickly entwined roots of the world's climate challenge and the likely complex and multi-faceted nature of its ultimate solutions are a strong argument for the continued relevance and value of liberal arts education. Far better for your students to complete a traditional liberal arts curriculum or a career preparatory degree program than an enticing but almost certainly disappointing boutique major in climate change.

On the other hand, a well-distributed climate minor could have significant value. Students with a strong interest in climate change who are receiving a solid grounding in their major -- be it Chemistry, Economics, History, or what have you – could benefit significantly from integrating half a dozen or so courses from across several departments. Such a minor could also include a relevant internship or work-study experience.

Increase the amount and quality of on-campus programming related to climate change. You have a wide range of possibilities: Talks by climate activists, authors, and political leaders. Presentations by academic experts doing important work. Panel presentations, symposia, debates, art installations, student poster exhibits, film showings and discussions. Commencement speakers. Honorary Degree recipients. Scholars and Practitioners in Residence.

Your university is already doing a lot of these things? Great. But task someone with doing a detailed inventory, take a fresh look – you will probably find opportunities to do more. Your community can sense the difference between a university that is deeply engaged in our climate challenge and one that is simply going through the motions.

Again, climate is not the only important topic you can prioritize for programming. But the climate issue is one that touches on practically every corner of your university. And your students, in particular, are embarking on lives in which climate concerns will be increasingly urgent.

Also, living out your university's climate commitment in this way will be noticed and can bolster multiple salutary purposes, including boosting applications, attracting faculty, increasing alumni and foundation funding, and so on.

For Faculty Members

You are at the heart of your university. You conduct, review, collect, and disseminate the research that advances humanity's store of knowledge. You open up the world to undergraduate students with your teaching and guide graduate students as they advance toward becoming professionals, whether inside the academy or in the world beyond. You have the opportunity to leverage your knowledge and concern about climate change across dozen or even hundreds of young people each year. And you can be the kind of professor who genuinely inspires and guides your students, in addition to helping them grow in knowledge.

You can make a significant impact whether your research is geared toward mitigation or adaptation. Some years ago, many environmentalists – including the author of this chapter – tended to disparage adaptation work as defeatist, as accepting climate change as inevitable, rather than finding ways to stop it. Sadly, humanity has dithered and stalled and avoided the issue for so long that our activities have already caused the climate to change. Significant adaptation is absolutely necessary to enable many millions of us to carry on and for all of humanity to become more resilient in the face of these changes.

Also, even though the creation or discovery of knowledge is central to the academic enterprise, new knowledge is not the only knowledge the world needs to combat climate change. If your research is the kind that Stokes labelled use-inspired, i.e., consciously geared toward having an impact on people and communities, consider that frequently the worlds of climate *practice* – be it in public policy, finance, engineering, or what have you – can sometimes make better, more rapid use of knowledge that is *not* novel as the academy typically defines it. Knowledge that is *new to them*, even if already established in your academic field, may be just the thing they can put into use now.

Also, keep in mind that the people who work on the front lines of climate change – public officials, corporate decision makers, planners, developers, and many more – are mostly not sitting in their offices waiting for the latest university research to tell them what to do. You will likely need to use means of disseminating your research insights other than the conventional academic journals and conferences to have the impact you desire more rapidly. Specifically, you may want to reach out directly to people and institutions you believe are well-positioned to make use of your data, methods, insights, and conclusions. In particular, reaching out to audiences in your own city or state may prove fruitful. Your university may have more credibility in your region than more distant universities that might be viewed as less in touch with the concerns of local communities.

Your work can play a part in moving humanity toward solutions to climate change, whatever your field of study. Your potential impact may be more obvious if you are in the sciences or social sciences, but a humanistic understanding of the climate crisis, enriched by the study of history, literature, philosophy, the arts, is also vitally important. People around the world, and young people in particular, are struggling to make sense of what it means to be living in this moment in time, when issues of threat, loss, meaning, lack of control, and the extinction of species are all around us. One of my former institutions, Clark University, with the support of several foundations, pioneered a program called "A new Earth conversation" which has been described as "a collective process of inquiry, listening and discernment," (Walsh, 2019). Might you collaborate with your university colleagues to organize something similar, geared to your particular circumstances?

Help your students see the genuine complexity of the climate challenge. Most of today's university students have grown up knowing that climate change is a real, serious, and growing threat. Many of them want to engage actively in addressing it, whether in their professional lives or in their role as citizens. Indeed, it will likely be the work of today's students and those who come after them that ultimately ends this dangerous warming.

But even though human choices brought about climate change and altering humankind's activities will be necessary to stop it, getting to that outcome is proving very difficult. What makes this problem so hard is that it is driven by literally billions of choices people make about things they want and like: houses that are warm in the winter and cool in the summer, automobiles and airplanes, cheeseburgers.

Scholars have much to say about these choices and the contexts within which they are made. But I worry that academic approaches to the problem sometimes may not be advancing us toward solutions. A case in point: economists have been arguing for years that imposing a carbon tax on fuels would "put a price on pollution" and incentivize the world to shift away from oil, gas, and coal to renewable energy sources that don't emit greenhouse gases. Markets work this way in theory, but in practice it has proven hard to achieve, because most people don't want to pay more for oil, gas, and coal and threaten to vote out of office any politicians who vote for or even suggest enacting such a new tax.

Or consider sociologists' fondness for intersectionality analysis, a relative new kid on the academic block but a popular one. It is certainly the case that the corporate and governmental entities that enabled and profited from the development and use of fossil fuels are and have been dominated by white, well-off men from the Northern Hemisphere. But if the world really can't decarbonize the global economy unless and until we end racism, capitalism, and patriarchy then we are truly done for.

Now that I have offended both the right and the left, let me simply urge that we all think critically about our own discipline's tools and analyses. Being in possession of sophisticated methods or deep insights may not bring the world any closer to solving the problem. And solving the problem is what our students – and everyone else – ultimately want and need to happen.

Also, if you are engaged in climate- or energy-related research, who is funding your work? Does some or all of your support come from fossil fuel companies?

These two things are inconveniently true: The world urgently needs a transition away from carbon emitting fossil fuels to halt climate change. But the world will also continue to need and use fossil fuels for the next several decades.

Much of the demand for fossil fuels until mid-century or so will likely come from the developing world, whose people have already waited too long to be able to turn on the lights or the air conditioning. Until humans figure out how to ramp up currently available carbon reduction technologies and develop new ones to fuel the parts of the economy that are not easily decarbonized, it really doesn't make sense to treat all fossil fuel companies as pariahs – they are satisfying a demand that the world's billions of people will continue to have until we figure out how to make the transition to a net-zero carbon world.

However, fossil fuel companies are *not* all acting the same with respect to this needed transition. They exhibit a spectrum of behavior that runs from making substantial investments in alternative technologies such as solar, wind, nuclear, and hydro power to actively working to increase future demand for oil and gas and coal and downplaying the need for the world to make the transition away from fossil fuels. These foot-dragging companies are eager to point to their research partnerships with respected institutions like universities while they get their remaining oil and gas reserves out of the ground for a few decades more.

If you are receiving support from fossil fuel companies, ask yourself: what is the evidence that your funders share your commitment to help the world solve the problem of climate change? My advice: Don't help the climate laggard companies greenwash their reputations by associating with you. Easy for me to say, but their money isn't worth it.

Advising is your hidden superpower. In addition to research and teaching, many faculty serve as advisors to students, both formally and informally. Often this work is tedious: signing off on dozens of students' course selections, writing anodyne recommendation letters for countless graduate school or job applications, and so on. But some students want and need more of what you have to offer, even if they don't know it yet. I was such a student, and I was fortunate to have an undergraduate advisor who went the extra mile. He sparked my desire to become a lawyer through his teaching of constitutional law, recruited me into a legislative internship program he directed, tipped me off about an off-campus work opportunity that became my first professional-level job, and wrote detailed, personalized recommendations for my law and graduate school applications. He became a mentor and lifelong friend. You can't provide this level of help for all of your students, but the potential is there for you to change lives and have an outsized impact beyond what you do in the lab or classroom.

I haven't distinguished between tenured and untenured faculty in these recommendations, and that is intentional. I urge you to begin pursuing your passions and interests now, whether you are just getting started in academia or are already an eminent scholar. The challenge of climate change is too urgent to bide one's time. You can do this work and still progress as an academic. If the work you most want to do would get in the way of receiving tenure, it may be that you're at the wrong institution, or even that you need to consider changing your career path.

For College Students

The first time I heard about the possibility of human-caused climate change was in 1975. I was a senior in college talking with a recent graduate who had heard the "greenhouse theory" described by one of his professors. I had no idea at the time that I would spend a significant part of my career engaged in environmental advocacy and policy, including several years working exclusively on climate change.

I wish my generation had made more progress on the issue, but here we are. So it will fall to you to help solve the problem of climate change, a problem that, as I described earlier, has been literally centuries in the making. If you have picked up this book, you likely have more than the average amount of interest in the topic, so allow me to share just a few thoughts.

Your college years offer you the chance to refine, or even reinvent, yourself. The experience of being a college student doesn't transform everyone, but it can, especially if you approach it in that spirit. If you want to help move the planet toward climate sustainability, what you choose to do in college can help you prepare yourself.

Your chances of making a positive difference will increase if you follow the advice given by, or at least often attributed to, Abraham Lincoln: w*hatever you are, be a good one*. Don't settle for just "getting the gist" of things that are important to you. You wouldn't want the services of a dentist or auto mechanic or athletic trainer who was just "decent" or "got the hang of it." Making change happen in the world is hard, and you will make more progress if you deeply know and understand what it is you are trying to change. This applies to other subjects too, of course, not just climate.

However, even if you are deeply concerned about climate and think it may be a major part of your life's work, you don't need to commit to that choice just yet. We all need to figure out the breadth of our interests and concerns, the place of work in our lives, how much of our time we dedicate to civic activism, and so on. You don't even need to major in a climate-related field for your future self to have an impact on climate issues. Neither Jodie Foster nor Mick Jagger majored in the fields in which they would later make their marks.

Also, your college years should be a time when you discover, if you hadn't already, that people hold a lot of different opinions on pretty much everything. Even if it seems self-evident to you that humanity must quickly move away from burning fossil fuels, many people are more concerned about the price of filling their gas tanks or the effects on the local economy of a fracking ban than they are about climate change. These differences of opinion are why we have *politics*, to give us a way of working out our disagreements peacefully and democratically. Such disagreements are your invitation to speak out, join an environmental organization, volunteer in a political campaign, and vote. And you can absolutely begin doing these things now, while you are still in school.

Martin Luther King Jr. famously said that the arc of the moral universe was long but that it bent toward justice. I don't think he meant that the arc bent by itself, rather, by the efforts of people who were engaged in the to work necessary to bring about change. You don't have the option of just, as the song says, "waiting for the world to change." You need to be the ones to change it, whether through your professional careers, your civic and political engagement, or some combination.

I am hopeful that your generation will succeed in the work that earlier generations have begun and that, through your efforts, you will help the world write a favorable ending to the climate change story.

CLOSING THOUGHTS

I hope you have found some useful information or advice in these pages. I will close with a few points that strike me as important but didn't find a home in the organization of this chapter:

- Climate change is, sadly, not the only major sustainability challenge our planet and its inhabitants are facing. The loss of forest lands, the decline in the health of oceans, the persistence of chemicals in the environment and their health effects on humans and other living things, and the accelerating extinction of species, among others, are each a cause for concern, research, and action.

- These issues and others are in urgent need of governments engaging in serious problem solving, both within their own borders and in cooperation with other nations. However, the trends in many parts of the globe toward reflexive negative partisanship, disinformation, nativism, and suspicion and distrust directed toward experts may make the world's environmental challenges much more difficult to overcome.
- An adage known as Stein's Law (named for the economist Herbert Stein) holds that, "If something cannot go on forever, it will stop"[6]. When applied to climate change, this is either comforting – or not – depending on what it is that stops the increase in atmospheric carbon and the attendant warming of the atmosphere, and when. Over the last couple of decades, as I witnessed the disappointing results of both domestic and international efforts to reduce greenhouse gas emissions, I became more pessimistic about the likelihood that our political processes were adequate to the task of reducing emissions before the climate effects were horrific. Nevertheless, I am now cautiously optimistic that the combination of new technologies (including the application of Artificial Intelligence) and their favorable economics will dramatically reduce greenhouse gas emissions sometime later in this century. However…
- I will likely not witness how the climate change issue is resolved. But perhaps, before this century gives way to the next, my children will tell their children and grandchildren how the world ultimately made the transition to zero carbon energy. I expect that universities will play an important part in producing that result, and that there will someday be books written and courses taught about how humans overcame the climate crisis. My hope is that the lessons of this struggle will inform future generations of researchers, teachers, students, and citizens in dealing with other, future challenges we can hardly imagine.

REFERENCES

Brenan, M. (2024, July 15). U.S. confidence in institutions mostly flat, but police up. *Gallup News*.https://news.gallup.com/poll/647303/confidence-institutions-mostly-flat-police.aspx

Bush, V. (1945). *Science, the endless frontier* (Reprinted 1960 by the National Science Foundation). https://archive.org/details/scienceendlessfr00unit/page/n3/mode/2up

Carson, R. (1962). *Silent spring*. Houghton Mifflin.

Emanuel, K. (1998). *What we know about climate change*. MIT Press.

Gore, A. (2006). *An inconvenient truth*. Rodale.

Howe, R. (2023, April 21). No more Silent Springs [Excerpt]. *richardhowe.com*. https://richardhowe.com/2023/04/21/no-more-silent-springs (Accessed September 1, 2024)

Intergovernmental Panel on Climate Change (IPCC). (2023). Summary for policymakers. In Lee, H., & Romero, J. (Eds.), *Climate change 2023: Synthesis report. Contribution of Working Groups I, II and III to the Sixth Assessment Report of the Intergovernmental Panel on Climate Change* (pp. 1–34). IPCC., DOI: 10.59327/IPCC/AR6-9789291691647

Jacobellis v. Ohio, 378 U.S. 184, 197 (Stewart, J., concurring) (1964).

Kurland, Z., Hafner, K., Feder, E., ... (2023, November 9). The woman who demonstrated the greenhouse effect. *Scientific American*.https://www.scientificamerican.com/article/the-woman-who-demonstrated-the-greenhouse-effect/

Lindsey, R. (2024, April 9). Climate change: Atmospheric carbon dioxide. *NOAA Climate.gov*.https://www.climate.gov/news-features/understanding-climate/climate-change-atmospheric-carbon-dioxide

National Geographic Society. (n.d.). *Mauna Loa Observatory*.https://education.nationalgeographic.org/resource/mauna-loa-observatory

National Oceanic and Atmospheric Administration. (n.d.). *Trends in atmospheric carbon dioxide*.https://gml.noaa.gov/ccgg/trends/data.html

Shabecoff, P. (1988, June 24). Global warming has begun, expert tells Senate. *The New York Times*.https://www.nytimes.com/1988/06/24/us/global-warming-has-begun-expert-tells-senate.html

Statista. (n.d.). Global carbon dioxide emissions. https://www.statista.com/statistics/276629/global-co2-emissions/

Stokes, D. E. (1997). *Pasteur's quadrant: Basic science and technological innovation*. The Brookings Institution.

Thompson, A. (2021, September 13). Here's how much food contributes to climate change. *Scientific American*. https://www.scientificamerican.com/article/heres-how-much-food-contributes-to-climate-change/

United Nations Department of Economic and Social Affairs. Population Division. (2008). *World population prospects: The 2008 revision* (only medium variants shown). https://web.archive.org

U.S. Congress. (1970–1976). *Clean Air Amendments of1970, Safe Drinking Water Act of 1974, Toxic Substances Control Act of 1976, & Resource Conservation and Recovery Act of 1976*. Public Laws 91-604, 93-523, 94-469, and 94-580.

Walsh, D. C. (2019, December 3). It's time for higher education to "get real" about the climate crisis. *Medium*. https://medium.com/@dcdubs/its-time-for-higher-education-to-get-real-about-the-climate-crisis-d0247b99d969

World Commission on Environment and Development. (1987). *Our common future*. Oxford University Press.

ENDNOTES

1 For example, see *Higher Ed Climate Action Plan,* Aspen Institute (2024) at https://www.thisisplaneted.org/img/HigherEdClimateActionPlan.pdf; *Race to net-zero: why universities must walk the walk on emissions,* Bothwell, Ellie (October 28, 2021) at https://www.weforum.org/stories/2021/10/race-to-net-zero-universities-emissions-cop26/; *This is how universities can lead climate action,* Mocatta, Gabi and White, Rob (2020) at https://theconversation.com/this-is-how-universities-can-lead-climate-action-147191.

2 After completing degrees in Political Science (B.A.), Public Policy (M.P.P.), and Law (J.D.), the author practiced law, worked as a political staff person, served as Massachusetts' Undersecretary of Environmental Affairs, was President and CEO of the Environmental League of Massachusetts, directed Clark University's Mosakowski Institute for Public Enterprise, and was Senior Advisor (Climate Change) to MIT's Vice President for Research.

3 Mancur Olson Jr.'s *The Logic of Collective Action* (1965; Harvard University Press) is the classic work on how individuals' shared interests often do not produce desired outcomes in the absence of governmental mandates. See also Hardin, Garrett, *The Tragedy of the Commons*, *Science* 162 (1968): 1243–1248.

4 For a more detailed description of Scope 1, 2, and 3 emissions, see https://www.epa.gov/climateleadership/scopes-1-2-and-3-emissions-inventorying-and-guidance.

5 For an eclectic survey of organizations' quest to produce excellence and build reputation, see Peters, Tom and Austin, Nancy (1985) *A Passion for Excellence*. Random House.

6 Washington Post. See https://www.washingtonpost.com/opinions/robert-j-samuelson-is-steins-law-real/2013/05/30/716942f2-c942-11e2-8da7-d274bc611a47_story.html

Chapter 3

Climate Action and the Ethics of Student Engagement

Steven Knapp
Carnegie Museums of Pittsburgh, USA

ABSTRACT

Critics of activism on university campuses argue that academic institutions should stick to their core missions of teaching and research. While there is some merit to this argument and to the related view that teachers should avoid using their academic authority to impose their political views on their students, universities should not shy away from engaging in research and teaching on subjects with policy implications or from basing their own policies on the best available science. Nor should they refrain from supporting extracurricular activities that enrich their students' education with real-world engagement. The complexity and civilizational implications of the climate crisis make it a uniquely rich subject for interdisciplinary study that fosters students' intellectual growth, as well as an appropriate focus of extracurricular activities that complement students' work in the classroom. The chapter proceeds from these theoretical considerations to concrete examples of university programs that effectively support student involvement in climate-related activities.

INTRODUCTION

In 2008, the ingenious and provocative literary and legal scholar Stanley Fish published a book with the (ingenious and provocative) title *Save the World on Your Own Time*. It was the same year in which, as the still-new president of The George Washington University in Washington, DC, I joined what were then nearly

DOI: 10.4018/979-8-3693-5454-4.ch003

300 colleagues in signing the American College & University Presidents' Climate Commitment, in which each signatory institution pledged to undertake a range of climate-related actions, setting a target date for achieving carbon neutrality. I had been (and still am) a follower as well as an occasional critic of Stanley Fish, so I was eager to find out what he might have to say about the role—or rather, in his view, the absence of a role—for universities in addressing global challenges. Nothing in the book argued against a university's deciding to reduce its carbon footprint, but Fish mounted a forceful argument, with his usual energy and eloquence, against bringing such issues into the classroom.

Although he did not present it in ethical terms per se, the challenge Fish was posing to activist faculty and administrators was essentially an ethical challenge. He was decrying what he regarded as a dereliction or diversion of institutional duty. And at the heart of his accusations was the implication that, by veering away from its core academic mission, a college or university was cheating its students, depriving them in some measure of the education the institution was ethically bound to provide.

Fish's argument, when I encountered it as a new university president, gave me pause, because the truth is that what led me to prioritize environmental sustainability was in large part my discovery, within weeks of arriving on campus, of how important the issue was to George Washington students. I wasn't under pressure from students to take any particular kind of climate action (that came later), but I was intrigued and inspired by the well-informed activism I encountered. I set up a faculty and staff Task Force on Sustainability and appointed as its coordinator a very able graduate student named Josh Lasky who, 16 years later and seven years after I left the presidency, is back at GW, now serving as director of the university's Office of Sustainability. It was this Task Force that recommended I sign the presidents' climate pledge and set up the university's first sustainability office.

Given the role of students as leaders of the climate movement on campus, it was obvious to me from the start that it would be impossible to draw a bright line between their extracurricular involvement (what they did "on their own time") and their academic interests and pursuits. Admittedly, George Washington is unusual among research universities in the degree to which the lines between learning and service are blurred. Many students are drawn to a Washington, DC university in the first place precisely because they want to connect their educational experiences with their commitment to "changing the world." For the same reason, the university tends to attract scholars and teachers who are more comfortable with applied research and service learning than many of their peers at other places. When service-oriented students and faculty come together, the kind of separation Fish advocates becomes very hard to enforce—even assuming one wanted to enforce it! And in fact, the Task Force presented a number of academic recommendations, summed up in the commendation that the university "approach sustainability as a central part of its academic mission," (Sustainability.gwu, 2008).[1]

What does it mean, however, to make sustainability—or, more in keeping with where the emphasis currently and, I believe, rightly falls, to make climate action—a central part of an institution's *academic* mission? Does it mean expanding the curriculum to include more courses on relevant topics? Hiring more faculty in fields like environmental engineering or policy studies? *Pace* Stanley Fish, such actions are entirely consistent with a university's mission as traditionally understood. The choice of what fields a university ought to invest in has always been driven by a combination "pure" and "applied" research functions, and the latter have always reflected choices about what kinds of application matter to society. Of course what matters to society is also always a subject of debate, and university administrations don't always agree with boards of trustees or, in the case of public institutions, boards of regents; with state or federal legislators; with alumni; with parents; or indeed with faculty members themselves on where such investments should be made.

So the subjects taught, the faculty hired, the departments invested in don't raise the kind of ethical issue I want to focus on in this chapter. What does raise that issue is the question of how far faculty and administrators ought to go in supporting, guiding, and even perhaps *enlisting* students in the climate movement. Do such activities violate the unwritten compact between teacher and student? Do they contradict the institution's commitment to open inquiry? Or are they, or at least some them, legitimate extensions or enhancements of the institution's educational role? And if it's OK for institutions to encourage, support, and/or promote their students when they advocate policies aimed at countering climate change—including, for instance, the investment policies of the institution itself—does that imply an obligation to support the activism of students who believe, for instance, that the social and economic costs of reducing the use of fossil fuels outweigh its benefits? Does it also imply an institutional duty to act similarly when it comes to student responses to painfully divisive issues like the Israel-Hamas conflict in Gaza?[1] Or whenever there is a political contest between two opposing parties?

The "slippery slope" argument implicit in such questions is always tempting when it comes to the formulation of institutional policies on complex and controversial matters. If we do X, the argument goes, we can't justify a refusal to do Y and Z. And we can only resist such slippage if we can show that X is an exceptional case.

The slippery slope argument might apply to an institution's decision to address climate change if the issue was correctly understood in partisan political terms; in other words, if the reason the institution chose to take this position was its allegiance to a particular political party or faction. And indeed, there is a concerted effort on the part of some powerful interests to characterize climate action in precisely that way. But colleges and universities, as well as the staff and teachers they employ, have every right, and indeed every responsibility, to base their decisions—both operational and academic—on the best information they have about things that matter to their educational mission, their service to the communities (national as well as local) on which they depend, and their own long-term sustainability. They should not allow themselves to be deterred by malicious or misguided accusations of partisanship, (Harvard.edu, 2024)[2]

The fact that, in my view, colleges and universities have every right and responsibility to respond to the best available information about issues affecting their missions and their long-term sustainability—and in the case of the climate emergency, that means *following the science*—does not tell us how institutions should or should not engage their *students* in the formulation or promotion of that response. In what follows, I will offer suggestions about where the lines ought to be drawn. I will also point to some of the practical ways in which engaging students in climate action can proceed once those lines are properly drawn. And between my discussions of those two topics, I will explore some of the ways in climate really does seem to me to be a special case, one with very broad implications for the mission of higher education.

So, to start with the negative: I share the view of those who think it's an abuse of academic power for teachers to use their authority as an instrument of political persuasion. In addition to its explicit or implicit coerciveness, such advocacy is at odds with the educational goal of enabling students to think for themselves. Even if we reject Fish's comparatively narrow view of higher education's purposes, we can agree that fostering students' intellectual growth is one of its core purposes and that anything that threatens to stunt that growth by shutting down the consideration of opposing views is incompatible with that purpose. And if we take a broader view of higher education's purpose—allowing it to include, for instance, the formation of autonomous, responsible citizens—then the abusive nature of partisan instruction becomes all the more obvious.

Avoiding that kind of abuse requires a good deal of discipline and discernment, especially in the case of an issue like climate change, where at least some of the policy implications of the scientific consensus are so glaringly obvious. Explicit advocacy of a partisan position is one thing, but how can one talk about, for instance, the effects of methane emissions from leaking oil and gas wells without implying the need for regulations that are bound to be politically controversial (Crownhart, 2024)?[2]

The answer, I believe, is that one can't, and the only way to avoid some degree of political influence on one's students in a class on such subjects would be to stay away from those subjects altogether. For reasons indicated above, I regard that kind of demand as illegitimately trespassing on the rights and responsibilities of academic institutions—and the teachers they employ—to address the subjects they regard as important, and to do so in a way that reflects the best information available to them.

All that said, courses on climate change might seem to be a particularly complex case, both because the policy implications of climate science are often so inescapable and because it is hard to imagine a richer subject of interdisciplinary study—a subject on which the natural and social sciences—and, for that matter, the humanistic disciplines of history, literature, philosophy, and religion - more readily converge. The very features of climate change that make it so hard to teach without crossing the boundary between academic analysis and political advocacy are the features that make it an ideal subject for studying the differences between, as well as the complementary contributions of, the various disciplines that a serious engagement with the subject compels one to engage. The potential intellectual benefits of providing students with opportunities to think across the disciplines while learning to appreciate the unique perspective of each separate discipline amply justify institutions in taking the risk that instruction may occasionally slip into the wrong kind of advocacy.

The educational significance of teaching and studying climate change, however, goes beyond the usual benefits of cross-disciplinary work. Climate change presents a unique set of challenges and opportunities that, when their depth, breadth, and urgency are fully appreciated, have profound implications for the very role and meaning of higher education in our moment. There is no way to say this without the risk of hyperbole, but the ramifications of climate change reach a level of complexity and importance that is nothing less than *civilizational.* They touch on every aspect of the way humans live together with each other and interact with the resources (including countless other species) on which they depend. And the response they demand is nothing less than shifting the bases of human civilization—or, if one prefers, human civilizations from modes of competition and extraction to modes of cooperation, renewability, and regeneration embodying a full appreciation of the interdependence of human and natural systems. The ultimate expression of that shift would be the achievement of what truly could be called an "ecological civilization" (or civilizations).

What exactly would constitute an ecological civilization, and how would it be different from the kinds of civilizations in which the lives of human beings are currently imbedded and organized? Here is a convenient definition, offered by Philip Clayton and Wm. Andrew Schwartz (keeping in mind that the kind of civilization they define is one that does not yet and may never exist):

> *What most clearly distinguishes an ecological civilization from other civilizations is the relation to the natural world. Whereas a defining characteristic of other civilizations is the manipulation of the environment for the sake of the people, an ecological civilization modifies the environment in sustainable and symbiotic ways, considering the well-being of nature as well as people. This is not a romantic ideal of a return to the pristine purity of nature; it represents, after all, a*

form of civilization. Not simply humans living in harmony with nature, ecological civilization also entails humans living in peaceful ways with other humans, to promote the flourishing of all life. (Clayton & Schwartz, 2019)

To which I would add the following helpful statement by Jeremy Lent:

In some ways it's preferable to define the term "ecological civilization" by the principles that underlie it, rather than by specific policies and actions. I generally define "ecological civilization" by simply saying that it involves the recognition that our current civilization itself needs to be transformed. Prior civilizations were based on wealth accumulation, extraction, and exploitation. It's that basic feature that needs to be transformed. An ecological civilization is a civilization that strives to achieve the conditions for all people to thrive on a regenerated earth. (Lent, 2024)

Any talk of a global transformation is bound to raise a suspicion of utopian thinking, and there is no reason to deny that a certain utopian impulse is present in all attempts to envision a just and sustainable human future. But an appreciation of the need for civilizational change is based on what is in fact an honest appraisal of the global situation. Thus, according to Clayton and Schwartz, the vision of an ecological civilization

begins with a harsh realism: the goal of long-term sustainability cannot be achieved within the context of global civilization as it currently exists... The pursuit of an ecological civilization then requires a second and equally demanding form of realism. It mandates concrete reflection on specific economic and social systems: forms of housing and transportation, of production and consumption, of government and business, even of religion and art, that will be genuinely sustainable over the long term. (Clayton & Schwartz, 2019)

In short, it is the realism not the utopianism of the idea that requires it to embrace so many areas of reflection and, in light of that reflection, to affect a vast range of both personal and collective policies. We might not go all the way with Kate Aronoff's recent bold statement that "All policy... – from housing to trade to financial regulation – is climate policy" (Aranoff, 2024). But it is true that the effects of climate change are so pervasive that an adequate response to those effects, certainly over the longer term, is likely to leave no area of policy untouched.

The pervasiveness and the civilizational scale of its implications are what makes climate change an unusually and perhaps uniquely powerful subject of cross-disciplinary courses and curricula. (Keeping in mind, of course, that the health of cross-disciplinary work depends on the health of the individual disciplines it brings into relation with each other.) And it's the urgency of responding to climate change that, in my view, gives institutions of higher education a responsibility to place the study of climate change at the core of their missions. The joke in Stanley Fish's title–*Save the World on Your Own Time*–lay partly in the apparent absurdity (as recently as 2008!) of supposing that saving the world was a sensible pursuit, in or out of the classroom. After all, it was not quite so obvious, even that recently, that there was anything to save the world *from.* But now the threat is obvious indeed, and none of us have the luxury any more of regarding the world as too big to save. If the purpose of higher education is such that it excludes participation in efforts to ensure the flourishing and even the survival of human civilization, what exactly is that purpose?

However challenging it may be to implement consistently and without crossing an ethical line, the distinction between partisan political advocacy and classroom teaching on subjects with sometimes inescapable policy implications is clear enough in principle. I have argued that teaching subjects with inescapable implications for policy is justified in general, and I have gone on to argue that teaching and learning about the global climate crisis are exceptional and perhaps unique in the range and exigency of their policy implications. With the partial exception of programs offered entirely online, however, an institution's interaction with students extends beyond classroom instruction and academic assignments. What place, if any, does climate activism have in the extracurricular opportunities an institution offers?

Countless studies have shown that, within certain limits, student engagement and academic performance are enhanced by participation in extracurricular activities. To cite one example, the most recent biennial survey on the subject administered by the University of California's office of Institutional Research and Academic Planning found that extracurricular participation was "positively related to higher sense of belonging, better academic satisfaction, higher GPA, and higher first-year retention," (Universtiy of California, 2023).

If I am right about the legitimacy and urgency of formal coursework on climate change, I would apply a version of the same argument to providing students with opportunities to participate in climate-related activities outside the classroom. In my view, colleges and universities should not be deterred from supporting, encouraging, and indeed from guiding—that is, essentially, *coaching*—student participation in activities ranging from global advocacy to community organizing around issues like environmental justice. The kind of cross-disciplinary exploration that enriches a student's intellectual life in the classroom can find its extracurricular parallel when a student works with a local community on, for instance, uncovering the history of federal and local housing policies, architectural designs, transportation planning, and entrenched poverty that have turned certain neighborhoods in so many American cities into treeless and lead-polluted "heat islands," where the too-often deadly effects of climate change are concentrated.

Does this mean that institutions have an equal responsibility to support student actions *against* climate activism, or *against* the pursuit of climate justice? Here I return to my earlier argument that an institution of higher education—or any institution, for that matter—has not only a right but a responsibility to act on the basis of the best information it has. And that right extends, in my opinion, to the selection of what kinds of extracurricular activity an institution chooses to support.

Now I can't imagine an institution actually forbidding a student to join a climate-denying organization, or to serve it as a volunteer or even as a paid intern. I suppose an institution that offered academic credit for, say, written work emanating from an internship might wonder whether to give such credit to a student who had interned with an organization it had reason to regard as scientifically illiterate, or as deliberately and knowingly engaging in the promotion of misinformation about what is raising sea levels, driving migration from increasingly drought-prone regions, or increasing the frequency of devastating weather events. But I would hope that such decisions would be made in an academically responsible way, weighing the academic merit of the student's work, independent of the student's personal allegiances or of the organizational context from which the work emerged.

Such instances aside, I strongly affirm the right of an institution to choose the extracurricular activities it is willing to endorse, and to supply those activities with material as well as coaching, advice, and support. And, given what I argued above about the civilizational implications of climate change, I would go further in the case of student participation in work addressing the challenges of climate change as they arise concretely and practically in the communities in which institutions and their students find themselves.

I have argued that addressing the climate crisis through courses and curricula that bring students into the project of civilizational change is a valid focus of higher education. Equally valid in my view are efforts to supplement the offering of courses on climate issues with extracurricular activities through which students participate in the efforts of institutions and their surrounding communities to reduce their carbon dependency, to enhance their resilience to the effects of climate change, or to mitigate the effects from which they may already be suffering.

In many cases, including all of those of which I am personally aware, institutional support for extracurricular activities related to climate change is imbedded in broader sustainability initiatives focused on such issues as food waste reduction, water conservation, and urban agriculture—all of which, however, sooner or later have climate-related implications. Addressing some of them reduces a campus's or a community's carbon footprint, while addressing others increases resilience to the damaging effects of climate change. Reducing the amount of waste that ends up in landfills reduces $CO2$ and methane emissions; conserving water mitigates drought; growing vegetables on or near a campus reduces the need for transportation from remote locations.

The results of all such efforts are likely to be modest in comparison to the global scale of the challenges in question, but, as students see and celebrate the measurable impact of their efforts, their achievements play an important role in motivating participation in the wider climate movement. They also serve an educational purpose, both by providing students with concrete illustrations of how the various causes and effects of climate change are interconnected and by enhancing students' engagement in ways that, as mentioned above, contribute to their academic success as measured by factors like GPA and retention.

Furthermore, there is no law requiring a rigid separation of student activities within and outside the classroom. Institutions can also make deliberate and explicit connections between the two through service learning and community-based research and teaching. It's important to note that the value of connecting academic research and teaching with community-based action is not confined to the subject of climate change. There are numerous precedents in academic responses to other concerns, for instance in the field of public health, where community-based participatory research has long played a significant role.

California State University Northridge (CSUN) offers an impressive example of an institution can has successfully imbedded extracurricular climate activities in a broad approach to sustainability while also forging explicit connections between classroom and community. Under the leadership of President Emerita Dianne Harrison, who signed the American College and University Presidents' Climate Commitment in 2013, and with the ongoing support of current President Erika Beck, CSUN has developed a wide range of sustainability programs, supported by three distinct but interrelated departments:

Facilities Planning, Design, and Construction works on infrastructure and energy management, the Institute for Sustainability leads academic and community education efforts, and Associated Students Sustainability organizes student-centered events and activities. Together, they form a comprehensive approach to fostering sustainability across campus, (University of California, 2023).

One example of how climate action is imbedded in CSUN's sustainability programs is the university's Zero Waste Program, which is

> *dedicated to diverting at least 90% of campus waste from landfills through recycling, composting, and waste reduction efforts. We actively engage students in this initiative through tabling events, class presentations, campus events, tours, and hands-on waste audits. Additionally, students participate in class projects where they monitor campus waste streams using innovative technologies, gaining practical experience in waste management and sustainability. These activities not only raise awareness but also empower students to play an active role in achieving our zero waste goals, making sustainability a collaborative effort across campus*[3].

The program encompasses an annual event, Matador Move Out, in which student and community volunteers collect "clothing, non-perishable food, and other miscellaneous items" donated by students vacating their dormitory rooms and distribute them to a variety of campus partners and community service organizations. According to the university, in 2024, the donated items amounted to 10,850 pounds of material, diverting it from landfill and preventing 2,712 pounds of climate-warming CO2 emissions[4]. Again, a small impact when measured again the tens of billions of tons emitted globally every year, but that in itself is an educational lesson in the magnitude of the challenge! In any event, the strictly extracurricular activity of collecting and diverting potential waste provides a meaningful hands-on complement to students' classroom work on monitoring waste streams, etc.

Programs like the ones developed by CSUN once again raise the question of how far institutions of higher education can or should go in integrating classroom work with efforts to address the challenges faced by the communities in which they reside, as well as (if only indirectly or symbolically) the larger challenges faced by human civilization in general. To my knowledge, the organization that is most ambitiously and systematically pressing that question, within and beyond the United States, is Educational Partnerships for Innovations in Communities—Networks (EPIC-N), (Canva.com, 2024).

EPIC-N traces its routes to a partnership launched in 2009 by two professors at the University of Oregon, who asked the government of the nearby city of Salem "to underwrite the costs of coordinating interested university faculty—and their classes—in a concentrated, time-limited effort to work on many of Salem's long list of identified problems."[3] According to its website, the network now supports more than 70 programs, has established a "fledgling EPIC-Africa Network" and has plans to expand into Asia, Latin America, and the Caribbean, (epicn.org).

Among the impressive features of "the EPIC model" are its reliance on existing courses, which reduces the friction and overhead involved in creating and seeking approvals for new curricula, and its intentional brokering of connections between the real-world needs of city managers and faculty members who, if they wish, can choose to focus their own expertise and the efforts of students in their classes on developing actionable responses to those needs.[5]

The Epic model provides a timely approach to applying the expertise of faculty members and the intellectual energies of their students to the challenges of urban planning. As one article on the EPIC approach puts it, "We are entering a prime time for planning—climate change demands our cities be organized differently, both to reduce our carbon emissions and to be resilient as natural disasters and seawater rise accelerates [sic]," (epicn.org). I would add to that list the need for planners to find ways of mitigating the extra severity of climate effects on marginalized communities. I would also add that connecting classroom learning with the daunting complexities of real-world problem-solving is not a diversion from serious academic work but a way of acknowledging what it means to be serious, academically and otherwise, in a civilizational crisis like the one we are undeniably, facing.

CONCLUSION

I opened this chapter by citing a prominent critic who regarded academic efforts to "save the world" as a diversion from the true mission of higher education—a dereliction of duty and even a kind of betrayal. This is not a criticism to take lightly, but, having acknowledged that burden, I replied to the criticism in two ways.

First, I pointed out that some subjects cannot be taught at all without raising inescapable policy implications. It's not clear that teaching those subjects is any less consistent with a traditional conception of the mission of higher education than teaching subjects that lack such implications. But I then moved on to a more radical as well as sobering argument: namely, that we are living in a moment—but a moment that could well last several centuries!—that compels us to reconsider higher education's mission itself. A question, however, perhaps lingers: how far is it appropriate for a college or university to go in recruiting students into efforts that, some would argue, are not directly related to their primary role as students?

I still remember how moved I was, many years ago, when I was admitted to the university of my choice with words welcoming me "to this community of scholars." As quaint as that phrase may sound at a time when the functions of a university have expanded in so many different directions, it provided the initial motivation of what became a career of some five decades of teaching and administration. What it meant to me then, and continues to mean now, is that a student's education, really at all levels, is not just an opportunity to accumulate knowledge or develop one's intellectual skills but an opportunity to participate in a shared enterprise—an enterprise that, ideally, is both a symbol and an instrument of the larger enterprise of human civilization as a whole. I see no reason to place a limit on how far students should be encouraged and enabled to participate in the work of advancing what now needs to be done to preserve that civilization and, moreover, to do so by transforming it into a civilization that, again in the words of Jeremy Lent, "strives to achieve the conditions for *all* people to thrive on a regenerated earth."

REFERENCES

Aronoff, K. (2024, September 30). Climate change does not care about your borders. *The New Republic.*https://newrepublic.com/article/186526/hurricane-helene-asheville-climate-havens

Clayton, P., & Schwartz, W. A. (2019). *What is ecological civilization: Crisis, hope, and the future of the planet.* Process Century Press.

Crownhart, C. (2024, March 13). Methane leaks in the US are worse than we thought. *MIT Technology Review.*https://www.technologyreview.com/2024/03/13/1089725/methane-leaks-oil-gas/

Educational Partnerships for Innovation in Communities Network. (n.d.). *EPIC-N.*https://www.epicn.org/

George Washington University. (2008, June 1). *GW Task Force on Sustainability: Final report and recommendations* (p. 6). https://sustainability.gwu.edu/sites/g/files/zaxdzs5666/files/downloads/taskforcereportandrecommendations%20%281%29.pdf

Harvard University. (2024, May). *Report on institutional voice in the university* (p. 1). https://provost.harvard.edu/sites/hwpi.harvard.edu/files/provost/files/institutional_voice_may_2024.pdf

Lent, J. (2024, September 5). Revised transcription of a comment recorded (with permission).

Matador Move Out. (2024). *Impact report.*https://www.canva.com/design/DAGF_QjU7bE/STfXaHA5UcqU29ivA53p1Q/edit

University of California, Institutional Research and Academic Planning. (2023, August). *Extra-curricular participation and student success.*https://www.ucop.edu/institutional-research-academic-planning/_files/extra-curricular-participation-and-student-success.pdf

ENDNOTES

1 In response to conflicting pressures on this issue, Harvard University has followed the example of several other institutions in announcing a policy based on the principle that "The university and its leaders should not... issue official statements about public matters that do not directly affect the university's core function." At the same time, [T]he university has a responsibility to speak out

to protect and promote its core function. Its leaders must communicate the value of the university's central activities. They must defend the university's autonomy and academic freedom when threatened – if, for example, outside forces seek to determine what students the university can admit, what subjects it can teach, or which research it supports. And they must speak out on issues directly relevant to the university's operation.

2 Of course they may be deterred by other things—for instance, legislation if they are public institutions.

3 From a summary of CSUN's sustainability programs by Aspen Coty, Austen Eriksson, and Sarah Johnson, provided to the author by courtesy CSUN President Erika D. Beck.

4 Ibid. See also the video on CSUN's Zero Waste webpage, https://www.csun.edu/zero-waste/pledge/zero-waste.

5 See https://www.epicn.org/about/.

Chapter 4

Advancing Climate Health Through Campus Civil Discourse and Course Integration:

Using the College Campus as a Catalyst for Promoting Holistic Sustainability

Mark Fincher
Mississippi State University, USA

ABSTRACT

Colleges and universities in the United States have a unique capability of convening important discussions about difficult topics. Nowhere else in modern society can people discuss challenging issues in an open forum in a civil manner. This capability is based in both the history of higher education, and in the rights and traditions of free speech in the United States. It is supported by promotion of thought and creativity being a major function of higher education.

HIGHER EDUCATION INSTITUTIONS AS CATALYSTS FOR CHANGE

The value of an advanced education arises from more than just learning facts. It also entails developing the ability to operationalize the acquired knowledge so that it can become valuable to the learner and to society as a whole (Dewey, 1916). A

DOI: 10.4018/979-8-3693-5454-4.ch004

foundational component of the American system of higher education is collegiality. Collegiality literally means a sharing of authority and a comradery. Learning to create, share ideas, gain understanding, and work together are cornerstones of the American collegiate experience (Rudolph, 1962). Students who gain these capabilities are helpful to society and highly valued by employers (VanDerziel, et.al., 2024). Colleges and universities should always be looking for opportunities to educate students beyond the classroom. This is what separates a truly valuable collegiate experience from a limited one (Cohen & Kisker, 2010).

Higher Education and Sustainability

Colleges and universities seek to equip students in creativity, communication, and civil discourse (Kisker, et.al., 2023). These characteristics have contributed to wholesale progress in the condition of mankind. The impacts that these institutions can have are both short-term and long-term. Colleges and universities in the United States have provided, through generating both human capital and new knowledge, the foundation for the modern world. This progress has been centered in the developed world but radiates throughout the planet.

Despite this progress, there is no topic facing mankind that is more difficult or in need of civil discourse and collective action than climate sustainability. The health of the environment is complex and global in nature (Nguyen, 2023). Any solution to the challenges that are currently being faced will be so as well.

While college and university faculty and administrations should be looked to for leadership, scholarship and innovation, the driving force behind environmental action on U.S. campuses must be the student body. Where do college students stand on the issue? Whether it is current college students, recent applicants or future generational cohorts, young Americans care deeply about the environment.

A study conducted between July and November 2023 surveyed individuals aged 16-25 across all 50 U.S. states. Findings revealed that 85% are at least moderately worried, and 57% are very or extremely worried about the climate crisis (The Guardian, 2024).

As for high school students applying to college, a 2023 survey by The Princeton Review found that 66% of responding prospective students would value information about a college's commitment to environmental issues (The Princeton Review, 2023).

And college students want to participate in constructive dialog on pressing issues such as the environment. One recent survey showed that 94% of students believe their schools should adopt programs promoting civil discourse and active engagement (Flaherty, C., 2024).

How about the next generation of college students? The first cohort of "Generation Alpha" is three years away from starting college. A 2024 nationwide survey by the Museum of Science in Boston found that 72% of middle and high school students believe climate change is already impacting their lives, and 53% expect it to be a major problem during their lifetime. Notably, 72% also believe their generation will find solutions to address climate change, and they desire careers that contribute to environmental preservation (PR Newswire, 2024).

Based on the data, there appears to be significant interest by current, incoming, and future college students around the issue of environmental sustainability. Equally important is a very strong interest in using their time on college campuses to discuss and take direct action surrounding those central issues.

The good news is that campus administrators are in lockstep with their students. On the issue of environmental stewardship, a significant number of college presidents have made a commitment to reducing carbon emissions, a key metric in addressing climate change.

One higher education network devoted to mitigating climate change is Second Nature, whose Climate Leadership Network is made up of more than 400 higher education institutions committed to acting on climate change. "Members of the Climate Leadership Network gain access to a diverse range of institutions working on various climate solutions, including operational decarbonization, climate justice, community engagement, workforce development, and climate research." (Second Nature, 2025) Second Nature has also taken over management of the American College & University Presidents Climate Commitment, which has been signed by well over 600 presidents (Cortese, 2010). A key metric in the climate commitment is to reduce institutional carbon emissions to a net zero level (Medlin, 2008).

One element of a global interest in lowering carbon emissions is the impact of gas-powered automobiles. The transportation sector accounts for approximately 21% of global CO_2 emissions, second only to the power generation industry. Passenger vehicles make up half of that figure (Fleck, A., 2023).

One response to the fossil fuel issue is the electric vehicle market, which is predicted to continue to advance rapidly in the coming decades. One source suggests that within five years, up to 45% of new vehicles sold globally will be electric, with 15% of all vehicles on the road being electric. Within 20 years, that figure will be 95% of all new car sales.

While for years the acceptance of climate change as a national priority and carbon emissions from energy production and transportation as a primary cause is generally stated as a consensus within the academic community, public opinion on environmental sustainability and climate change is mixed. For instance, a recent national poll showed that of 21 perceived possible priorities for Congress, climate change was in the bottom five. (Pew Research Center, 2023). Although the general issue of climate change appears to be correlated to partisan politics (59% of Democrats consider it a national priority while only 13% of Republicans do), individual strategies are less partisan in gaining public approval. For instance, improving transportation, improving the energy system, and protecting the environment in general were rated as a priority by more Americans, with improving the energy system being a priority of over half the respondents. Fully 67% of respondents favor renewable energy sources such as wind or solar over fossil fuels. And young adults (under 30) are most likely to support environmental sustainability initiatives (Pew Research Center, 2023).

Data cited earlier in this chapter makes clear that current, incoming and future student cohorts want to attend college at an institution that is committed to and working to positively impact climate change and environmental sustainability. Fortunately, colleges and universities already possess the missions, mechanisms and resources to promote a more holistic and practical approach to this challenge (Fullan, 2005). The goals of both the institutions and their students can be served through an emphasis on promoting the overall health of the planet. Colleges and universities are expected to produce graduates who can perform advanced intellectual functions and contribute to society. Employers routinely look for people who have technical knowledge and skills, but can also communicate, create, and work in groups (NACE, 2014). A knowledge of international issues and diverse cultures is also appreciated. Promoting discussion of the health of the environment can support these goals. This is a critical aspect of anything that is done on campus, as students need to clearly see the value of what is being offered and required by the institution.

Unlike with generations past, the current students or their families pay the great majority of the cost of their education. Consequently, merely becoming a better-informed global citizen may not be a sufficient return on their sizable investment (Fincher & Katsinas, 2017). They have every right to be discerning about what they pay for and resist supporting activities that do not benefit them directly. However, becoming a better citizen while gaining skills and knowledge that last a lifetime can be very attractive to many students. This can be achieved by getting students involved in the protection of the environment from the ground up (Cairo, 2023).

Communicating, creating, cooperating, and becoming culturally competent can cause a student to become much better able to negotiate the modern work world (NACE, 2014).

The mission of most colleges and universities is not to tell students what to think or do. That is not the prerogative of higher education. It is instead to prepare them to be formidable and valuable citizens and equip them to learn to think and to discern what to do (Campbell, 1995). The purpose of this chapter is to equip leaders of colleges and universities with actionable ideas on how to develop better and more capable college graduates who can build a healthier world.

Sparking the Interest Throughout

Few Americans of any age understand their role in the condition of the natural environment. Access to food, water, air, and power are critical for civilization, yet most Americans do not know how they get these resources or how they impact the environment (Cohen, 2014). While many strategies and tactics related to the health of the environment are issue specific, some radiate throughout the entire institution (Fullan, 2005). Most colleges and universities have faculty members with great expertise on related issues and subjects, and the potential interest of the students is almost universal. Exposing students to the knowledge of faculty members on the origin of what they consume and the place/people that produce it can help them become interested in protecting these resources. This can be largely furthered by promoting a holistic examination of the health and sustainability of the environment.

Environmental sustainability has many aspects, most of which get little consideration. Emissions from automobiles get the most attention in the United States, but many researchers consider other issues to be greater threats to sustainability of the environment and the corresponding wellness of mankind. This chapter shows higher education leaders how to develop students by exposing them to critical issues threatening environmental sustainability.

Maintaining The Classical Elements: Earth, Wind, Fire, and Water

To improve the health of the natural environment through the actions of colleges and universities, it is necessary to define specific challenges to environmental health and develop strategies and tactics to involve students. As this chapter is intended to give practical guidance about how to effectively engage students in holistically improving the health of the environment, this can be best done through showing specific environmental challenges and practical tactics that colleges and universities can use to facilitate their students making a difference on these issues.

Exploring the Health of the Environment Through the Classical Elements

Colleges and universities have long examined the natural world. One of the earliest ways of viewing nature was through the lens of the four classical elements. It is fitting that these environmental challenges can be broken into the four classical elemental categories of Earth, Wind, Fire, and Water (Ball, 2004). For each of those categories, addressing a major environmental challenge with institutional strategies and tactics could create a gameplan for making a difference in the health of the environment. Just as the classical elements interact, the critical issues impacting the health of the environment also interact. Consequently, the greatest impact of these activities may be achieved when institutions address the health of the environment in multiple ways. For Earth, a related issue could be consumer waste. For Wind, deforestation. For Fire, power generation. For Water, microplastic and chemical pollution.

Each of these elements (Wind, Fire, Water, Earth) could frame many other important environmental issues to address. Each campus and each faculty member must decide what makes sense—given their geographical location, mission, student body and other factors—to address.

Earth: Consumer Waste

Consumer waste is a problem that everyone can directly influence. You can reduce waste from to-go containers by not ordering portions that are too large, going to fewer fast-food restaurants, or going to places that use environmentally friendly takeout containers that are recyclable or biodegradable. Another popular approach being used on college campuses are trayless dining halls, where students only eat what they can carry to the dining table. This cuts down on food waste, which amounts to 24% of all landfill waste, according to the EPA, generating most of the methane gas that is emitted into the air. (EPA, 2023)

Using products that are easy to recycle is another way to cut down on waste. This can be a very helpful approach with clothing. When purchasing clothing, choosing natural materials like cotton or wool can be far more friendly to the environment. Most synthetic clothing materials tend to break down overtime and may end up in landfills or burned. When materials like polyester are burned or placed in landfills, toxic gases can be released as well as carbon dioxide (The Sustainable Fashion Forum, 2023). Cotton may be more expensive, but it lasts much longer, so clothes don't have to be replaced as frequently. The materials are also easy to reuse after the clothing eventually does wear out or get damaged.

While some farming practices can be unhealthy for the environment, cotton can be very advantageous to the environment if it is produced sustainably. Deforestation occurs to clear away trees to make farmland and this can have dire consequences. Sustainable farming, however, is farming without causing too much damage to the land or exhausting the soil (USDA, 2024). That way, people do not have to continually cut down and burn down forests. Organic farming is similar, but it also avoids the use of synthetic chemicals and GMOs to lessen pollution. Food products that are organically farmed or made with organically farmed ingredients are labeled and certified as organic. Looking for those labels and buying food from stores that often carry organic foods are good ways to make better choices when grocery shopping.

While one person cannot impact the environment as a consumer of goods, an abundance of people can act together with dramatic effect. The purchasing habits of consumers drive many manufacturing decisions and processes. Mass manufacturing is a boon to mankind, but some approaches and processes are much more supportive of a healthy environment than others. It helps to avoid using too much power because it can decrease the need for fossil fuels and help save money. Disposable single use items, like bottles, shopping bags, and takeout containers, are discarded in the trash and put in landfills (Vasarhelyi, 2023).

This waste can cause substantial damage, in addition to filling the limited capacity of landfills. Many of these items are made of materials like certain Styrofoam, plastic, and other materials that create a lot of waste. Some materials can be recycled, like metals, glass, paper, and certain plastics (EPA, 2023). This leads to a commercially viable industry for recycling. Others cannot be recycled, like Styrofoam, and non-recyclable plastics. Some of the recyclable plastics aren't as easy or profitable to recycle and often just end up in landfills or as pollution (Fire Dawgs Junk Removal, 2019). Metal cans are easier to recycle than plastic bottles, so using them can help reduce waste. It helps to use reusable goods, like cloth bags, reusable water bottles, and Tupperware containers in place of single use items. Many products have different wastes from use or production. Some have little waste and are reusable, but disposable single use items, like bottles, shopping bags, and takeout containers, are discarded in the trash and put in landfills (UNEP, 2024). Much of this single-use material can cause substantial harm to the environment beyond the Earth and its landfills. The production and disposal of single-use plastic materials, for example are projected to produce almost 20% of the Earth's airborne carbon by 2040 (UNEP, 2024).

Some of the recyclable plastics aren't easy to recycle and often just end up in landfills or as pollution (Cho, 2020), but this could be changed by better organization and collection procedures. This could also be changed by using plastic materials that could be cost-effectively recycled and avoiding using the materials could not (Cho, 2020). Not only can recycling materials improve the quality of the soil, but it can also save money reduce the need for power that would be used by manufacturing entirely new products. It helps to use reusable goods, like cloth bags, reusable water bottles, and Tupperware containers in place of single use items. You can reduce waste from to-go containers by not ordering portions that are too large, going to fewer fast-food restaurants, or exclusively going to places that use more environmentally friendly takeout containers that are recyclable or biodegradable. One strategy gaining popularity among college students is the availability of water refill stations across campus so that reusable water bottles replace the use of disposable plastic bottles. Since 2013, the University of Vermont has banned bottled water on campus. The school installed more than 100 water bottle refill stations to support reusable bottle use. Other universities with similar initiatives include Washington University in St. Louis, University of California system campuses, and Harvard University, though UVM is lauded for fully banning bottled water sales (Wakefield, J. 2012).

Using end products, rather than just materials or packaging, that are easy to reuse or recycle is another way to cut down on waste. This can be a very helpful approach with clothing (Trzepacz, et.al, 2023). When purchasing clothing, choose natural materials like cotton or wool. Most synthetic clothing materials tend to break down overtime and may end up in landfills or burned. On college campuses, one popular approach is to create a "thrift shop" on campus where students can swap "gently used" clothing and cut down on new clothing purchases.

When burning materials like polyester, toxic gases can be released as well as carbon dioxide (Le, 2020). Cotton may be more expensive, but it lasts longer, so clothes don't have to be replaced as frequently (Trzepacz, et.al, 2023). While not all farming practices are healthy for the environment, even to the point of using deforestation to clear trees to make farmland, farming can be done in a way that is even beneficial to the environment. Sustainable farming is farming without causing damage to the land or exhausting the soil (Boyle, 2024). This can protect the Earth by better managing chemical exposure and nutrient levels. This can remove the need for more farmland through deforestation by reviving and rejuvenating the existing farmland. Organic farming is similar to effective soil management, but it also avoids the use of synthetic chemicals and GMOs to lessen pollution. Food products that are organically farmed or made with organically farmed ingredients are labeled and certified as organic. Consumers that purchase organic foods not only experience potential personal health benefits, but they also influence consumers to introduce less toxic waste into the environment.

Wind: Deforestation

The gradual destruction of the Amazon Rainforest is considered by many researchers to be the greatest single threat to global environmental sustainability (Baker & Spracklen, 2019; Rain-Tree, 2024). On the surface, this seems unlikely since it is an isolated region with a relatively small population. In practice, it has an outsized impact on the entire planet. This can easily be recognized through the multiple nicknames given the rainforest. Some climate scientists refer to it as the air conditioner of the world, as it consumes around 30% of the carbon dioxide that is absorbed from the atmosphere (Zimmer, 2019). It is also known to some as the lungs of the Earth as it produces around 9% of the planet's oxygen (Georgiou, 2019). The Amazon Rainforest is even known as the pharmacy of the Earth (Holland, 2021) as it is the most prolific source of medicines in the world. This striking global impact is largely driven by the trees and vegetation and the year-round growth and photosynthesis that occurs in the region.

Unfortunately, these positive impacts may have diminished by a third since 1990 due to mass clear cutting of the rainforest (Haywood, 2024). These continuing losses are exacerbated by the fact that the lost rainforest is largely burned, releasing perhaps a billion tons of carbon dioxide into the atmosphere. When combining both output and reduced use, the Amazon Rainforest has gone from being the largest net reducer of airborne carbon dioxide in the world to being the larger producer (Gatti, et.al., 2021). The carbon dioxide addition to the atmosphere is largely a one-time event that could be stopped, but the loss of carbon dioxide consuming rainforest is permanent without slow and costly intervention (Spring, 2023).

The Amazon Rainforest does not exist in a vacuum. It has inhabitants that are citizens of several countries with differing cultures, economies, and governments. It receives little to no foreign direct investment and gains virtually no benefits from industrialization. This complex and challenging cultural and economic reality creates a situation where natural resources with tremendous value to the world but little for those living there could be logically replaced with low quality pastureland that provides some value to the inhabitants (Baker & Spracklen, 2019). The annual massive reduction of the rainforest also appears to have a major impact on the economy and livability of faraway northern Africa. There is a complex relationship between the rainforest and the Sahara Desert (WWF, 2024). The downstream effects of reducing the size and impact of the Amazon Rainforest could directly impact people of different countries and cultures in Africa in ways that are yet to be fully understood. When taken in combination, these complex factors and the critical nature of the outcome, the Amazon Rainforest makes a great topic for exploration and discussion by college or university students and faculty.

Fire: Generating and Conserving the Power That Drives Mankind

The generation and use of power is critical to the prosperity of mankind. Most advances of modern civilization are built on a foundation of consistently and widely available power. This availability of power is critical for quality of life throughout the world. Much of this power is delivered in the form of electricity. Electricity can be produced in different ways, some of which can cause more air pollution or hazardous waste than others. A critical factor in the availability of electricity is its usage and conservation. Inefficiency and waste of electricity drastically increases the need for power production and the corresponding pollution that it would produce.

Fire, or combustion, is a requirement for power generation with modern needs and technologies. Combustion is the base power generation method in almost all countries. This is partially because burning fuel can be used to predictably generate electricity 24 hours a day, 7 days a week (EPA, 2024). Many ideas exist for emission free power generation, but yet none of them are practical on a large scale. With the exceptions of nuclear energy, geothermal energy, and hydroelectric energy, power cannot be reliably generated without combustion. At present, geothermal and hydroelectric power are at near practical maximums of production and the political and societal will for mass expansion of nuclear energy production is lacking. When producing energy through combustion, many fuels can be burned to produce the same electricity, but the impact on the environment varies sharply by the type of fuel that is burned.

The burning of coal has had a profound positive impact on mankind. In many ways coal is a perfect fuel source for generating electricity. It is very simple to use and is extremely effective at producing electricity on demand. Coal-fired energy production was critical for the creation of the modern world. Unfortunately, coal fired energy production places a massive burden on the health of the environment. Other than burning wood, coal produces more carbon emissions than any other type of fuel relative to energy output (EIA, 2023). It is possible that new methods could be developed to burn coal in a more environmentally friendly manner, but efforts do so have had very limited success. Carbon capture technology exists that can drastically reduce carbon emissions, but at present this is too costly to be a large-scale solution. Coal, despite all its advantages, remains a terribly flawed power source for generating electricity.

With current technology, other fuels are far more friendly to the health of the environment. Natural gas, by its chemical makeup, produces cost-effective electricity with little more than half the emissions of coal. It also has the capabilities of producing energy in large amounts at any time and being quickly started when needed (IEA, 2023). Consequently, needed electricity can be produced using natural gas consumption with far less carbon emissions than by burning coal. This is far from a zero-emission power source, but it produces much lower emissions than the dominant coal burning methods (IEA, 2023). Other technologies and fuel sources show some promise, such as the incineration of solid waste. Unfortunately, current technologies lack either cost-effectiveness or sufficient reduction in carbon emissions.

It would be advantageous if mankind could function without burning fuel for power generation, but that is simply not possible with current needs and technologies. Eliminating combustion would require mass rationing and disruption of power availability. This would greatly reduce the production of energy critical goods, and this impact would largely be felt by the less affluent (IEA, 2024). If given no alternative to going without dependable and affordable electricity, countries will likely expand the number of people using the least clean energy sources, such as burning coal or wood (IEA, 2024). Consequently, any attempts to improve the health of the environment through the management of energy production must be based in providing viable and affordable power generation that meets the needs of all countries.

The demand for and uses of electricity have increased dramatically throughout the 21st century. In developing countries, industrialization and increased quality of life have led to rapid rises in the demand for electricity. The globalization of the economy has led to additional needs for both power and electricity. This is driven by both production of goods and their transport. In economically developed countries, modern conveniences use massive amounts of electricity in ways that would not have previously been imagined. Every smart device contains a computer that uses electricity. Artificial intelligence (AI) is the most extreme example of this in that it replaces human thought using massive amounts of computer power and electricity. Cryptocurrency generation is a new phenomenon that also uses massive amounts of electricity (Wilson & Zimmerman, 2023). It exists in digital form and is mined through a completely digital and therefore electrical process. Finally, air conditioning homes and businesses and the use of electric vehicles has grown rapidly with no sign of slowing down (I.E.A., 2018).

With so many expanding uses of power, there are also many opportunities to reduce its negative impact on the environment and its usage. Every process of power creation or usage can potentially be made more efficient. Tasks that are more frequently done by computer can instead by done by the human mind. Products and materials can be moved locally or regionally instead of globally. New technologies can be developed to make massive zero-emission power generation possible or the burning of fuels more friendly to the environment. Many options exist for improving the health of the Earth's atmosphere, but this will not happen without effort.

Water: Microplastic and Chemical Pollution

Water quality is an important part of protecting the environment because every living organism needs clean water to survive. When water gets polluted, it can be unsafe for the plants and animals that consume or live in it. When rainwater lands on the ground it can seep into the soil, or flow into bodies of water. Pollutants that dissolve in water, such as chemicals and microplastics can contaminate soil and poison plants and animals (Pelton, 2014). Runoff can pick up all kinds of pollutants such as trash, sludge, fertilizers, loose soil, and chemicals. Trash and chemicals can poison or injure organisms. Loose soil and fertilizers can cause eutrophication, murkiness, and excessive microbes.

While these problems are very concerning, the greater worries for water quality are both literally and figuratively downstream. Water contamination is an immediate challenge, but it may have much greater long-term effects. Lakes, oceans, and topsoil may become compromised, and their ability to support animal and plant life may be diminished. New challenges in water contamination due to new innovations have led to additional worries about the health of the environment. The most disturbing of these sources of potential damage may be microplastics.

Microplastics. Plastics can break down into tiny plastic particles that are far too small for people to notice (UNEP, 2015). These are known as microplastics. Microplastics can mix into water and soil and enter the bodies of organisms. They have toxic chemicals in them that can not only contaminate any organism that consumes them but also poison whatever eats the contaminated organism. Microplastics can harm people and animals by causing damage to DNA, metabolic disorders, disrupting organs, and many other ways (Barber, 2024). Microplastics that contaminate water can mix into the soil or get washed into bodies of water through runoff. A lot of microplastics can be found in the great Pacific garbage patch. Microplastics have even been found in human breast milk and waste, which shows someone has either drunk contaminated water or eaten contaminated food.

A particularly concerning aspect of microplastics is that it is unknown how these particles can impact the human body. It is possible that these particles could impact human functions as diverse as mental health and fertility. The great increase in microplastic exposure seems to have coincided with both an increase in mental disorders and a decline in fertility (Longman, 2004). The decline in fertility may be particularly concerning. Beyond the global trend of people wanting fewer children, there is a substantial drop in the physical ability of young people to father or bear children (Freeborn, 2024). It is unknown exactly why this occurred or if it will continue. The potential universality of a growing amount of microplastics in the water supply creates the possibility of whatever impact these particles have being very widespread.

Pseudo-estrogens. In addition to microplastics, it has also been suggested that these problems with fertility could be related to another type of water pollutant that can travel in water: pseudo-estrogens. These are chemicals with a variety of valuable commercial uses that have a similar chemical makeup to estrogen, or other hormones found in the healthy human body (Bitner, et.al, 2014). It is hypothesized that the human body may not be able to effectively distinguish between its own hormones and similar manmade pseudo-estrogens, such as those classified as xenobiotic chemicals. A similar phenomenon has occurred when people take performance enhancing drugs along the lines of anabolic steroids. These artificial hormones influence the otherwise healthy body to both increase muscle production and reduce natural hormone output (Hackney, 2018). Common symptoms of this hormone change include the taking on of secondary sex characteristics of the opposite gender. The long-term effects of these substances can be the reduction of fertility in both men and women. It is hypothesized that pseudo-estrogen chemicals could impact the reproductive capabilities of people in the same way.

Residual Runoff. The use of chemical fertilizer and pest control products has been part of great expansions in food production and stability around the world. Their overuse, however, has also been associated with some disturbing results related to water quality. Chemicals used to suppress insect populations and enhance crop growth in fields are related to environmental changes in water as much as a half a continent away. High chemical levels in the fields are associated with chemical infused runoff, which flows into tributaries, which in turn flows into major rivers (Pelton, 2014). One of several examples of the problems that this can cause is the creation of dead zones. An imbalance of algae fueled by chemical runoff can create a low oxygen expanse known as a dead zone. Little or no life can thrive in these areas, and it is unknown how this impacts wildlife and fisheries. The largest known example of these zones is in the Gulf of Mexico. What had previously been teeming with life throughout now contains a dead zone roughly the size of the state of New Jersey (NOAA, 2024). As with other runoff issues, the potential impact of this phenomenon is yet to be fully understood.

Aquaculture as a Positive Force. Commercial activities have an even greater potential for environmental benefit as for environmental harm. This may be seen on the micro level as well as the macro. Researchers and practitioners are finding ways to enhance the environment through commercial activities, and this has the potential to profoundly impact the health of the environment. An example of this can be found in the response to a serious problem. Green crabs are an invasive species of crab that originally came from Europe to coasts near the United States (Lutto, 2022). Due to a lack of natural predators near the United States, their populations grow out of control. They crowd out native crabs and damage the coastal banks and plant life through excessive digging. One good way to control their population is to eat them as seafood and use them in commercial products. Researchers and small businesspeople are doing this in the field. By harvesting green crabs, people can control their populations and make money by selling them as food for people to cook and eat. The local profit motive is driving these people, and it is done with a mind of environmental sustainability. This makes money rather than costs money. It helps the environment and provides food for people. Plus, by eating green crabs, people are eating fewer native crabs, which gives them a chance to repopulate. This is sustainable, not only for the environment but also for the local economy (Skerry, 2024).

This same principle of aligning economic and environmental sustainability can be found in commercial kelp farming. Kelp farming is another way for people to earn money while benefiting the environment. Kelp spores are woven into a string that is wound around a rope underwater for two or more months. It is easy and inexpensive to grow and quite profitable (Skerry, 2024). Kelp can also help reduce global warming and ocean acidification. As it grows it absorbs carbon dioxide to lessen global warming and ocean acidification. Kelp is used in many foods and products all over the world. For example, when mixed into animal feeds for livestock (Roque, et.al., 2019), it can reduce the methane released from cow belches and manure (Roque, et.al., 2021). Just taking kelp from kelp forests can damage ecosystems through habitat loss. When people grow it intentionally, it can meet demands without exhausting important resources. Kelp farmers can also raise mussels alongside kelp and increase the kelp production. When mussel larvae are put on the ropes, the kelp has been shown to absorb enough carbon dioxide to protect mussel shells from ocean acidification (Pessarrodona, 2023). In addition, the mussels produce waste that can fertilize kelp. Thus, not only does the kelp grow better, but there can be bountiful mussel crops. The mussels consume whatever organic matter or tiny organisms float by, so it takes no extra expense to feed them. It's profitable with a low impact (Skerry, 2024). This principle of joining economic sustainability with environmental sustainability is very important for any long-term efforts to promote a healthy environment.

The sum of these concerns is that water that has been polluted in ways that were not previously possible could be causing health concerns that have not previously been seen. Water pollution has long been a problem with bad consequences. It is now, however, dangerous in new ways that may not subside over time.

Addressing Environmental Challenges as a Whole: The Four Elements in Unison

Just as with the interaction of the four classical elements, the different environmental crises that mankind faces interact with one another. Optimally, the methods that a college or university uses to enhance the health of the environment will interact as well. Consumer waste first pollutes the *earth* and then turns into microplastics which pollute the *water*. Deforestation weakens the *air* and the atmosphere which makes the atmosphere unable to manage the *fire* of energy production. Even beyond the interaction of these challenges to the classical elements, in combination they lessen the quality of life for all of mankind. While this combination of problems will be felt sooner and more painfully by those not fortunate enough to be born in affluent countries, it will be felt by all. Concentrating on a specific issue can be an effective entry into building the capability to support the health of the environment, but it necessarily leads to promoting a holistic approach. This can be approached in several ways.

Principles of Promoting Campus Civil Discourse and Engagement

Each college or university that chooses to incorporate a holistic view of promoting the health of the environment must chart its own path. No two institutions are alike so no single plan will be the most effective for all. Since colleges and universities are primarily made of people and these people largely determine what they will do, promoting civil discourse and individual engagement are critical to this enterprise.

Much has been written on how civil discourse and engagement can be promoted on campus. While every college or university will most benefit from a unique approach to reflect its unique makeup, certain approaches can be useful in most cases. A report from the American Association of University Professors (AAUP) recommended the promotion of classroom civil discourse by building it into lessons with defined expectations, promoting free expression of all ideas, assessment of learning, and neutral feedback (Schwartz & Ritter, 2019). The promotion of civil discourse also goes beyond the classroom.

The overall role of the institution includes seeking and disseminating truth. It also includes producing responsible citizens. Efforts have been taken to collect best practices from multiple institutions that are actively working to produce civil discourse and engagement on their campuses (Drury, et.al., 2024). Some of these promising methods include focusing discourse on learning and collaboration while accepting divergent viewpoints, the use of self-reflection and guided discussion, the promotion of individual humility, and the framing of questions in a way that promotes the acceptance of divergent viewpoints (Drury, et.al., 2024). This can be furthered by an institutional focus on pursuing the truth regardless of what it is and to model for students that they should go beyond determining who has the right to say something and explore what is a responsible action (Schwartz & Ritter, 2019). Academic freedom does not come without responsibility and modeling this as an institution can move beyond simple arguments to truly productive discourse (Tight, 1988).

PRACTICAL STEPS AND STRATEGIC THINKING

What follows are both practical and strategic steps colleges and universities can take in responding to sustainability issues while engaging in rigorous discourse in classrooms and beyond.

Practical Steps

Individual faculty, students, academic departments and administrative offices can effectively work independently on singular aspects of sustainability with positive outcomes, especially in terms of student learning. Below is a brief set of ideas that can be used to jump start or advance individual courses, student projects, and other sustainability initiatives. Some, as described, relate to the four "elemental" topics previously covered, but they can certainly be used on other sustainability projects. There are also some general practical steps that can be taken.

Waste Reduction

Everyone on campus is a consumer and therefore producer of waste products. Simple yet measurable opportunities for waste reduction on campus include:

1. Reduction in the use of bottled water. (The University of Vermont installed more than 100 water bottle refill stations and abolished the sale of bottled water in 2013.)
2. Trayless dining in dining halls. Students only eat what they can carry.
3. Diversion of unused food from dining halls to local soup kitchens/food pantries.
4. Partner with local organizations on a large-scale, local composting program.
5. Competition among residence halls to recycle whatever possible. Make this an active not passive program, with students' behavior being informed by their learning, i.e. the impact of waste reduction is explored in academic settings beyond the Housing Office. Promotion of reusable and recyclable materials and products should be part of campus life.
6. Recycle students' furniture and other excess belongings during Move Out days.

Alternative Energy/Energy Conservation

1. Depending on where your campus is located, promote internships, field trips, and other direct experiences for students to expose them to solar, wind and other energy alternatives.
2. Invite your local utility to come to campus and talk about what it is doing to develop alternative, green sources of energy.
3. Build a solar array, a reflector oven, or another form of solar-powered device/battery system.
4. Promote electric vehicles through outreach and the installation of charging stations.

5. Have your class visit a wind farm, solar installation, or other green electricity generating utility.
6. Develop internships at your local utility company.
7. Promote energy conservation on campus (dormitory contests); Install LED lighting and promote energy conservation by turning off printers, computers and lighting on the weekend.
8. Promote faculty and student research around alternative energy. There is currently no viable technology that can replace or even substantially reduce the reliance on burning fossil fuels for power generation. A new generation of research and researchers can change this. Many standard technologies were complete impossibilities when creative students were inspired to conquer them 30 years before. This is the sort of long-term solution that colleges and universities are best equipped to facilitate. The development of human capability is at the core of what institutions of higher education provide and this sort of effort can provide ripple effects with many benefits. Even in small ways, campuses can move toward solar, wind or hydro-electricity projects to reduce their carbon footprint. No college or university can achieve zero emissions overnight, but every movement forward gets an institution closer to its goal.

Clean, Available Water

More than 2 billion people live in countries with water scarcity. In our own country, the states of California, Arizona, Nevada, Utah, Colorado, New Mexico and Texas are all dealing with water shortages. Causes range from rapid urbanization to drought, limited surface water, groundwater depletion, and agricultural demand. Colleges and universities in those states can serve as leaders in discovering solutions and working with local authorities and organizations to manage water shortages. Water conservation, like energy conservation, would appear to be a topic that would spur much interest and attention on the part of college students. Water pollution is another serious problem and can be explored regardless of your campus location.

The health of the oceans is also an area worth exploring. The University of Miami's Rosenstiel School of Marine, Atmospheric, and Earth Science offers internships and undergraduate and graduate level research through its Coral Reef Futures Lab at sites throughout the Caribbean to study the impact of ocean warming and acidity on the health of coral reefs. In addition to field sites, the university has a close partnership with the National Oceanic and Atmospheric Administration. (University of Miami, 2025)

Deforestation and the Amazon Rainforest

American colleges and universities are not geographically positioned to directly impact the deforestation of the Amazon rainforest. However, there are still opportunities to study the issue. For instance, the effect of the rainforest on the Sahara Desert in Africa is a reminder that we all share this fragile planet. In addition, 40% of the lumber harvested in the Amazon ends up in China, the United States, and Europe. The economics again remind us of a global economy and global society. It would be an interesting assignment to have an ad hoc faculty task force examine the rain forest issue to see how many different academic departments on campus could find a reasonable, feasible, and meaningful learning experience for students, with or without an interdisciplinary approach.

General Practical Steps

In addition to the specific recommendations above that are tied to the four "elemental" topics covered earlier in this chapter, there are several practical steps faculty, and students can take regardless of the sustainability topic being pursued.

1. Adopt a sustainability problem for a semester and "work it" across campus. For instance, on the matter of "hunger" (a UN Sustainable Development Goal), you could create an on-campus food pantry, volunteer at a local soup kitchen, sponsor a "hunger marathon," raise funds and food for homeless shelters, and otherwise impact the issue.
2. Seek synergies across sustainability topics, e.g., hunger and waste reduction, poverty and job creation. Assign to specific student clubs and faculty sponsors.
3. Engage the campus.
    - Sponsor frequent contests and competitions, e.g., energy usage in the dormitories
    - Generate, evaluate and articulate data on various sustainability issues. For instance, you could ask every faculty member, staff person and student to evaluate their own energy and food consumption and pledge to reduce waste and consumption.
    - Challenge each student to develop a sustainability project that aligns with the campus's sustainability plan or the UN's 17 Sustainable Development Goals.
    - Sponsor a community garden and support students in growing their own food.
4. Bring in the outside world.

- Book guest speakers, especially local or regional partners and suppliers, e.g. local utility companies, fellow universities, nonprofits working in sustainability fields.
- Sponsor frequent (annual, semi-annual) symposiums on issues surrounding local sustainability. Invite utilities, local elected officials, NPOs, and other organizations to engage in meaningful conversations about real issues facing local communities. Make sure students attend and be ready with questions. Have classes write papers based on symposium discussions.
- Create and manage partnerships with appropriate organizations in your community. These could range from a homeless shelter to a community garden. Invest time and energy in these partnerships so that the college/university is seen as a change agent and friend. And students and faculty can develop internships and class projects through the partnerships.

5. Carry on an ongoing dialog.

Discourse can be two people talking about their electric cars. Or it can be an entire campus engaged in real discussions of actions being taken daily on campus to reduce waste, conserve energy and impact local, regional, national and international sustainability goals. (See "Strategic Thinking" for more on this.) There are many issues in society today that are divisive, partisan and debated at length without any seeming progress. The ongoing conflict in the Middle East and immigration here at home are two examples. At the same time, 75% of Americans believe we have experienced global warming in the past 100 years (Kosnick, J., and MacInnis, B., 2024). Fully 77% of Americans support renewable energy research. (Leiserowitz, A., et. al., 2024). And 70-75% of Americans support expansion of wind and solar farms. (Pew Research Center, Environment and Energy, 2023). In terms of demographics, younger Americans skew higher on these initiatives. All this seems to make a good case that students on college campuses are eager to talk about and act upon issues of sustainability—locally, nationally, and internationally. Classroom discussions, organized campus forums, student-led blogs and podcasts . . . this is a well that won't run dry any time soon.

6. Track success.
   - Develop energy dashboards for each dormitory and major campus building. Track energy usage and share across the campus. Award students who conserve the most energy.
   - Develop a sustainability website and dashboard to document courses with sustainability connections, as well as campus operational projects

aimed at waste reduction, energy conservation, and other sustainability goals.

- Publish, disseminate and discuss an annual sustainability report.
- Champion sustainability leaders on campus.
- Showcase green projects with an annual "Green" day, either during Earth Week or another appropriate time.

Strategic Thinking

While the author understands the interest of individual faculty and students in personally responding to the sustainability challenge, the fact is, each individual project that can be envisioned will be amplified and empowered when set in the context of an institution-wide sustainability culture. Funding and other resources can be shared, campus dialog around the multiplicity of issues related to sustainability becomes a daily discussion, and the campus itself can become a powerful laboratory for student growth and learning.

While individual faculty, solitary student campus organizations and other singular players on campus learn about the environment and develop meaningful strategies to respond to climate change, there is another much more impactful opportunity available. With college administrators and faculty committed to environmental sustainability and the student learning and growth that can occur in the process of galvanizing environmental programming on campus, many college presidents are making an "all in" effort to create sustainability cultures on their campuses. In such a model, every student on campus can learn about their own carbon footprint and how to minimize it. The entire curriculum can be examined for opportunities to incorporate sustainability into courses in most disciplines. In the process, students can gain valuable applied knowledge, impact sustainability on their own campus and in their local community and contribute to what is becoming a green movement across higher education.

The opportunity to impact one's campus, local community and more was greatly expanded in 2015 when the United Nations adopted the Sustainable Development Goals. In addition to environmental stewardship and equity, the UN added goals that address social and economic equity across the globe. This means several things for college campuses. It has broadened the alignment between college and university curriculum and other campus operations and this newer definition of sustainability. As more academic departments are involved, the creation of a sustainability culture will gain additional momentum on any campus attempting to ramp up their efforts and taps into further interest among students.

As much as current and future students care about the environment, they are perhaps even more focused on social equity. A United Way NCA survey found that nearly 40% of college-aged Gen Z individuals engage in activism or social justice work. (United Way of the National Capital Area, 2024)

Other data shows Generation Alpha also takes a strong position when it comes to social equity. A 2019 study by Hotwire and C+R Research found that 96% of Generation Alpha members believe people should be treated fairly, regardless of their appearance. This contrasts with 79% of Millennials and 58% of Baby Boomers who held similar views at the same age. Additionally, 93% of Gen Alpha respondents emphasized the importance of being accepted for who they are. (Gingerich, J., 2019)

Similar research findings show this newest generation of young adults believe in other measures of social equity such as food security, housing, and meeting other basic needs. (Fernandez, M., 2025) This newest generation of college students will be actively engaged in talking about and engaging the pressing issues of social equity and environmental sustainability. Even in middle and high school, Generation Alpha places a high stake in engaging with the world. A Junior Voices survey by The Week Junior and YouGov found that 83% of Generation Alpha think it's important to learn about current events, with 79% discussing these topics with friends and family at least weekly. (Garcia, C., 2021)

With a commitment to build a sustainability culture across campus, and a global framework to guide that work (United Nations, 2015), what would be the key features of a college- or university-wide approach to this broad definition of sustainability?

1. **Develop a comprehensive plan.** Creating a sustainability plan for your campus must be as inclusive and empowering as the implementation strategies it contains. That means campus-wide participation in developing the plan: faculty, students, administrators, local partners, industry partners. It needs to reflect curriculum needs, co-curricular activities and campus operations. Achievable goals — both short-term and long-term — must be developed and reasonable benchmarks established. While global issues such as oceans rising and the Amazon Rainforest can and should be addressed, it is critical that a campus focus on its own environmental circle of influence as well as its local community's needs. Those local and regional conditions are what your graduates are likely to be working with as community citizens after graduation. This context and relevance will not only provide applied learning opportunities to students, but it will also ground the campus's internal activities.
2. **The plan must carry the weight of authority.** That means the college/university president must be the loudest advocate for sustainability on campus, both in launching the initiative and in articulating subsequent major announcements. A sustainability plan should not stand alone but should be included as an es-

sential element of the college/university's overall strategic plan. It should also be elevated to the appropriate level during budget deliberations.

3. **Designate a Faculty Champion.** As much as sustainability must become a campus culture, owned by everyone, appointing a full-time faculty advocate is essential to maintain momentum. This individual does not have to be an environmental scientist but needs to have some direct relationship to achieving sustainability goals. Such an individual will facilitate groups of faculty and students in fulfilling their portion of the plan, providing ongoing communication updates, sponsor sustainability events such as Earth Day, and otherwise maintain the progress and movement essential in such a campus-wide enterprise.
4. **Create a standing committee.** An interdisciplinary committee made up of faculty, campus operations, students, administrators, and even alumni representatives from the local community can sponsor projects, bring in guest speakers, disseminate reports and otherwise serve as a clearinghouse and a cheering squad for a campus sustainability program. However, this committee and the designated faculty champion cannot be perceived as the drivers of sustainability on campus. The development of curriculum, growing faculty expertise, the creation of co-curricular clubs, and more must occur within the existing structures of the college/university. Sustainability becomes a culture when everyone on campus shares the same goals, the same language, and practices environmental stewardship together.
5. **Focus resources on faculty development on the issue of sustainability.** "How do I incorporate sustainability into my course?" "Is my course already impacting our sustainability goals?" "What changes in classroom pedagogy must I consider?" "How do I engage the local and regional community in creating internships and other applied learning opportunities for students?" These are the types of questions that faculty will ask. Answering these and other questions with a faculty development plan that considers the goals of the campus sustainability program will be a critical element in achieving the program's goals and maximizing student learning in the process. Another critical question that faculty will ask is, "How can I integrate sustainability into my existing course(s)?" Faculty should not have to address this question on their own. With the UN's 17 Sustainable Development Goals in play, most if not all academic disciplines can find opportunities to include sustainability learning plans in their courses. Faculty workshops and seminars, release time, and other tools needed to develop sustainability course components must be provided.
6. **Seek external funding.** The reality is, when a campus seeks to reduce its dependence on fossil fuels by conserving electricity or by buying green electricity, it must do so largely with institutional funds. However, at the same time, cost savings that are realized by energy efficiency can be reinvested into the

college/university's sustainability program. As importantly, various external sources—both public and private—of funding support should be aggressively pursued. Often these are to start up innovative technology-centered projects, such as solar installations. Public utility rebates are also a possibility.

7. **Network with local and regional entities.** This could be everything from social service agencies whose clients face environmental, social and economic inequity to local industries that can assist by providing internships to students. Municipalities must also address sustainability issues similar to those found on your campus and could become strong partners in waste management and other mutual concerns.
8. **Build experiential learning into sustainability curriculum.** Every college and university have its own eco-system, including local and regional environmental conditions as well as social structures. Whether it is interning at a veteran's center, taking field trips with faculty to witness a wind farm in action, exploring sea surge erosion along the southern shores of New England, or conducting undergraduate research on clean energy solutions, the more applied the learning, the more effective it is in providing students with measurable competencies and marketable job skills.
9. **Create manageable student projects.** No single classroom, no one university can single-handedly solve hunger, reduce carbon emissions to zero on campus, or magically make the ocean waters stop warming. And some issues—reversing the impact of deforestation in the Amazon rainforest or eliminating the impact of microplastics—are beyond the reach of most American colleges and universities. However, lesson plans and syllabi can be designed to allow students to explore in depth the issues surrounding such global problems, gaining valuable insight into the interconnectedness of global eco-systems. In addition, projects and courses with more local implications — partnering with local organizations to establish a composting center or monitoring energy usage on campus — are doable, measurable, and effective in teaching students' important green economy skills.
10. **Build assessment into your sustainability plan.** Monitoring the use of electricity in a campus building or across an entire campus is not difficult. Assessing the use of gasoline by all members of the campus community is more difficult. Other elements of a sustainability plan may not easily be assessed. The point here is that you should be building in assessment tools and methods whenever possible. One important assessment strategy to pursue is to participate in the Sustainability Tracking, Assessment & Rating System (STARS) of the Association for the Advancement of Sustainability in Higher Education (AASHE). This self-reporting system allows colleges and universities to evaluate their progress along several criteria spectrums and earn relative recognition from AASHE

(bronze to platinum). Those spectrums include Academics (Curriculum, Research); Engagement (Campus Engagement, Public Engagement) ; Operations (Buildings & Grounds, Energy & Climate, Food & Dining, Procurement & Waste, Transportation); Planning & Administration (Coordination & Planning, Investment Social Equity, Wellbeing & Work; Innovation & Leadership.

CHAPTER SUMMARY

College campuses are recognized as safe havens for meaningful dialog, innovation, and action surrounding a broad array of societal issues. While climate changes and global warming are "litmus test" issues with partisan political implications, the American public largely believes that protecting the environment is a priority that should be addressed. The vast majority of college students—current and upcoming—want to proactively engage in discourse and activism around issues of environmental and social sustainability. Colleges and universities are well positioned to be sustainability leaders, with student/faculty interest, research capacity, local partnerships, and other resources serving as a solid foundation on which to make progress. In addressing sustainability, college students and their professors can embrace their role as society's "marketplace of ideas," where important topics can be carefully yet vigorously considered, tested, and in the case of sustainability, acted upon.

A discussion around the classical "elements" of fire, wind, water, and earth is useful in identifying potential environmental sustainability problems to address, including water pollution, deforestation, solid waste management, and generating renewable energy sources as alternatives to fossil fuels.

There are many practical steps that individual faculty and students can take, as well as single academic departments and student groups, to learn about sustainability and impact on-campus issues such as recycling, waste reduction, energy conservation, and other important sustainability issues.

If colleges and universities take a broader approach to sustainability that includes social and economic sustainability, in addition to environmental stewardship (aligning with the UN SDGs), they can create a campus-wide sustainability culture. In such a culture, every faculty member, every student, every staff member can participate in working toward a more sustainable future for their campus, for their local community and region, for our nation, and for the global community.

REFERENCES

Baker, J., & Spracklen, D. (2019). Climate benefits of intact Amazon forests and the biophysical consequences of disturbance. *Frontiers in Forests and Global Change*, *2*(47), 47. DOI: 10.3389/ffgc.2019.00047

Ball, P. (2004). *The Elements: A Very Short Introduction. Very Short Introductions*. OUP Oxford. DOI: 10.1093/actrade/9780192840998.001.0001

Barber, C. (2024). America has a $250 billion problem: Microplastics have invaded our bloodstreams and may increase the risk of heart attack and stroke. *Fortune*. May5th, 2024. https://fortune.com/well/article/microplastics-health-effects/

Bittner, G. D., Denison, M. S., Yang, C. Z., Stoner, M. A., & He, G. (2014, December 4). Chemicals having estrogenic activity can be released from some bisphenol A-free, hard and clear, thermoplastic resins. *Environmental Health : A Global Access Science Source*, *13*(1), 103. DOI: 10.1186/1476-069X-13-103 PMID: 25477047

Boyle, K. (2024). The importance of sustainable agriculture: The future of farming. *Lead The Change*.https://leadthechange.bard.edu/blog/the-importance-of-sustainable - agriculture-the-future-of-farming

Cairo, A. (2011). Student engagement in campus sustainability. *Facilities Manager*, March/April. https://files.eric.ed.gov/fulltext/EJ920384.pdf

Campbell, J. (1995). *Understanding John Dewey: Nature and cooperative intelligence*. Open Court.

Cho, R. (2020). Recycling in the U.S. is broken: How do we fix it? *State of the Planet*.https://news.climate.columbia.edu/2020/03/13/fix-recycling-america/

Cohen, A., & Kisker, C. (2010). *The shaping of American higher education: emergence and growth of the contemporary system* (2nd ed.). Jossey-Bass.

Cohen, S. (2014). Understanding how Americans view the environment. *Huffpost*, May 24. https://www.huffpost.com/entry/understanding-how-america_b_5020669

Cortese, A. (2010). Higher education's purpose: A healthy, just, and sustainable society. *Inspired Economics*, 1/28/2010.

Dewey, J. (1916). *Democracy and education: An introduction to the philosophy of education*. Macmillan Publishing.

Drury, S., Briscoe-Smith, A., Longo, N., Napoli, L., Winslow, R., Farley, M., & Weaver, L. (2024). *Better Discourse: A Guide for Bridging Divides in Challenging Times*. Campus Compact.

EIA. (2023). *Carbon Dioxide Emissions Coefficients*. U.S. Energy Information Administration. Washington: DC. https://www.eia.gov/environment/emissions/co2_vol_mass.php

Ekins, E. (2017). *The state of free speech in America: Attitudes about free speech, campus speech, religious liberty, and tolerance of political expression*. Cato Institute, 10/31/2017

EPA. (2023). *Recycling Basics and Benefits*. Environmental Protection Agency. Washington: DC. https://www.epa.gov/recycle/recycling-basics-and-benefits

EPA. (2023). *Quantifying Methane Emissions in Landfilled Food Waste*, Environmental Protection Agency. Washington: DC. October 2023, https://www.epa.gov/land-research/quantifying-methane-emissions-landfilled-food-waste

EPA. (2024). *Electric Power Sector Basics*. Environmental Protection Agency. Washington: DC. https://www.epa.gov/power-sector/electric-power-sector-basics

Fernandez, M. (2025). *Generation Alpha Characteristics: Innovators or Trading Humanity for Technology?* Softonix.com, January 30, 2025; https://en.softonic.com/articles/generation-alpha-characteristics-stats

Fincher, M., & Katsinas, S. (2017). Testing the limits of the price elasticity of potential students at colleges and universities: Has the increased direct cost to the student begun to drive own higher education enrollment? *Journal of Higher Education Policy and Management*, *39*(1), 31–39. DOI: 10.1080/1360080X.2016.1211975

Firedogs Jumk Removal. (2019). *What Does Firedogs Do with Junk*. Indianapolis: IN. https://firedawgsjunkremoval.com/what-does-fire-dawgs-do-with-junk/

Flaherty, C. (2024). *How Students Feel About Campus Speech, in 5 Graphics,* Inside Higher Education, December 13, 2024; https://www.insidehighered.com/news/student-success/academic-life/2024/12/13/survey-students-thoughts-campus-speech-five-graphics

Fleck, A. (2023). *Cars Cause Biggest Share of Transportation CO2 emissions*, September 22, 2023, in Statista.com.

Freeborn, J. (2024). Microplastics in testicles may play a role in male infertility, study suggests. *Medical News Today*. May 24, 2024 https://www.medicalnewstoday.com/articles/microplastics-in-testicles-may-play-a-role-in-male-infertility-study-suggests

Fullan, M. (2005). *Leadership & sustainability: System thinkers in action*. Corwin Press.

Garcia, C. (2021). *Week Junior survey finds Generation Alpha wants adults to listen more,* The Week Junior (With YouGov), November 10, 2021

Gatti, L. V., Basso, L. S., Miller, J. B., Gloor, M., Gatti Domingues, L., Cassol, H. L. G., Tejada, G., Aragão, L. E. O. C., Nobre, C., Peters, W., Marani, L., Arai, E., Sanches, A. H., Corrêa, S. M., Anderson, L., Von Randow, C., Correia, C. S. C., Crispim, S. P., & Neves, R. A. L. (2021). Amazonia as a carbon source linked to deforestation and climate change. *Nature*, *595*(7867), 388–393. DOI: 10.1038/s41586-021-03629-6 PMID: 34262208

Georgiou, A. (2019). How much oxygen does the Amazon rainforest provide? *Newsweek,* August 27. https://www.newsweek.com/how-much-oxygen-amazon-rain-forest-1456274

Gingerich, J. (2019). *Fairness, Sustainability Defines Alpha Generation*, O'Dwyers PR, December 9, 2019, https://www.odwyerpr.com/story/public/13499/2019-12-09/fairness-sustainability-defines-alpha-generation.html

Graham, E. (2014). Mandated trigger warnings threaten academic freedom. *Reclaiming the Narrative*, (November-December), 2014.

Hackney, A. (2018). *Emerging Issues in Analytical Chemistry, Doping, Performance Enhancing Drugs, and Hormones in Sport*. Elsevier., DOI: 10.1016/B978-0-12-813442-9.00002-X

Haywood, A. (2024). Amazon Rainforest deforestation and climate change. *National Geographic,* March 11. https://education.nationalgeographic.org/resource/amazon - deforestation-and-climate-change/

Holland, J. (2021). Nature's pharmacy: The remarkable plants of the Amazon Rainforest and what they cure. *The Telegraph*, February 19. https://www.telegraph.co.uk/travel/cruises/articles/how-to-be-a-botanical-buff/

IEA. (2018). *The Future of Cooling: Opportunities for energy-efficient air conditioning*. IEA., DOI: 10.1787/9789264301993-

IEA. (2023). *Prospects for EV Development,* October 2023; www.iea.org)

Kisker, C., Cohen, A., & Brawer, F. (2023). *The American Community College* (7th ed.). Jossey- Bass. (Original work published 2023), Retrieved from https://www.perlego.com/book/4213866

Kosnick, J., and MacInnis, B. (2024). *American Understanding of Climate Change*, published by Resources for the Future and Stanford University, July 15, 2024.

Le, N. (2020). *The Impact of Fast Fashion on the Environment*. Princeton Student Climate Initiative. Princeton University. Princeton: NJ. https://psci.princeton.edu/tips/2020/7/20/the-impact-of-fast-fashion-on-the-environment

Lederman, D. (2023). Showcasing solutions for better campus discourse. *Insider Higher Education,* October 23rd. https://www.insidehighered.com/news/students/free- speech/2023/10/10/showcasing-solutions-promote-civil-discourse-campuses

Leiserowitz, A., Maibach, E., Rosenthal, S., Kotcher, J., Goddard, E., Carman, J., Myers, T., Verner, M., Marlon, J., Goldberg, M., Ettinger, J., Fine, J., & Thier, K. (2024). Climate Change in the American Mind: Politics & Policy, Fall 2024. Yale University and George Mason University. New Haven, CT: Yale Program on Climate Change Communication.

Longman, P. (2004). *The Empty Cradle : How Falling Birthrates Threaten World Prosperity and What to Do About It.* Basic Books.

Lutto, A. (2022). *Crabby About Invasive Crabs*. U.S. Fish & Wildlife Service. https://www.fws.gov/story/2022-10/crabby-about-invasive-green-crabs

Medlin, E. (2008). Sustainability and the APUCC. *Facilities Manager*, (March/April), 10–11.

MIT Free Speech Alliance. (2024). Debate: Is Decarbonization Worth the Cost? Massachusetts Institute of Technology. November 14, 2024.

N.A. (2024). *Boston Museum of Science Survey Finds 'Generation Alpha' Taking Up the Challenge to Adapt and Act on Climate*, PR Newswire, April 22, 2024; https://www.prnewswire.com/news-releases/boston-museum-of-science-survey-finds-generation-alpha-taking-up-the-challenge-to-adapt-and-act-on-climate-302123223.html

N.A. (2024). *Overwhelming majority of young Americans worry about climate crisis*, The Guardian, October 17, 2024; referring a study by Avaaz published in The Lancet Planetary Health. https://www.theguardian.com/environment/2024/oct/17/young-americans-climate-change

NACE. (2014). *Career Readiness Competencies: Employer Survey Results*. National Association of Colleges and Employers. https://www.naceweb.org/career- readiness/competencies/career-readiness-competencies-employer-survey-results

Nature, S. (2025). *Climate Leadership Network*; https://secondnature.org/climate-leadership-network/

Nguyen, T., Nguyen, T., Grote, U., Neubacher, F., Rahut, D., Do, M., & Paudel, G. (2023). Security risks from climate change and environmental degradation: Implications for sustainable land use transformation in the Global South. *Current Opinion in Environmental Sustainability*, *63*, 101322. Advance online publication. DOI: 10.1016/j.cosust.2023.101322

NOAA. (2023). *Gulf of Mexico 'dead zone' larger than average, scientists find.* National Oceanic and Atmospheric Association. Washington: DC. https://www.noaa.gov/news- release/gulf-of-mexico-dead-zone-larger-than-average-scientists-find

Pearce, F. (2024). Mind the gaps: How the U.N. climate plan fails to follow the science. *Yale Environment 360*, 1/9/2024.

Pelton, T. (2014). *Polluted Runoff: How Investing in Runoff Pollution Control Systems Improves the Chesapeake Bay Region's Ecology, Economy, and Health.* Chesapeake Bay Foundation. Washington: D.C.https://www.cbf.org/document-library/cbf-reports/2014-Polluted-Runoff- Report-compressed20b2.pdf

Pessarrodona, A., Franco-Santos, R. M., Wright, L. S., Vanderklift, M. A., Howard, J., Pidgeon, E., Wernberg, T., & Filbee-Dexter, K. (2023). Carbon sequestration and climate change mitigation using macroalgae: A state of knowledge review. *Biological Reviews of the Cambridge Philosophical Society*, *98*(6), 1945–1971. DOI: 10.1111/brv.12990 PMID: 37437379

Peterson, R. (2014). Permission to Speak: College Presidents, the Israeli Boycott, and Climate Change, *National Association of Scholars*, January 22, 2014.

Pew Research Center (2023). *Economy Remains the Public's Top Policy Priority; COVID-19 Concerns Decline Again*, February 2023.

Raintree (2024). *About The Rainforest.* https://rain-tree.com/facts.htm#:~:text=Unbelievably%2C%20more%20than%20200%2C000%20acr es,threatened%20as%20the%20destruction%20continues.

Review, P. (2023). *College Hopes & Worries Survey*, March 15, 2023; https://www.princetonreview.com/press/college-hopes-worries-press-release-2023

Roque, B., Brooke, C., Ladau, J., Polley, T., Marsh, L., Najafi, N., Pandey, P., Singh, L., Kinley, R., Salwen, J. K., Eloe-Fadrosh, E., Kebreab, E., & Hess, M. (2019). Effect of the macroalgae Asparagopsis Taxiformis on methane production and rumen microbiome assemblage. *Animal Microbiome*, *1*(1), 3. DOI: 10.1186/s42523-019-0004-4 PMID: 33499933

Roque, B., Venegas, M., Kinley, R. de Nys, R., Duarte, T., Yang, X. & Kebreab, E. (2021). Red seaweed (Asparagopsis Taxiformis) supplementation reduces enteric methane by over 80 percent in beef steers. *PLoS One*, Mar 17;16(3).

Rudolph, F. (1962). *The American College and University: A History*. Alfred A. Knopf.

Schwartz, L. & Ritter, D. (2019). Civil discourse in the classroom: Simple approaches to tough conversations. *New Culture Wars*, November 2019.

Skerry, B. (2024). The Gulf of Maine is warming fast: What does this mean for lobsters-and everything else? *National Geographic*. June 2024. https://www.nationalgeographic.com/environment/article/gulf-of-maine-climate-change- photographer

Spencer, A., Ross, S., & Tyson, S. (2023). *How Americans view electric vehicles*. Pew Research Center.

Spring, J. (2023). Replanting the Amazon could help save the world's climate. Here's why it's so hard to do. *Reuters*. June3rd. https://www.reuters.com/investigates/special-report/brazil- environment-reforestation/

The Sustainable Fashion Forum. (2023). *Is Polyester *Really* That Bad?* October 11, 2023. https://www.thesustainablefashionforum.com/pages/quick-question-what-is-polyester- and-is-it-really-that-bad-for-the-environment

Tight, Malcolm & Society for Research into Higher Education. (1988). *Academic freedom and responsibility / edited by Malcolm Tight*. Milton Keynes [Buckinghamshire] ; Philadelphia, Pa.: The Society for Research into Higher Education and Open University Press

Trzepacz, S., Lingås, D., Asscherickx, L., Peeters, K., van Duijn, H., & Akerboom, M. (2023). *LCA-based assessment of the management of European used textiles*. Norion Consult. https://euric.org/images/Position-papers/lca-based-assessment-of-the-management-of- european-used-textiles_corrected.pdf

UNEP. (2015). *Plastics in Our Environment: Are We Polluting the Environment Through Our Personal Care?* United Nations Environment Programme. New York, NY. https://www.unep.org/resources/report/plastic-cosmetics-are-we-polluting-environment- through-our-personal-care

UNEP. (2024). *Our Planet Is Choking in Plastic*. United Nations Environment Programme. New York, NY. https://www.unep.org/interactives/beat-plastic-pollution/

United Nations (2015) Transforming Our World: The 2030 Agenda for Sustainable Development.

United Way of the National Capital Area. (2024). *The Gen Z Activism Survey,* March 5, 2024. https://unitedwaynca.org/blog/gen-z-activism-survey

University of Miami. (2025). *Rosenstiel School of Marine, Atmospheric, and Earth Science Coral Reef Futures Lab*https://coralreeffutures.earth.miami.edu/research/index.html

USDA. (2024). *Sustainable Agriculture*. The United States Department of Agriculture. Washington: DC. https://www.nifa.usda.gov/topics/sustainable-agriculture

VanDerziel, S., Gatta, M., Jahn, J., & Koncz, A. (2024). Job Outlook 2025. National Association of Colleges and Employers, November 2024.

Vasarhely, C. (2024). *The Impact of Plastic on Climate Change*. University of Colorado. Boulder: CO. https://www.colorado.edu/ecenter/2023/12/15/impact-plastic-climate- change

Wakefield, J. (2012). *UVM Celebrates End of Bottled Water Sales With Bottled Water 'Retirement Party,* December 5, 2012, https://www.uvm.edu/uvmnews/news/uvm-celebrates-end-bottled-water-sales-bottled-water-retirement-party

Wilson, J., & Zimmerman, Z. (2023). *The End of Flat Power Demand Is Over*. Grid Strategies. https://gridstrategiesllc.com/wp-content/uploads/2023/12/National-Load-Growth-Report-2023.pdf

WWF. (2024). *Climate Change in the Amazon*. World Wildlife Federation. https://wwf.panda.org/discover/knowledge_hub/where_we_work/amazon/amazon_threats/climate_change_amazon/

Zimmer, K. (2019). Why the Amazon doesn't really produce 20% of the world's oxygen. *National Geographic,* August, 28. https://www.nationalgeographic.com/environment/article/why-amazon-doesnt-produce- 20-percent-worlds-oxygen

Chapter 5
Scaling Up Faculty Engagement in Education for Sustainability Across the Curriculum:
Five Strategies for Amplifying Impact

Rebecca A. Watts Hull
https://orcid.org/0000-0002-5277-8725
Georgia Institute of Technology, USA

ABSTRACT

Incorporating sustainability across the curriculum is a growing component of college and university sustainability commitments. Education for Sustainability aligns with institutional commitments to advance solutions to the world's most pressing challenges. Most sustainability curricular change initiatives take a "bottom up" approach, engaging interested faculty through small learning communities. While effective, this approach is slow and many instructors choose not to engage. This chapter draws on the experience of a technologically focused university to suggest strategies for "scaling up" sustainability across the curriculum. The following strategies supported significant expansion of sustainability course integration: laying the groundwork through a major initiative to enhance undergraduate learning; leveraging strategic planning to win high-level commitment and resources; connecting sustainability education with professional development in teaching and learning; empowering faculty to co-design initiatives; and supporting wellbeing through on-going learning communities.

DOI: 10.4018/979-8-3693-5454-4.ch005

INTRODUCTION

What strategies can colleges and universities use to scale up sustainability curricular integration, while respecting faculty autonomy? This chapter describes several strategies that enabled a comprehensive, public university to significantly *scale up* faculty participation in a course redesign program that brings sustainability and the United Nations Sustainable Development Goals (SDGs) into diverse courses across many disciplines. Between 2020 and the present, sustainability education stakeholders collaborated to embed sustainability learning into the university's strategic plan, formally connected sustainability education with the Center for Teaching and Learning, expanded faculty *agency* and *buy-in* through a representative committee to oversee sustainability education, and administered a course redesign grant program that *meets faculty where they are* and *engages them as co-learners rather than trainees*. These strategies resulted in an expansion of sustainability and SDG course integration that reaches virtually every undergraduate student at Georgia Tech. These strategies described in this chapter built upon a solid foundation established through Georgia Tech's 2015-2020 Quality Enhancement Plan, the Center for Serve-Learn-Sustain (SLS).

BACKGROUND

Over the past several decades, the number of colleges and universities advancing sustainability and climate action through campus operations, research, and education has expanded significantly (AASHE, 2024). With respect to the education for sustainability (EfS) portion of this growth, case studies and published reviews demonstrate a diversity of approaches adopted by institutions to advance sustainability learning across disciplines, in the face of significant challenges to curricular change (Barlett and Chase 2013; Leal Filho et al., 2019; Weiss et al., 2021a). Many adopt a "bottom up" approach, inviting interested faculty to workshops and faculty learning communities and slowly growing the number of instructors engaged in teaching a course that incorporates sustainability (Barth & Rieckmann, 2011; Barlett & Chase, 2012). Inviting faculty to opt into workshops and other professional development opportunities in sustainability education is consistent with norms around academic freedom—respecting instructor autonomy—and recognition that policy directives about instructional strategies are rarely effective (Henderson et al., 2011).

While these professional development programs have expanded sustainability content and pedagogies in college curricula and facilitated "social learning" in EfS (Barlett & Rappaport, 2009; Barth & Rieckmann, 2011), many programs struggle to reach more than a modest proportion of faculty members. Structural barriers, competing demands, disciplinary differences, and perceived risk can constrain instructors from taking advantage of professional development opportunities (Armstrong & Stewart-Gambino, 2016). As a result, while sustainability content and instructional approaches slowly gain traction at many institutions, meaningful sustainability integration across disciplines remains limited among colleges and universities, including technological institutions like Georgia Tech (Weiss et al., 2021b; Menon & Paretti, 2024).

This chapter will describe several key elements of Georgia Tech's sustainability and climate across the curriculum strategic approach: 1) laying the groundwork through the Center for Serve-Learn-Sustain, Georgia Tech's 2015-2020 Quality Enhancement Plan; 2) institutionalizing high-level commitment through the strategic planning process; 3) building strategic institutional partnerships to embed sustainability teaching and learning within evidence-based teaching and learning professional development; 4) creating a structure for ongoing faculty leadership; and 5) "expanding the tent" to meet diverse faculty where they are and engage them in co-directing their own professional development. A final chapter section will discuss the Georgia Tech case within the context of other published case studies and suggest ways in which sustainability education leaders working within a different institutional context may need to adapt the strategies detailed here to align with their specific circumstances. The chapter draws on scholarship of organizational change generally and sustainability leadership and faculty development in higher education specifically to interpret the impacts of these strategic initiatives and explore their relevance for other institutions. The Georgia Tech experience is consistent with other case studies and meta-analyses demonstrating the importance of administrative commitment, strong faculty/staff leadership and professional development, alignment of sustainability integration across research, education and operations, and meaningful engagement of all stakeholders (Mcmillan & Dyball, 2009; Atkins et al., 2019; Ramísio, 2019; Weiss et al., 2021b).

FIVE STRATEGIES FOR SCALING UP EDUCATION FOR SUSTAINABILITY

Laying the groundwork

Georgia Tech's first formal initiative to advance sustainability across the curriculum was in 2015 with the creation of the Center for Serve-Learn-Sustain (SLS). SLS was the Institute's second Quality Enhancement Plan (QEP), a blueprint for enhancing undergraduate learning that is a requirement of the Southern Association of Colleges and Schools Commission on Colleges (SACSOC). The Center's charge was to "equip students with the knowledge and capabilities to effectively address sustainability challenges and inter-related community-level societal needs in their professions and their civic lives." It advanced this aim by integrating sustainability-focused service-learning projects with community partners into the undergraduate curriculum (Georgia Institute of Technology, 2021).

SLS fostered structures, relationships, and learning that laid the groundwork for the initiatives and outcomes described in this chapter. The "Affiliated Courses Program" was particularly significant for engaging faculty in sustainability course integration. Over five years, this program included more than 100 courses with one of four broad student learning outcomes related to community sustainability. The voluntary program introduced faculty and students to the benefits offered by a sustainability-across-the-curriculum model. To support the work of participating faculty members, SLS staff offered faculty fellows programs and learning communities that provided faculty across disciplines with a shared interest in community sustainability the opportunity to develop relationships with one another. In addition, an online teaching toolkit was developed to help instructors incorporate sustainability content outside their areas of disciplinary expertise. For example, a popular teaching tool called "Technology in Social Context" helped faculty in the College of Engineering incorporate social science perspectives related to the impacts of various technologies on community sustainability.

The SLS faculty development programs, as well as student learning through the Affiliated Courses and other programs, supported development of a common language for sustainability learning at Georgia Tech. Most notably, the program's decision to center equity in its sustainability work nurtured a cultural shift toward sustainability discourse that includes all "three Es"—equity, economy, and environment. In addition, support for community-engaged sustainability fostered exploration of transdisciplinary perspectives: learning that not only transcends disciplinary boundaries but also includes perspectives and expertise from outside academia (Mokski et al., 2023). By supporting the development of equitable partnerships with community-based organizations, SLS helped instructors and students see leaders outside Georgia Tech as collaborators, innovators, and co-educators (Hirsch et al., 2023). These approaches and regular opportunities for faculty to engage with "big ideas" through challenging cross-disciplinary conversations nurtured a broader understanding of education for sustainability on campus (Georgia Institute of Technology, 2021).

Strong networks that engage all stakeholders, including faculty and staff, are often key to advancing sustainability initiatives at colleges and universities (Novy et al., 2021; Weiss et al., 2021b). By cultivating a network of faculty interested in engaging with each other and their students around the equity, environmental, and economic dimensions of community sustainability, and developing structures for sustainability integration across the curriculum, SLS laid the groundwork for "scaling up" education for sustainability and the SDGs at Georgia Tech. As the program was preparing its impact report and planning for institutionalization, a growing number of colleges and universities were exploring how to use the United Nations Sustainable Development Goals (SDGs) to support sustainability integration across the curriculum (Serafini et al., 2022). SDG education requires teaching approaches that address diverse social, economic and environmental challenges, their complex interlinkages, coping with uncertainty, and exploring conflicting values and perspectives (Leal Filho et al., 2019). The broad, multi-dimensional nature of the SDGs aligned well with the holistic understanding of sustainability learning nurtured by SLS. With the arrival of a new president deeply engaged in the role of universities in advancing the SDGs, the next phase of Georgia Tech's sustainability education efforts leveraged the strong foundation provided by SLS and an opportunity to expand engagement by using the SDGs as a common framework for sustainability education.

Institutionalizing Georgia Tech's Commitment

Colleges and universities periodically revise and rewrite their institutional strategic plans, and these processes provide opportunities for sustainability champions to institutionalize high-level commitment and win resources. In 2019 Georgia Tech welcomed a new president, Dr. Ángel Cabrera, who had demonstrated leadership in advancing the role of higher education in advancing the SDGs. Dr. Cabrera is a co-founder of the University Global Coalition (UGC), a global network of universities working in partnership with the United Nations in support of the SDGs. UGC is "...a call to action to universities around the world to engage, to lead, and to collaborate. It [is] a movement, a platform for collective action, and a mechanism to amplify the voice of higher education and bring it to the table in the most relevant global efforts" (UGC 2022). Soon after Dr. Cabrera joined Georgia Tech, the Institute began developing a new strategic plan to guide its progress from 2020-2030. The groundwork laid by SLS and a network of campus sustainability stakeholders enabled strong representation by individuals interested in advancing sustainability in Georgia Tech's curriculum, research, and operations across all the strategic planning teams.

As a result of administrative commitment and widespread participation of sustainability champions, Georgia Tech's 2020-2030 Institute Strategic Plan (ISP) included a bold commitment to transformative teaching and learning that enables students to advance the SDGs. President Cabrera's public message accompanying the ISP on the Institute's web site articulated the plan's alignment with the SDGs:

"*...Our world today is wrought with complex challenges that will define the future of our communities, our nation, and our species on this planet: how to power a growing world economy without heating our planet beyond repair and without* *harming the ecosystems our lives depend on; how to provide access to food, clean water, education, and healthcare to a world population that has doubled during my own lifetime and will add another billion by the end of this decade; how to combat social inequities, racial injustice, and gender discrimination; how to strengthen our democracy, resolve conflict, maintain peace, and defend ourselves against emerging security threats; how to prepare the world to fend off the next pandemic..." (Cabrera, 2024)*

After the ISP was published, a team was formed to develop a detailed plan for all the sustainability and climate-related goals in the plan. In 2023 Georgia Tech published the *Sustainability Next Plan*, which included a blueprint for expanding sustainability education across the curriculum (Georgia Institute of Technology, 2024a), and a *Climate Action Plan* with goals for advancing climate education throughout the curriculum (Georgia Institute of Technology, 2024b). The ISP and these two plans have been key for the dedication of resources to scaling up sustainability and climate education across Georgia Tech's curriculum.

Many case studies have identified "high-level commitment" as one of several factors that contribute to the efficacy of efforts to incorporate sustainability across the curriculum at HEIs (see, for example, Tilbury, 2011; Serafini et al., 2022). Strong leadership, including strategic planning for sustainability, the provision of incentives, and coordination are all associated with more comprehensive integration of sustainability into courses and curricula (Weiss et al., 2021b). One of the benefits of high-level commitment, such as a clear commitment to sustainability in an institution's strategic plan, is the ability to secure human and financial resources to support education for sustainability initiatives. At Georgia Tech, the team charged with implementing the *Sustainability Next Plan* was invited to submit a proposed budget to support its initiatives, including an ambitious plan to significantly "scale up" sustainability across the curriculum. Committing resources to support institutional change initiatives is critical to their success. As Armstrong & Stewart-Gambino argue, "assuming that institution-wide change can be achieved through enticing individual instructors to change their courses one by one runs all the classic risks of volunteer efforts – feelings of exhaustion, frustration, and even exploitation among leaders, uneven change confined to the 'believers,' and institutional narratives of progress that gloss over problem areas where no change has taken place" (2016, p. 115). Institutional commitment to advancing sustainability across the curriculum was supported with a budget that Sustainability Next leaders could use to fund seed grants for faculty interested in making a significant time investment to incorporate sustainability and the SDGs into the classes they teach. As of summer 2024, ISP resources had supported 45 course design and redesign projects resulting in new sustainability content in all six colleges at Georgia Tech. See Table 1 for a description of several of the grant-funded projects.

Seed grant examples

A project in the **Ivan Allen College of Liberal Arts** redesigned Introduction to Sociology to incorporate the SDGs. This course usually is at capacity with close to 200 students across diverse majors, as it satisfies a general education requirement. The instructor used a case study approach to connect the SDGs to local sustainability challenges through in-depth examination of Atlanta's Beltline project. The Beltline is a multi-use path with the City of Atlanta that has sparked significant debate regarding transportation infrastructure, affordable housing, and gentrification.

In the **College of Computing**, faculty initiated a pilot project evaluating the effect of integrating global sustainability goals into the curriculum on students' motivation to enhance the efficiency of their code. Case studies presented the impact of energy use in computation and the effects of computational efficiency on energy use. Students learned techniques to optimize their code, emphasizing the role of efficient coding in conserving natural resources. Results suggested many students were inclined to invest extra effort in code optimization when provided with a socially relevant rationale. Furthermore, these students demonstrated a keen interest in continuing to develop their optimization skills.

A project in the **College of Design** engaged eight members of the building construction faculty in workshops that mapped the SDGs onto the major's curriculum and the courses selected for a new minor in Sustainable Development and Construction. The school's chair spearheaded the project that aims to establish Georgia Tech's Construction Management and Real Estate Development programs as construction industry leaders in reducing global resource consumption and greenhouse gas emissions attributed to the built environment.

A pilot project in the **College of Sciences** evaluated the impacts of incorporating the Global Goals into Cellular and Molecular Biology. Case-based assignments included an essay on cholera that required explanations of the mode of bacterial transmission and mechanism by which V. cholerae's toxin causes illness as well as reflection on how studying V. cholerae biology addresses Global Goals including *SDG 3: Good Health & Wellbeing*. Instructors evaluated essays for emergent themes. They compared survey data from students in the Global Goals (GG) course with those in a regular course and found that students in the GG course reported increased in understanding of the role of their discipline in solving global problems and how current events related to their discipline. This pilot suggests the feasibility and benefits of incorporating GGs into a course normally focused on cellular mechanisms and pathways. A paper describing results has been accepted for publication (Hull et al., 2024, in press).

In the **College of Engineering**, a Materials Processing Lab incorporated Global Goals throughout the semester. As students were introduced to processes for producing polymers, ceramics, aluminum and more. Students used lifecycle analysis to explore environmental, economic, and social equity connections. In lab, students worked in small groups to relate aspects of materials processing to Goal 12: Responsible Consumption and Production. The instructor reported that her participation in the newly launched Community of Practice facilitated her ability to incorporate social and economic aspects of Global Goals into her teaching.

In the **Scheller College of Business**, a seed grant supported the development of a new course in sustainability reporting and accounting, equipping students to interpret and evaluate sustainability reports. The instructor adopted a case-based, experiential approach that incorporated business industry guest speakers. The course is the first of its kind at Georgia Tech and the instructor plans to gradually expand its scope to include diverse, global case examples to prepare students to engage effectively within diverse regulatory and sustainability reporting contexts.

Building Strategic Institutional Partnerships (CTL integration)

Sustainability across the curriculum programs in high education often have been initiated by sustainability office staff and/or faculty "champions" working in collaboration with staff in the campus sustainability office. As a result, on many campuses sustainability curricular integration initiatives are not directly tied to teaching and learning centers, even though they include faculty professional development in sustainability course design and pedagogies. And yet, robust and ongoing support for faculty in course design and evidence-based teaching and learning strategies is essential to advancing EfS.

Instructors face many barriers to incorporating sustainability and the SDGs across disciplinary courses. Demands on HEI instructors have never been higher, with faculty burnout a growing concern (Pope-Ruark, 2022; Zahneis, 2022). In addition to competing demands, barriers to engaging faculty in teaching with the SDGs revealed through survey research include lack of or inadequate training and resources, perceived lack of relevance to certain courses and fields, and a perceived lack of power to modify courses (Leal Filho et al., 2019). Faculty also report challenges associated with content-packed courses and large courses, which can make inquiry-based pedagogies harder to incorporate (Natkin & Kolbe, 2015). As described above, high-level commitment can help address barriers related to resources by providing financial support to release faculty from other commitments to work on course design and for educational development programming and staff. When provided sufficient resources, strategic partnerships with teaching and learning professional development units can facilitate initiatives that address other barriers instructors face, including unfamiliarity with education for the SDGs and its relevance to their discipline, navigating institutional norms and rules around course redesign, and student-centered pedagogies effective for teaching with "wicked problems," such as problem-based learning and other experiential pedagogies.

At Georgia Tech, the decision to formally link faculty development in education for sustainability with faculty development for teaching and learning broadly contributed to the Institute's ability to address these challenges and "scale up" teaching and learning with sustainability and the SDGs across the curriculum. In 2022, a formal partnership was established between the unit charged with expanding sustainability across the curriculum—Sustainability Next's education team (SNE)—and the Center for Teaching and Learning (CTL), whose mission is "to enhance the learning and teaching environment at Georgia Tech by encouraging a fully engaged, sharing community with communication networks, resources, and innovative programs for faculty, postdoctoral scholars, and graduate students" (Georgia Institute of Technology, 2024). A new position in CTL, assistant director, faculty development for sustainability education initiatives (ADSE), established a formal and ongoing structure for leveraging CTL's faculty development in teaching and learning resources and expertise in support of SNE's mission. The new position also formalized a mechanism for enhancing CTLs programming through the inclusion of EfS approaches and resources into "signature programs" such as CTL's regularly offered Course Design Studio.

Developing a direct and transparent relationship between an institution's education for sustainability program and the unit supporting teaching and learning on campus offers several benefits. First, it can significantly expand the depth of expertise engaged in designing faculty workshops and other professional development opportunities. Faculty development support is essential to capacity-building among instructors to facilitate the integration of education for sustainability across the curriculum (Ryan and Cotton, 2013; Mulá et al., 2017; Weiss et al., 2021). Incorporating complex real-world learning into courses and curricula is challenging and benefits from the support of educational developers who can provide training in course redesign, sustainability competencies, and pedagogical approaches relevant to EfS (Lozano et al., 2017; Cross and Congreve, 2020; Expósito & Sánchez, 2020; Natkin and Hill, 2020).

The current structure linking CTL with EfS faculty development at Georgia Tech built upon earlier collaboration, beginning in 2017 with consultations between SLS and CTL on the development of community sustainability learning outcomes and other aspects of the SLS program. Over several years, SLS and CTL partnered on occasional workshops and then deepened the relationship through a 2021 faculty learning community on teaching with the SDGs that integrated course design and pedagogy "best practices" with education for sustainability instructional approaches. The resources made available through Sustainability Next, described above, later provided funding for the new ADSE position within CTL. As part of CTL's Faculty

Teaching and Learning team, creation of the new role enabled the author to draw on the deep expertise in teaching and learning in the Center to enrich education for sustainability offerings while also slowly incorporating elements of EfS into CTL programming.

A second benefit of explicitly connecting sustainability education with evidence-based course design and teaching strategies is that it can mitigate the response among faculty to new professional development offerings that sustainability education is just another "add-on" or fad that will come and go. Teaching core content within the context of urgent societal challenges students care about can enhance teaching and learning generally (Brundiers et al., 2010; NRC, 2012; Dabbagh, 2019). Establishing formal connections between sustainability education offerings and the university's teaching and learning center communicates that synergy to instructors.

A third benefit for sustainability education programs offered by a formal alliance with campus teaching and learning units, including global learning and service-learning offices as well as CTLs, is the rationalization of faculty development offerings related to teaching and learning for instructors. As discussed above, the ability of instructors to identify and take advantage of opportunities and resources that will enhance their teaching is constrained by the many demands placed on them. Because instructors have limited time and attention to dedicate to professional learning, campus units that support effective teaching should collaborate to integrate and streamline their offerings. Many faculty members who choose to incorporate EfS into their teaching have some experience with course design frameworks and High-Impact Practices, or HIPs (Kuh, 2008). Integrating EfS approaches with teaching frameworks familiar to faculty can both support their comfort with EfS approaches and underscore the ways in which EfS aligns with and support effective teaching and learning generally.

Creating a Structure for Formal Faculty Input

Scaling up sustainability across the curriculum requires collaboration across many units, and building and expanding campus networks is essential to supporting collaborative initiatives in HEIs (Kezar, 2006). However, large universities often struggle with the challenges of coordination and cooperation across units, and Georgia Tech is no exception. SLS was housed in the Office of Undergraduate Education (OUE), which is separate from the colleges and schools (the latter are referred to as "departments" at most universities). SLS, like other units within OUE, was charged with reaching out to faculty in all six colleges and all schools within them[1]. Adding to the challenge of that broad scope, some engaged faculty report that some of their colleagues view initiatives and opportunities housed outside their school as a lower priority than requests and opportunities shared with them by the school chair or college dean. Even though the SLS initiative resulted from faculty input, once it was established SLS operated outside the schools and colleges. This section of the chapter describes the strategy we employed to address the challenge of separation between administrative initiatives and academic units in order to increase faculty "buy-in." The steps described below expanded the strategic engagement of diverse faculty in scaling up sustainability and the SDGs across the Georgia Tech curriculum.

Once the strategic plan was completed and the implementation plan for sustainability education had been published, the leaders of Sustainability Next formed a new standing committee, the Undergraduate Sustainability Education Committee (USEC), to foster school-level faculty engagement and increase the agency of faculty in the strategic direction of sustainability across the curriculum initiatives at Georgia Tech. We contacted deans and chairs to request nominations and were able to recruit representatives from twenty of the twenty-eight schools at the Institute. The committee structure was designed to significantly expand diverse faculty engagement in decision-making about sustainability across the curriculum approaches and initiatives.

USEC's initial charge was to shape and manage an Undergraduate Sustainability Education Innovation (USEI) grant program, funded by Sustainability Next because of the commitment to EfS in the Institute Strategic Plan. In its first year, USEC funded thirty projects across all six colleges and 18 schools that redesigned courses with sustainability and the SDGs. A grants subcommittee crafted the call for proposals and evaluation rubric, supported marketing, and organized the proposal review process, with support from Sustainability Next coordinators and the ADSE. In addition, a curriculum subcommittee completed an in-depth assessment of sustainability content within courses, to identify assets and gaps in sustainability across the curriculum. The committee raised the profile of Sustainability Next's education agenda on campus through meetings with President Cabrera and key stakeholders on campus. USEC's communications subcommittee collaborated with Sustainability Next's program and portfolio manager and the ADSE to develop an online presence that elevated committee initiatives and the innovative teaching approaches of the grantees.

Cultivating Faculty Leadership

The importance of giving faculty agency in any major curricular change initiative seems intuitive, and yet these initiatives often are advanced by units outside the colleges and schools. Advancing changes in undergraduate education requires engagement at all levels of the university, not just select units (Smith et al., 2024). Faculty members may perceive opportunities to enhance their courses that are communicated from outside their units as "top-down" impositions. Faculty agency and engagement in strategic decisions is essential to the effective implementation and long-term sustainability of curricular change initiatives (Vandenbussche, 2021, Weiss et al., 2021a). For example, Weiss et al. (2021) found that sustainability curricular integration was more robust at institutions where faculty had been formally engaged in change processes. A review of ESD initiatives similarly concluded that "Professional learning…should connect staff learning in ESD with sector initiatives and institutional change processes" (Mulá et al., 2017, p, 813). The creation of a standing committee composed mostly of faculty, with a faculty chair, communicated the Sustainability Next education team's commitment to privileging faculty voices in shaping the next iteration of sustainability across the curriculum initiatives.

Bringing in Diverse Perspectives

Henderson et al.'s 2011 review of literature on professional development initiatives designed to improve STEM instructional practices found that educational developers, STEM education researchers, faculty development researchers, and higher education researchers approached the question of how to advance change in instructional practices in very different ways. One of their recommendations was for stronger coordination and communication among these groups (Henderson et al., 2011). Similarly, faculty within and across schools at Georgia Tech who are interested in EfS hold diverse views about the best ways to expand and enhance sustainability learning in the curriculum. USEC served not only to give faculty greater agency but also to expand the kinds of views, perspectives, and expertise engaged in strategic planning to advance EfS. Expanding the group of decision makers resulted in calls for proposals that appealed to faculty who had not yet participated in EfS offerings and "expanded the tent" of engaged faculty, as described below. In addition, recruiting participants with relevant "local" knowledge and networks enhances a group's strategic capacity, or ability to develop plans well-suited for their context (Ganz, 2009). By recruiting committee members with deep knowledge of their schools and with networks helpful to spreading the word about the grant opportunity, we strengthened our ability to devise effective outreach strategies.

Leveraging Peer Learning and Peer Networks to Expand Faculty Engagement

As discussed above, creating a formal relationship between CTL and Sustainability Next facilitated Georgia Tech's ability to leverage faculty development expertise in teaching and learning with the expertise of sustainability leaders. The formation of USEC further strengthened the diversity of expertise shaping faculty support in sustainability/SDG course integration by creating a subcommittee focused on faculty development. This team of diverse faculty meet several times each year and provide input to the ADSE on faculty development offerings and resources, including the community of practice described below.

Meeting Faculty Where They Are and Building Community

The final strategy described in this chapter is how and why we deliberately broadened the scope of sustainability education to attract faculty not previously engaged by "meeting them where they are." To maintain engagement and leverage faculty leadership to further expand participation in sustainability course integration across the Institute, we initiated a new structure for long-term peer- and near-peer learning: a Community of Practice on Transformative Teaching with the SDGs (CoP).

Broadening the Scope of EfS to "Expand the Tent"

As described above, SLS focused on community-engaged, equity-centered sustainability curricular and co-curricular innovation. Its approach succeeded in shifting campus discourse and understanding of the term "sustainability" to be more holistic, with attention to equity and what it means to meaningfully incorporate community-based perspectives into sustainability learning. The Center's initiatives also resulted in the expansion of community-engaged learning and the incorporation of social aspects of sustainability among faculty with a sustainability interest. Following completion of the Strategic Plan and the associated *Sustainability Next Plan*, sustainability education leaders developed the "seed grant" program described above. The creation of the Undergraduate Sustainability Education Committee (USEC) facilitated faculty input from diverse disciplines in the development of the calls for proposals. At the beginning of the 2024-25 academic year USEC was renamed to become the Sustainability Education and Curriculum Committee (SECC) to reflect its work to expand graduate, as well as undergraduate, teaching and learning in sustainability and climate solutions.

The USEC subcommittee guiding the process decided to "expand the tent" by specifically targeting faculty teaching large, core courses who had not previously been engaged in teaching with sustainability. The call for proposals was broad, open to proposals focused on incorporating at least two of the three pillars of sustainability (equity, economics, environment) and/or at least one of the U.N. Sustainable Development Goals (SDGs). The call required instructors to describe what evidence-based, High-Impact Practices they would employ but did not require them to adopt specific pedagogies. While this change from the equity and community partnership focus of SLS was a source of concern for some, the diverse faculty guiding the call was convinced a more inclusive description of sustainability, along with prioritization of large, core courses, was needed to "scale up" sustainability across the curriculum. This "big tent" decision did, indeed, significantly expand faculty participation, including computing courses taken by virtually every undergraduate student at Georgia Tech.

The creation of USEC, later SECC, had additional outcomes supporting the expansion of sustainability across the curriculum. With the transition of Serve-Learn-Sustain to the Center for Sustianable Communities Research and Education (SCoRE), the SLS Affiliated courses program that helped students identify courses with sustainability and community engagement came to an end. USEC formed a curriculum subcommittee that began work to build a comprehensive database of courses with sustainability or SDG content at Georgia Tech and a dashboard that enables students to search for courses. A communications subcommittee developed new web content to increase visibility of courses redesigned with support from the seed grants. In addition, USEC/SECC members leverages networks and communications platforms within their home departments to publicize grant opportunities and the outreach events offered by members of the new community of practice, described below.

A New Faculty Community of Practice Facilitates Peer Learning and Outreach

USEC's formation was the first step to bring more faculty into strategic planning for EfS. As we began distributing grants, we engaged a subset of USEC to plan professional development opportunities for the grantees. "Expanding the tent" meant that we had quite a few instructor grantees with relatively limited experience teaching with sustainability and the SDGs. We knew that ongoing faculty development support would be essential for effective implementation of sustainability across the curriculum initiatives (Henderson et al., 2011; Beach et al., 2016). A second, unanticipated benefit of the new faculty group we developed was that many participants also were eager to support outreach strategies for for reaching more faculty and students, complementing the efforts of USEC.

Engaging faculty in the infusion of sustainability, or any effort to spark change throughout the curriculum of an institution, requires consideration of "structural elements that shape institutional, disciplinary, and individual power dynamics" (Amstrong & Stewart-Gambino, 2016, p. 115). For example, at Georgia Tech teaching faculty may be tenure-track professors, academic professionals, or lecturers, and each category has different performance expectations and promotion processes. The ability of instructors to carve out time for a major course redesign project depends upon resources available, scheduling flexibility within their department, and the demands and reward structures associated with their role and status.

The Community of Practice on Transformative Teaching with the SDGs (CoP SDGs) launched in August 2023 with grantees and several additional instructors interested in incorporating the SDGs into their teaching. Faculty learning communities (FLCs) and communities of practice (CoPs) are groups of instructors that meet regularly to discuss teaching and learning and build community, usually guided by an educational developer from the teaching and learning center. Participation in FLCs and CoPs have been shown to have many benefits for faculty and students, including the adoption of multidimensional instructional strategies that improves learning outcomes, increased interdisciplinary content, and the development of faculty networks that support a sense of community (Cox, 2004, 2013; Chism et al., 2014; Lotti et al., 2022).

In general, longer-term faculty development approaches like FLCs and CoPs tend to be more effective at promoting sustained teaching improvements than "one-shot" faculty development approaches (Beach et al., 2016). Learning communities focused on sustainability integration can both expand faculty understanding of sustainability concepts and use of sustainability learning outcomes and pedagogies in their courses (Natkin & Kolbe, 2015) and nurture a community of instructors committed to advancing innovative EfS (Barlett & Rappaport, 2009). CoPs are a broader, more open-ended type of faculty learning group than FLCs, which usually have a discrete timeframe, include scholarship of teaching and learning goals, and often focus on producing a specific outcome, such as a publication (Cox, 2004, 2013). Our founding participants had highly varied levels of experience with high-impact pedagogies. For example, several had deep experience working with community-based organizations to help students explore local sustainability challenges, while others were brand-new to teaching with sustainability and had more limited experience with active or experiential teaching practices. We chose a flexible community of practice model to support emergent peer and near-peer learning and a culture of ongoing adaptation in response to the group's priorities. This aspect of the group's design supported ongoing interdisciplinary integration into the course design projects and peer learning about challenging pedagogies including community engaged teaching.

We began with a two-hour workshop to shape the objectives and structure of the group that balanced an introduction to EfS for those new to the approach with examples from experienced faculty and an inquiry process that invited participants to co-design what we would do together in the CoP. Participants at the initial workshop identified three kinds of goals for the group: 1) peer learning and cross-disciplinary learning by sharing and discussing their course redesign projects-in-process; 2) building community with like-minded peers through regular discussion; and 3) learning fundamentals of EfS and accessing resources that would enhance their course design or redesign projects. The group's priorities aligned well with elements demonstrated to enhance effectiveness through research on FLC impacts, including mentoring and peer learning, cross-disciplinary learning opportunities, forums for sharing works in process, and the provision of mutual support (Richlin & Cox, 2004).

Studies exploring faculty motivation suggest that intellectual engagement and a desire to make meaningful contributions to society are important to many (Pope-Ruark, 2022; Richard et al., 2022). Consistent with these kinds of intrinsic motivation for teaching, many CoP participants reported one of the most important benefits of participation was building community with and being inspired by colleagues who share their values and passion for positive social change. In the first year of the CoP, participants had the opportunity to provide feedback through two surveys, through a collaborative poster project, and through a two-hour "reflection and planning" session. The most often cited benefits of the group included enjoyment associated with the collegiality and collaborative spirit of the group, opportunities to engage across disciplines with peers they otherwise would unlikely ever speak with, and inspiration and hope resulting from regular discussions with peers about urgent societal challenges. As one participant noted: "The CoP is a powerful platform for interdisciplinary collaboration and sharing of ideas. There is not a 'one size fits all' solution for how to incorporate SDGs into the classroom, which makes this a wonderful learning journey for everyone involved" (anon., 2024, personal communication).

In the second semester of its operation, nine new grantees from the second round of seed grant funding were invited to join the CoP and six of the nine chose to participate (one additional faculty member was interested but was unable to meet at any of the times selected). Three cohorts were created to accommodate schedules, meeting five times over the courses of the semester. An unusual element emerged in the CoP as the result of democratic decision-making, rather than facilitator-guided objectives. While meetings during fall semester had prioritized peer sharing and discussion of course design or redesign projects in process, from the initial August workshop forward this group of instructors, most of whom were not part of USEC/SECC, expressed interested in advancing faculty and student engagement in sustainability learning at Georgia Tech—beyond the projects for which they had been funded. In other words, they were interested in volunteering to support the work of USEC/SECC to expand faculty engagement in EfS. This finding provides support for the hypothesis that many faculty are intrinsically motivated by a desire to advance positive change at their institutions and in the world.

The CoP meetings and priorities for spring 2024 shifted in response to the strategic interests of the group. Through a series of meetings during final exam reading days in December, we developed plans for several spring projects: a CoP-led faculty workshop on teaching the SDGs and a faculty showcase of SDG teaching examples to be offered during SDG Action and Awareness Week (March 4-8, 2024); an SDG scavenger hunt for students featuring a dozen posters highlighting SDG-intensive course opportunities, also during SDG Action and Awareness Week; and the development of a new module for Georgia Tech's learning management system, Canvas, designed to enable GT 1000 and 2000 instructors to easily incorporate introductory SDG knowledge into their courses.[2] CoP meetings were spent collaboratively planning and marketing these initiatives, with about 30 faculty members engaging in one or more projects voluntarily. In addition, during fall 2023 a CoP subgroup group collaborated on a proposal to present an interactive workshop on Teaching with the SDGs at the University System of Georgia (USG) 2024 Teaching & Learning Conference. That workshop was presented in April 2024, with 10 faculty members from the CoP co-facilitating the conference session in Athens, Georgia.

As with the creation and structure of USEC, the participatory decision-making approach adopted for the CoP was intentional, with the aim of facilitating a sense of ownership and agency on the part of the faculty. Had we approached the faculty development initiative for grantees with a stronger emphasis on instruction, with CTL educational developers leading a series of workshops or dedicating significant time each session to direct instruction, it is unlikely the group would have expanded its goals beyond their own course projects and volunteered significant time and effort to outreach, education, and marketing to advance EfS at Georgia Tech. Of course, there are trade-offs with any approach, and this "co-learner" approach to the CoP meant that faculty spent less time learning about transformative sustainability learning pedagogies and key sustainability competencies than they would have with a more facilitator-directed approach.

In April 2024 we closed the academic year with a two-hour reflection and planning session. Participants identified several elements of course design and EfS pedagogies they wanted to learn more about, and we co-developed a structure that would meet those needs for fall 2024. Most of the 2023-24 participants decided to continue in 2024-25, and in August 2024 we incorporated new members from the group of 15 round-three seed grant recipients. During the 2024-25 academic year, the CoP continued meeting regularly to explore diverse aspects of EfS and to plan outreach about EfS within and beyond the Institute. Inspired by what they appreciated about the CoP, in April 2025 six members of the group delivered a teaching and learning conference presentation entitled "Building an Effective Community of Practice: Design Principles that Support Teaching with Impact." The design principles the group identified are shared in Table 2, below.

Table 2. Design principles developed by the community of practice on transformative teaching with the SDGs

1. Deliberately cultivate a sense of community
    - Shared mission
    - Caring and camaraderie
    - Ongoing reflection and co-design, co-facilitation

2. Provide diverse resources and opportunities to support teaching
    - o Peer learning/cross-disciplinary examples
    - o Access to relevant networks and resources
    - o Support with challenging pedagogies (e.g., community-based learning, problem-based learning)
3. Offer structure with flexibility
    - o Regular and ongoing opportunities, resources and communication
    - o Members comfortably engage and disengage, as needed
    - o Opportunities to lead (outreach, engagement, peer learning)

These community of practice outcomes and the implementation of more than 45 course redesign projects completed suggest that the participatory structure adopted was effective in securing deep engagement and commitment from faculty. Just as identity, perspective-taking, and agency are as important to nurturing students as sustainability change agents (Novy et al., 2021), our experience suggests that educational development that nurtures faculty identity as sustainability change-leaders can support deep engagement of instructors in EfS.

Discussion and Considerations for Other Institutions

Georgia Tech's strategic approach to scaling up sustainability integration across the curriculum has been effective, in part, because of its alignment with the mission, values, and organizational norms of the institution. Theories of organizational and social change support the importance of that alignment (see, for example, Campbell, 2005). Efficacy of organizational change strategies is influenced by multiple aspects of regional and institutional context, include dominant sustainability challenges, organizational structure, decision-making structures, and mission alignment (Arthur, 2016; Stephens et al., 2008; Weiss et al., 2021a). As a result, the specific strategies that prove effective at one institution may not necessarily prove to be effective at another. This section will first explore ways in which this case study compares with other published examples of initiatives to expand sustainability integration across the curriculum. That comparison is followed by several suggestions for ways in which sustainability education leaders working in different environments might consider adapting the strategies described to align with their context.

Several aspects of Georgia Tech's strategic approach are consistent with experiences at other universities. For example, many published case studies highlight the importance of high-level commitment to the success of meaningful curricular integration of sustainability. At Royal Melbourne Institute of Technology in Australia, for example, high-level commitment was structured through a sustainability governing body at the highest levels of the university. This signaled high-level commitment and alignment with university priorities, which was important for engaging widespread support (Mori Junior et al., 2019). Georgia Tech's experience confirms the value of "top-down" commitment. Our experience also suggests that high-level commitment should be paired with "bottom-up" engagement that empowers diverse stakeholders as leaders. Other institutions have reported similar benefits of this "mixed approach." For example, a nine-year case study at University of Minho in Portugal revealed that a combination of high-level elements, including institutionalization through strategic and action plans, and grassroots engagement of stakeholders resulted in significant sustainability curricular integration that reflected the values and priorities of the university community (Ramísio et al., 2019).

Comparing Georgia Tech strategic approach to that of other universities, the element of our recent experience that is most distinct is the extent of faculty engagement and agency in co-directing initiatives to expand participation and faculty development opportunities in EfS. While many institutions report their use of governing bodies to provide strategic direction for sustainability, the author has not encountered other examples of a coordinating body charged specifically with guiding the strategic expansion and enhancement of sustainability and SDG teaching and learning throughout the curriculum, across disciplines. The structure of the community of practice that serves as a primary mechanism for faculty development in EfS reflects a similar commitment to faculty agency that appears to be relatively unusual when compared to other FLCs and CoPs in EfS. While we have not yet undertaken a comprehensive survey of all faculty engaged through one group or the other, preliminary and informal feedback suggests that giving groups of faculty greater decision-making responsibility for the nature and structure of initiatives to expand sustainability across the curriculum can increase engagement and cultivate a community of "faculty champions" who volunteer to collaborate with sustainability and CTL staff and faculty to advance EfS at their institution.

Another benefit of shared leadership in EfS has been robust peer learning and cross-disciplinary learning and collaboration. Sustainability challenges are "wicked problems" that are inherently interdisciplinary, requiring knowledge, skills, and perspectives from many disciplines as well as expertise outside of the academy (Rittel and Webber, 1973; Thomashow, 2020; Tomkinson and Tomkinson, 2020). However, the field of higher education has been slow to reduce barriers and advance new structures and processes to support interdisciplinary teaching and learning (Marchesi and Rolls, 2016; Turner et al., 2022;). Cross-disciplinary exchange of ideas through faculty CoPs offer a near-term opportunity to help faculty bring disciplinary content and perspectives into their courses, while institutions work to reduce structural barriers.

Shared leadership in our CoP also has provided mutual support and assistance with pedagogies that are new to some instructors. Community-based learning (CBL) is an essential component of education for sustainability as it supports the ability of students to explore diverse perspectives, develop an appreciation of different kinds of expertise, and engage with real-world initiatives (Hirsch et al., 2023; Aramburuzabala and Cerrillo, 2023). CBL also is labor-intensive and challenging to implement, making a supportive environment essential (Hou and Wilder, 2015). Our CoP has leveraged the deep experience of several members within the group to support instructors getting started with community partners for the first time. In our experience, with highly challenging pedagogies like CBL peer learning is essential for making the approach feel "doable" for faculty. Enhancing self-efficacy with respect to service-learning supports faculty motivation to engage with this high-impact practice (Darby and Willingham, 2022). When instructors hear stories from their peers detailing how they got started and how they addressed challenges that arose, they feel more confident they can be successful with CBL.

Like Georgia Tech, several other colleges and universities have integrated faculty development in Backward Design (Wiggins & McTighe, 2005), an evidence-based, widely used course design framework, and evidence-based active learning approaches with EfS professional development. At James Madison University, the Office of Environmental Stewardship collaborated with the Center for Faculty Innovation to create a year-long faculty development program, similar to an FLC, that helped faculty redesign course modules with sustainability content (Hurney et al., 2016). As initiatives to expand and enhance sustainability learning across the curriculum gain traction, we expect to see more examples of formal linkages between centers that support teaching and learning (which operate under a variety of names) and the sustainability office or the entity charged with advancing sustainability initiatives on campus. Our experience at Georgia Tech suggests that these partnerships, when combined with approaches that empower faculty leadership, can significantly expand and enhance participation in EfS faculty development.

While the examples provided above suggest many commonalities in effective change processes, strategies for instructional changes that works well in one setting are not necessarily effective in other contexts (see, for example, Henderson et al., 2011). One of barriers commonly cited to engaging more instructors in teaching with sustainability and climate is the financial resources required to offer incentives. As discussed above, our strategy of embedding sustainability commitments into the strategic plan was important for securing financial and human resources for scaling up sustainability across the curriculum at Georgia Tech. Other large or well-resourced institutions may be able to adopt this approach or a similar strategy, such as advocating a sustainability education commitment in a capital campaign or a major teaching and learning initiative required for accreditation. At smaller institutions, and those with fewer resources, sustainability educators can explore alternative to seed grants for addressing barriers that made it hard for instructors to participate in faculty development opportunities. For example, at University of Pennsylvania, relatively low-cost undergraduate research assistants were paired with faculty to facilitate their ability to redesign courses with sustainability content. Each student assistant was paired with two faculty members, and this approach facilitated relatively low-cost cross-disciplinary collaboration and student engagement (Dmochowski et al., 2014). For institutions with funding available for student positions, offering course credit for research that supports student learning as well as the instructor's course redesign work may be another low- to no-cost alternative.

CONCLUSION

Education for sustainability that advances the SDGs, including climate action, requires significant changes in the curriculum and robust support for instructors in course redesign and EfS pedagogies. Ideally, transformative sustainability learning includes both discipline-specific and cross-disciplinary knowledge and skills, sustainability competencies such as systems and strategic thinking, the ability to analyze these challenges from diverse perspectives, and the mindset and capacity to create positive change in their future professions and communities (Sterling, 2011; Mulá et al., 2017; SDSN, 2020). The curricular changes required are challenging and instructors have many demands on their time. The experience at Georgia Tech suggests the value of creating formal structures that give diverse faculty greater agency in advancing sustainability across the curriculum together with high-level commitment and the integration of EfS with evidence-based course design and pedagogies broadly. While institutions with different characteristics and fewer resources may need to adapt these strategies, many aspects can be modified to "scale up" sustainability learning across the curriculum. The "collaborative culture" fostered through this kind of EfS strategy may also support and enhance other campus priorities, while fulfilling institutional missions to advance social good (Wright et al., 2022).

ACKNOWLEDGEMENT

The initiatives described in this chapter are the product of dozens of committed staff, faculty, and administrators at Georgia Tech. They include the faculty and staff that built the Center for Serve-Learn-Sustain, those who gave countless hours of time to shape the Institute Strategic Plan, the members of the Sustainability Next leadership and implementation teams, the members of the Undergraduate Sustainability Education Committee, later renamed the Sustainability Education and Curriculum Committee, and the members of the Community of Practice for Transformative Teaching with the SDGs. The author has no conflicts of interest to declare.

REFERENCES

Akins, E., Giddens, E., Glassmeyer, D., Gruss, A., Kalamas Hedden, M., Slinger-Friedman, V., & Weand, M. (2019). Sustainability education and organizational change: A critical case study of barriers and change drivers at a higher education institution. *Sustainability*, *11*(2), 501. DOI: 10.3390/su11020501

Aramburuzabala, P., & Cerrillo, R. (2023). Service-learning as an approach to educating for sustainable development. *Sustainability (Basel)*, *15*(14), 11231. DOI: 10.3390/su151411231

Armstrong, M., & Stewart-Gambino, H. W. (2009). Building curricular diversity through a "social movement": How faculty networks support institutional change. *Journal of Curriculum Theorizing*, *25*(1). Advance online publication. DOI: 10.63997/jct.v31i1.512

Arthur, M. M. L. (2016). *Student activism and curricular change in higher education*. Routledge. DOI: 10.4324/9781315611129

Association for the Advancement of Sustainability in Higher Education (AASHE). (2024). Sustainability Tracking, Assessment & Rating System (STARS). Accessed July 20, 2024. https://reports.aashe.org/institutions/participants-and-reports/

Barlett, P. F., & Chase, G. W. (2012). Curricular innovation for sustainability: The Piedmont/Ponderosa model of faculty development. *Liberal Education*, *98*(4), 14–21.

Barlett, P. F., & Chase, G. W. (Eds.). (2013). *Sustainability in higher education: Stories and strategies for transformation*. MIT Press. DOI: 10.7551/mitpress/9418.001.0001

Barlett, P. F., & Rappaport, A. (2009). Long-term impacts of faculty development programs: The experience of Teli and Piedmont. *College Teaching*, *57*(2), 73–82. DOI: 10.3200/CTCH.57.2.73-82

Beach, A. L., Sorcinelli, M. D., Austin, A. E., & Rivard, J. K. (2016). *Faculty development in the age of evidence: Current practices, future imperatives*. Routledge.

Benbow, R. J., Lee, C., & Hora, M. T. (2021). Exploring college faculty development in 21st-century skill instruction: An analysis of teaching-focused personal networks. *Journal of Further and Higher Education*, *45*(6), 818–835. DOI: 10.1080/0309877X.2020.1826032

Brundiers, K., Wiek, A., & Redman, C. L. (2010). Real-world learning opportunities in sustainability: From classroom into the real world. *International Journal of Sustainability in Higher Education*, *11*(4), 308–324. DOI: 10.1108/14676371011077540

Cabrera, A. (2024). A Message from President Cabrera. Georgia Institute of Technology, https://strategicplan.gatech.edu/about/cabrera-message

Campbell, J. L. (2005). Where do we stand? Common mechanisms in organizations and social movements research. In *Social Movements and Organization Theory* (pp. 41–68). Cambridge University Press., DOI: 10.1017/CBO9780511791000.004

Chism, N. V. N., Holley, M., & Harris, C. J. (2012). 9: Researching the impact of educational development: Basis for informed practice. *To Improve the Academy*, *31*(1), 129–145. DOI: 10.1002/j.2334-4822.2012.tb00678.x

Collazo Expósito, L. M., & Granados Sánchez, J. (2020). Implementation of SDGs in university teaching: A course for professional development of teachers in education for sustainability for a transformative action. *Sustainability (Basel)*, *12*(19), 8267. DOI: 10.3390/su12198267

Coops, N. C., Marcus, J., Construt, I., Frank, E., Kellett, R., Mazzi, E., Munro, A., Nesbit, S., Riseman, A., Robinson, J., Schultz, A., & Sipos, Y. (2015). How an entry-level, interdisciplinary sustainability course revealed the benefits and challenges of a university-wide initiative for sustainability education. *International Journal of Sustainability in Higher Education*, *16*(5), 729–747. DOI: 10.1108/IJSHE-04-2014-0059

Cox, M. D. (2004). Introduction to faculty learning communities. *New Directions for Teaching and Learning*, *2004*(97), 5–23. DOI: 10.1002/tl.129

Cox, M. D. (2013). The impact of communities of practice in support of early-career academics. *The International Journal for Academic Development*, *18*(1), 18–30. DOI: 10.1080/1360144X.2011.599600

Cross, I. D., & Congreve, A. (2021). Teaching (super) wicked problems: Authentic learning about climate change. *Journal of Geography in Higher Education, 45*(4), 491–516. DOI: 10.1080/03098265.2020.1849066

Dabbagh, N. (2019). Effects of PBL on critical thinking skills. The Wiley Handbook of Problem-Based Learning, 135-156. DOI: 10.1002/9781119173243.ch6

Darby, A., & Willingham, L. (2022). Faculty motivation in service-learning based on expectancy x value theory. *Journal of Experiential Education, 45*(3), 337–353. DOI: 10.1177/10538259211058292

Dewey, J., Pautz, M. C., & Diede, M. K. (2024). How do we Address Faculty Burnout? Start by Exploring Faculty Motivation. *Innovative Higher Education, 49*(3), 521–539. DOI: 10.1007/s10755-023-09685-2

Diani, M., & McAdam, D. (Eds.). (2003). *Social Movements and Networks: Relational Approaches to Collective Action*. Oxford University Press. DOI: 10.1093/0199251789.001.0001

Dmochowski, J. E., Garofalo, D., Fisher, S., Greene, A., & Gambogi, D. (2016). Integrating sustainability across the university curriculum. *International Journal of Sustainability in Higher Education, 17*(5), 652–670. DOI: 10.1108/IJSHE-10-2014-0154

Ganz, M. (2009). *Why David sometimes wins: Leadership, organization, and strategy in the California farm worker movement*. Oxford University Press. DOI: 10.1093/acprof:oso/9780195162011.001.0001

Georgia Institute of Technology (2021). Georgia Institute of Technology QEP Impact Report, March 2021.

Georgia Institute of Technology. (2024a). Sustainability Next Plan. https://sustain.gatech.edu/sustainability-next-plan/

Georgia Institute of Technology. (2024b). Climate Action Plan. https://sustain.gatech.edu/georgia-tech-climate-action-plan/

Georgia Institute of Technology. (2024c). Center for teaching and learning. https://ctl.gatech.edu/

Henderson, C., Beach, A., & Finkelstein, N. (2011). Facilitating change in undergraduate STEM instructional practices: An analytic review of the literature. *Journal of Research in Science Teaching, 48*(8), 952–984. DOI: 10.1002/tea.20439

Hirsch, J., Yow, R., & Wu, Y. C. S. (2023). Teaching students to collaborate with communities: Expanding engineering education to create a sustainable future. *Engineering Studies*, *15*(1), 30–49. DOI: 10.1080/19378629.2023.2176767

Hou, S. I., & Wilder, S. (2015). Changing pedagogy: Faculty adoption of service-learning: Motivations, barriers, and strategies among service-learning faculty at a public research institution. *SAGE Open*, *5*(1), 2158244015572283. DOI: 10.1177/2158244015572283

Hurney, C. A., Nash, C., Hartman, C. J. B., & Brantmeier, E. J. (2016). Incorporating sustainability content and pedagogy through faculty development. *International Journal of Sustainability in Higher Education*, *17*(5), 582–600. DOI: 10.1108/IJSHE-12-2014-0180

Kezar, A. (2006). Redesigning for collaboration in learning initiatives: An examination of four highly collaborative campuses. *The Journal of Higher Education*, *77*(5), 804–838. DOI: 10.1080/00221546.2006.11778945

Kezar, A. (2012). Bottom-Up/Top-Down Leadership: Contradiction or Hidden Phenomenon. *The Journal of Higher Education*, *83*(5), 725–760. DOI: 10.1080/00221546.2012.11777264

Kotter, J. P. (2007). Leading change: Why transformation efforts fail. In *Museum management and marketing* (pp. 20–29). Routledge.

Kuh, G. D. (2008). High-impact educational practices. *Peer Review : Emerging Trends and Key Debates in Undergraduate Education*, *10*(4), 30–31.

Leal Filho, W., Shiel, C., Paço, A., Mifsud, M., Ávila, L. V., Brandli, L. L., Molthan-Hill, P., Pace, P., Azeiteiro, U. M., Vargas, V. R., & Caeiro, S. (2019). Sustainable Development Goals and sustainability teaching at universities: Falling behind or getting ahead of the pack? *Journal of Cleaner Production*, *232*, 285–294. DOI: 10.1016/j.jclepro.2019.05.309

Lozano, R., Merrill, M. Y., Sammalisto, K., Ceulemans, K., & Lozano, F. J. (2017). Connecting competences and pedagogical approaches for sustainable development in higher education: A literature review and framework proposal. *Sustainability (Basel)*, *9*(10), 1889. DOI: 10.3390/su9101889

McAdam, D., & Scott, W. R. (2005). Organizations and movements. Social movements and organization theory, 4.

Mcmillin, J., & Dyball, R. (2009). Developing a whole-of-university approach to educating for sustainability: Linking curriculum, research and sustainable campus operations. *Journal of Education for Sustainable Development*, *3*(1), 55–64. DOI: 10.1177/097340820900300113

Menon, M., & Paretti, M. (2024). Faculty Perspectives on Integrating Sustainable Development Into Engineering Education. *IEEE Transactions on Technology and Society*, *5*(3), 316–324. DOI: 10.1109/TTS.2024.3403482

Meyerson, D. E., & Scully, M. A. (1995). Tempered Radicalism and the Politics of Ambivalence and Change. *Organization Science*, *6*(5), 585–600. DOI: 10.1287/orsc.6.5.585

Mokski, E., Leal Filho, W., Sehnem, S., & Andrade Guerra, J. B. S. O. D. (2023). Education for sustainable development in higher education institutions: An approach for effective interdisciplinarity. *International Journal of Sustainability in Higher Education*, *24*(1), 96–117. DOI: 10.1108/IJSHE-07-2021-0306

Mori, R.Junior, Fien, J., & Horne, R. (2019). Implementing the UN SDGs in universities: Challenges, opportunities, and lessons learned. *Sustainability (New Rochelle, N.Y.)*, *12*(2), 129–133. DOI: 10.1089/sus.2019.0004

Mulà, I., Tilbury, D., Ryan, A., Mader, M., Dlouhá, J., Mader, C., Benayas, J., Dlouhý, J., & Alba, D. (2017). Catalysing change in higher education for sustainable development: A review of professional development initiatives for university educators. *International Journal of Sustainability in Higher Education*, *18*(5), 798–820. DOI: 10.1108/IJSHE-03-2017-0043

National Research Council (NRC). (2012). Discipline-Based Education Research: Understanding and improving undergraduate learning in science and engineering. Washington, D.C., the National Academies Press. https://nap.nationalacademies.org/catalog/13362/discipline-based-education-research-understanding-and-improving-learning-in-undergraduate

Natkin, L. W., & Hill, L. M. (2020). Faculty integration of sustainability learning outcomes into curriculum: A case study of a faculty learning community and teaching practices. *New Directions for Teaching and Learning*, *161*(161), 123–138. DOI: 10.1002/tl.20377

Natkin, L. W., & Kolbe, T. (2016). Enhancing sustainability curricula through faculty learning communities. *International Journal of Sustainability in Higher Education*, *17*(4), 540–558. DOI: 10.1108/IJSHE-02-2015-0024

Natkin, L. W., & Kolbe, T. (2016). Enhancing sustainability curricula through faculty learning communities. *International Journal of Sustainability in Higher Education, 17*(4), 540–558. DOI: 10.1108/IJSHE-02-2015-0024

Novy, J. W., Banerjee, B., & Matson, P. (2021). A core curriculum for sustainability leadership. *Sustainability (Basel), 13*(19), 10557. DOI: 10.3390/su131910557

Pope-Ruark, R. (2022). *Unraveling faculty burnout: Pathways to reckoning and renewal*. JHU Press. DOI: 10.56021/9781421445137

Ramísio, P. J., Pinto, L. M. C., Gouveia, N., Costa, H., & Arezes, D. (2019). Sustainability Strategy in Higher Education Institutions: Lessons learned from a nine-year case study. *Journal of Cleaner Production, 222*, 300–309. DOI: 10.1016/j.jclepro.2019.02.257

Richard, F. D., Berkey, B., & Burk, H. M. (2022). Motivation and Orientation: Faculty Perspectives on Development and Persistence in Service Learning and Community Engagement. *Journal of Community Engagement and Higher Education at Indiana State University, 14*(1), 12–23.

Richlin, L., & Cox, M. D. (2004). Developing scholarly teaching and the scholarship of teaching and learning through faculty learning communities. *New Directions for Teaching and Learning, 2004*(97), 127–135. DOI: 10.1002/tl.139

Rittel, H. W., & Webber, M. M. (1973). Dilemmas in a general theory of planning. *Policy Sciences, 4*(2), 155–169. DOI: 10.1007/BF01405730

Ryan, A., & Cotton, D. (2013). Times of change: shifting pedagogy and curricula for future sustainability. In *The sustainable university* (pp. 151–167). Routledge.

SDSN, F. (2020). Accelerating education for the SDGs in universities: a guide for universities, colleges, and tertiary and higher education institutions.

Serafini, P. G., de Moura, J. M., de Almeida, M. R., & de Rezende, J. F. D. (2022). Sustainable development goals in higher education institutions: A systematic literature review. *Journal of Cleaner Production, 370*, 133473. DOI: 10.1016/j.jclepro.2022.133473

Smith, D., Finkelstein, N., Horii, C. V., Miller, E., & York, T. T. (2024). *Levers for change: Enacting a national agenda for undergraduate STEM education* (Scherer, L., & Veague, T., Eds.). American Association for the Advancement of Science.

Snow, D. A., Rochford, J., Worden, S. K., & Benford, R. D. (1986). Frame Alignment Processes, Micromobilization, and Movement Participation. *American Sociological Review, 51*(4), 464–481. DOI: 10.2307/2095581

Stephens, J. C., Hernandez, M. E., Román, M., Graham, A. C., & Scholz, R. W. (2008). Higher education as a change agent for sustainability in different cultures and contexts. *International Journal of Sustainability in Higher Education*, *9*(3), 317–338. DOI: 10.1108/14676370810885916

Sterling, S. (2011). Transformative learning and sustainability: Sketching the conceptual ground. *Learning and teaching in higher education,* *5*(11), 17-33.

Strang, D., & Soule, S. E. (1998). Diffusion in Organizations and Social Movements: From Hybrid Corn to Poison Pills. *Annual Review of Sociology*, *24*(1), 265–290. DOI: 10.1146/annurev.soc.24.1.265

Stubbs, W., Ho, S. S., Abbonizio, J. K., Paxinos, S., & Bos, J. J. A. (2021). Addressing the SDGs through an integrated model of collaborative education. In *Handbook on teaching and learning for sustainable development* (pp. 252–271). Edward Elgar Publishing. DOI: 10.4337/9781839104657.00025

Thomashow, M. (2020). Postface: Why Environmental and Sustainability Learning Matters (Now More Than Ever). *New Directions for Teaching and Learning*, *2020*(161), 189–193. DOI: 10.1002/tl.20381

Tilbury, D. (2011). Higher education for sustainability: a global overview of commitment and progress. *Higher education in the world,* *4*(1), 18-28.

Tomkinson, B., & Tomkinson, R. (2020). An Interdisciplinary Problem-Based Approach to Education for Sustainable Development. In *Teaching and Learning Strategies for Sustainable Development* (Vol. 19, pp. 29–40). Emerald Publishing Limited. DOI: 10.1108/S2055-364120200000019006

University Global Coalition. (2022). About Us. https://universityglobalcoalition.org/about/

Vandenbussche, J., Ritter, L., Callahan, K. M., & Westlund, E. E. (2021). Using strand committees to build faculty support for departmental change. *PRIMUS (Terre Haute, Ind.)*, *31*(3-5), 330–342. DOI: 10.1080/10511970.2020.1772916

Weiss, M., Barth, M., & von Wehrden, H. (2021a). The patterns of curriculum change processes that embed sustainability in higher education institutions. *Sustainability Science*, *16*(5), 1579–1593. DOI: 10.1007/s11625-021-00984-1

Weiss, M., Barth, M., Wiek, A., & von Wehrden, H. (2021b). Drivers and Barriers of Implementing Sustainability Curricula in Higher Education—Assumptions and Evidence. *Higher Education Studies*, *11*(2), 42–64. DOI: 10.5539/hes.v11n2p42

Wenger, E., McDermott, R. A., & Snyder, W. (2002). *Cultivating communities of practice: A guide to managing knowledge.* Harvard business press.

Wiggins, G. P., & McTighe, J. (2005). *Understanding by design.* Ascd.

Wright, C., Ritter, L. J., & Wisse Gonzales, C. (2022). Cultivating a collaborative culture for ensuring sustainable development goals in higher education: An integrative case study. *Sustainability (Basel), 14*(3), 1273. DOI: 10.3390/su14031273

Zahneis, M. (2022). A rare survey of faculty morale shows that the pandemic's effects continue to ripple. Chronicle of Higher Education. https://www.chronicle.com/article/a-rare-survey-of-faculty-morale-shows-that-the-pandemics-effects-continue-to-ripple

ENDNOTES

1. In 2024 Georgia Tech added a seventh college and 29th school of Lifetime Learning.

2. GT 1000 (for incoming first-year students) and GT 2000 (for incoming transfer students) are one-credit courses designed to support the successful transition of students to the Georgia Tech learning environment.

Chapter 6
Africana Studies and the Environmental Humanities

Régine Michelle Jean-Charles
Northeastern University, USA

ABSTRACT

This essay explores the intersection of two robustly related fields--Africana Studies and the Environmental Humanities. Drawing on expertise in Black feminist literary studies, I argue that the current climate and higher education requires that we develop new pedagogical approaches to engage students about the twin impacts of racial and environmental injustice. Throughout the essay I refer to scholars of Africana Studies, Environmental Humanities, Black Feminism, Caribbean Studies and eco-feminism who have made notable interventions in the field and whose work helps us to understand the overlapping concerns in these interrelated areas of study. I argue that taking into consideration how students' understanding of the manifold connections between racial and environmental justice bolsters their commitment to both, demands that higher education account for and address their perspectives by taking up new pedagogical approaches that dissolve the boundaries between theory and praxis, analysis and experience, and even humans and nature. Of particular interest is how engaged pedagogy and Black feminist pedagogy offer an array of tools for teaching about environmental justice through an intersectional lens in the 21st century.

DOI: 10.4018/979-8-3693-5454-4.ch006

INTRODUCTION

Fighting against social ills and fighting for empowerment and sustainability are political tasks for which Africana and environmental studies are totally unapologetic and well suited, as they are for the task of advancing innovative and rigorous scholarship.

—*Greening Africana Studies*, Ruben Patterson

It was a bright but bitterly cold New England day in January of 2020 when I began teaching "Between Nature and Culture: Caribbean Literature and the Environment" for the second time. Though it was the first day of class, this advanced level cross-listed Africana Studies and French Literature seminar was already filled beyond capacity with both undergraduate and graduate students. I was proud of the syllabus I had painstakingly crafted selecting works from my current research and prior teaching. The cover page was adorned with gorgeous artwork from the Haitian artist Edouard Duval-Carrié's "Imagined Landscapes" series (see fig. 1). Entitled "After Church–Mystic Lagoon," the black and silver painting is a dark image of a lush forest encircling a brightly glimmering pond. Duval-Carrié's painting emblematized the attention to detail in representations of nature that I planned for the class to experience throughout the semester. On the syllabus I explain our focus on the natural world as an important site of Caribbean art and literature. Applying eco-criticism to the field of francophone Caribbean literature, our goal was to examine how fiction approaches the relationship between human activity and the environment. How does the novel inhabit Caribbean its unique ecologies and topographies? In what ways do Caribbean texts offer meditations on nature, culture, and history? As the earthquake in Haiti 2010 demonstrated with calamitous force, and the cycles of Caribbean hurricanes have shown over the years, so-called natural disaster is a political crisis. In view of this, we also consider the legacies of slavery and colonialism in relation to systemic inequality and injustice through an intersectional lens. (Steinmetz, 2020) With literature and visual art as our guides, we grappled with how novels, films, poetry, and essays attune us to the natural environment and portray elements from land to sea as well as the relationships between humans and nature.

Figure 1. "After Church–Mystic Lagoon" (Edouard Duval-Carrié (2013) mixed media on aluminum, 96" x 144)

My students were an environmentally conscious and politically minded group eager to embark on this new journey. As they introduced themselves on that first day, many shared that they were drawn to the class due to their own activism centered on environmental justice, though few had had the opportunity to pursue these interests in a university classroom. The class worked in the interest of those committed to thinking about the environmental humanities through the lens of racial justice. For example, several students used their oral presentations as an occasion to advocate for distinct environmental causes. One graduate student from Spain presented on the preponderance of trees in the novel *The Infamous Rosalie* by Évelyne Trouillot. She noted how the trees were more than the landscape—they became forms of refuge for the enslaved people on the plantation whether they were maroons seeking hiding places in the forest laden mountains, or children playing games in the brush. She provided a history of deforestation in the creation of the plantation economies and encouraged students to pay more attention to the trees that surround them (Paravisini-Gebert, 2011). For her presentation, another student from Colombia homed in on the unique ecology of the mangrove as a Caribbean entity and issued a call to action, reminding us that Environmental Studies is "oriented toward a better understanding of negative human impacts on the environment and ways to reduce them" (Patterson, 2015, p. 5). Our classroom environment was full of energy, with the students' burgeoning curiosity and attentive participation serving as fertile ground for our inquiries about how environmental justice and racial justice intersect within this unique context.

We had almost reached the halfway point of the semester when the class, the university, the country, and the world, were forever altered due to the unprecedented outbreak of the novel coronavirus. Unsure of what lay ahead, we sullenly met in person for the last time in a classroom on campus, feeling like we had had the wind knocked out of us. After pausing for three weeks while the college's administration determined how to move forward, we resumed the semester in the same way that educators all over the world were doing during the pandemic—virtually online via Zoom. We never reconvened in person after Spring break or for the remainder of the semester. Instead, we peered at one another through computer screens, technology both facilitating and inhibiting our interactions. With each student contained in their tiny Zoom boxes, the convivial atmosphere of our classroom community lost its luster. As a Black feminist scholar who takes seriously bell hooks' mandate to create a classroom environment in which students are invited to bring their entire selves and who has embraced the tenets of engaged pedagogy throughout my teaching career, I was challenged and perplexed by how to proceed in this new normal (hooks, 1994). Desperate for ways to connect with my students and attend to the concerns they had beyond the scope of the syllabus, I devised a new assignment. Instead of writing papers and delivering presentations I instructed each student to go on weekly, if not daily nature walks. It was a habit I had taken up with my own family to offer our four elementary school aged children a change of pace from their newly monotonous Zoom school lives. While on the walk, the students should take a picture of nature and then find a passage from one of our novels to connect to the image that they took. For many, the landscapes they were in were a far cry from the tropical environments in the literature we were reading, but the assignment allowed them to make connections across time and space. Even more, this activity allowed my students to contemplate and experience the natural world around them in a personalized way. At a time when the government forced people to remain cloistered indoors, being outdoors, reflecting on space beyond the walls that contained them, also provided a much-needed change of environment for my students.

At a loss for how to teach in this new normal while managing the needs of my own four children, my other strategy was to reach for the wisdom of the writers we had been learning from throughout the semester and to follow their examples. Jacques Roumain: "Je suis ça, je suis cette terre là. [I am this, I am the land.]" Zora Neale Hurston: "All geography is within me." Kettly Mars: "I am the water enclosed in the rocks." Maryse Condé: "How true! Life's problems are like trees. We see the trunk, we see the branches and the leaves. But we can't see the roots, hidden deep down under the ground. And yet it is their shape and nature and how far they dig into the slimy humus to search for water that we need to know. Then perhaps we would understand" (Condé, 1989). Inspired by these Black authors from throughout the Americas, I entreated my students to look to the land, to be with the land, and to take refuge in the environment not only for the sake of learning, but for their own wellbeing and self-actualization. By turning to nature and engaged pedagogy I found a way to care for, connect with, and continue educating my students in what was one of the difficult periods of my career as a professor (hooks, 1994).

TEACHING RACIAL AND ENVIRONMENTAL JUSTICE IN THE 21ST CENTURY

Reflecting on my experience teaching "Caribbean Literature and the Environment" during the unique time that was spring 2020, it strikes me that what we experienced together came at a critical juncture that is also emblematizes a new reality in higher education. Not only is teaching about environmental justice and racial justice together necessary and urgent in our current climate, but in the 21st century, we are serving a new generation of students passionate about environmental justice, enraged about the imperiled planet, and lamenting the deleterious effects of unsustainable life who are deeply committed to approaching these problems with an intersectional lens that is attentive to how race, class, gender, sexuality, and other locations of identity inter-imbricate our experience of the natural world. These students live in the context of a political landscape in which abortion and affirmative action have been struck down, diversity, equity, and inclusion (DEI) efforts are being systematically dismantled, the humanities are being defunded, and both African American history and critical race theory are under attack. Many of them are also racial justice warriors critical of how white supremacy, heterosexism, and capitalism are shaping the world, and they are eager *to do something about it.* In my view this new context is an invitation to leverage pedagogical approaches such as experiential learning, Black feminist pedagogy, and engaged pedagogy as praxis in order to teach about racial and environmental (in)justice (Freire, 1970; hooks, 1994). Engaged pedagogy is of particular relevance for how it "views students as whole and embodied beings; it destabilizes the mind/body split upon which traditional pedagogical practices are built and which positions knowledge as neutral and devoid of affect; it takes students seriously as knowers; it enables a process of conscientization." (Tack, 2023). Furthermore, because engaged pedagogy calls for "teaching students in a manner that respects and cares for their souls as opposed to a rote, assembly line approach" it is especially useful in times of crisis like what our students experienced during the pandemic (hooks, 1994, p. 13). I am also interested in how Black feminist scholar, professor, and author bell hooks advocates for education that goes beyond the classroom, because, in my view this is exactly what the environmental humanities must do (Florence, 1998). David Pellow's essay "Toward a Critical Environmental Justice Studies: Black Lives Matter as an Environmental Justice Challenge," makes the case for an environmental humanities approach that is multi-disciplinary, draws on diverse fields including critical race theory, ecofeminist theory, and political ecology, and places Black Lives Matter resolutely at the center. To do so he offers the framework of "Critical Environmental Justice Studies," that "brings greater attention to how multiple social categories of difference are entangled in the production of environmental injustice, from race, gender, sexuality, ability, and class to species, which would attend to the ways that both the human and more-than-human world are impacted by and respond to environmental injustice" (Pellow, 2016, p. 224). Following Pellow, in this chapter I explore how Africana Studies intersects with the environmental humanities as well as its attendant and adjacent fields of environmental justice, climate change, and sustainability. I argue that taking into consideration how students' understanding of the manifold connections between racial and environmental justice bolsters their commitment to both, demands that higher education account for and address their perspectives by taking up new pedagogical approaches that dissolve the boundaries between theory and praxis, analysis and experience, and even humans and nature. In my view, engaged pedagogy and Black feminist pedagogy offer an array of tools for us to deploy in teaching the Africana Studies and the environmental humanities together.

Defined as "the disproportionate burden of environmental harm facing these populations," environmental injustice has far-flung manifestations in Black communities throughout the diaspora (Natural Resources Defense Council, n.d.). Whether it is the lack of green spaces in highly surveilled urban centers with large Black populations, the inordinately negative effects of climate change on Caribbean islands and in cities throughout the African continent, or how ecological crises unjustly ensnare marginalized people, we know that environmental justice and racial justice are intimately entangled. Scholars such as bell hooks, Alexis Gumbs, Nathan Hare, Audre Lorde, and Fred Moten among others have championed the cause of the planet, the earth, the non-human environment, and the climate as pressing subjects essential to people of African descent who are severely impacted by environmental injustice. An understanding of what has been dubbed "Black Ecology," offers a salient point of departure for thinking critically about Africana Studies as an environmental humanities project. As a concept, an analytic, and a mode of being, "Black ecology…moves by way of an aesthetic that is visible and invisible, earthly and cosmos, phantasmic and imaginative" (Cerenvak, 2023, p. 7). In "Black Ecology," the seminal essay in which the concept is named for the first time, Nathan Hare delineates the impact of racism on the lived experience of Black people in environmental terms. The essay is purposefully experiential and activist minded, consistent with each field's intention to dissolve the boundaries between the academy and the community. As Sarah Cerevnak explains describing *Black Ecology,* "Hare's Black ecology is regarded as a key text that identifies and extends the need for environmental justice activism" (Cerenvak, 2023, p. 7). *Black Ecology* confirms one of the most important ways that Black Studies (or Africana Studies) and Environmental Studies intersect—each is conceived as a social justice project whose utility must also be applied outside of the academy. What is essential to understand about the term black ecology is how it opens up political possibilities and modes of resistance even while it attends to fraught histories and environmental harm. In *Dark Agoras,* J.T. Roane argues that Black ecologies are "a formulation describing not only Black communities relationship to climate change but also the relations of mutuality and resistance giving expression to possibilities beyond the delimitation of Blackness as living in vulnerability and death taking seriously various cultural mediums of Black expression as vehicles of insurgent and practical knowledge in the face of environmental degradation, geographic dislocation, and territorial violence" (Roane, 2023, p. 8). The responses to climate change and environmental harm are as important as the ability to identify and analyze how it operates in the lives of marginalized people.

Africana Studies and Environmental Studies are proximate fields with similar investments. Numerous scholars and studies, "have documented that people of color, people of lower socioeconomic status, indigenous and immigrant populations, and other marginalized communities are disproportionately affected by ecologically harmful infrastructures, such as landfills, mines, incinerators, polluting factories, and destructive transportation systems, as well as by the negative consequences of ecologically harmful practices, such as climate change/disruption and pesticide exposure" (Pellow, 2016, p. 222). As the authors of *Black Studies and Environmental Studies* have written, "we must recognize the two fields of study as complementary in their respective missions...both have an academic mission and an unapologetic social mission" Patterson, 2015, p. 2). The field of Environmental Studies includes studies of the environment that account environmental justice, and sustainability studies. Patterson is also careful to issue a point of caution writing that, "The notable successes in Africana and environmental studies should be kept in perspective, considering how much more work needs to be done in the areas of racial inequality and environmental degradation and its attendant climate chaos" (Patterson, 2015, p. 8). In other words, and as our students' activism continues to demonstrate, there is still much work to be done in the efforts to eradicate racial and environmental injustice. Defining the environmental justice movement, Pellow offers that it is, "composed of people from communities of color, indigenous communities, and working-class communities who are focused on combating environmental injustice—the disproportionate burden of environmental harm facing these populations. For the [environmental justice] movement, social justice is inseparable from environmental protection" (Pellow, 2016, p. 222). These studies help to illustrate the extent to which environmental justice and racial justice require activist and pedagogical approaches that not only acknowledge their proximity in the 21st century but allow students to think and act creatively to advance these causes together.

AT THE INTERSECTION OF AFRICANA STUDIES AND THE ENVIRONMENTAL HUMANITIES

The course that I taught in Spring 2020, "Caribbean Literature and the Environment" is but one example of how and intersection of Africana Studies and the environmental humanities can occur in the classroom, that, if done correctly will also call for students to move outside of its limits. In almost two decades of teaching Africana Studies classes I have encountered curricula with exciting examples for how to approach the environmental humanities like "Black and Green" focused on environmental justice in Black communities, "Race and Place" in which students examine how racial identity relates to the built environment, or "Visions of the Sea" in which students home in on the multiple meanings and manifold renderings of the sea in African diasporic history, literature, and culture. For the class that I taught, though we explored human-environment interactions in general, we also paid attention to how the environment functions in the literary texts and in the lives of Black people in the Caribbean. I introduce my students to critical frameworks like eco-criticism and ecofeminism as well as concepts such as environmental justice, and representations of nature in literature by authors from the French-speaking Caribbean. We explore numerous topics such as the effects of slavery and colonialism on the natural environment, how the enslaved used nature to facilitate flight and power their resistance, parallels between gender-based violence and violence towards the environment, the effects of natural disasters, and the disproportionate impact of climate change on Caribbean islands. We read captivating novels that span across centuries: Jacques Roumain's *Gouverneurs de la rosée/ Masters of the Dew* (1945, Haiti), Simone Schwarz-Bart *Pluie et Vent sur Télumée Miracle/ The Bridge of Beyond* (1972, Guadeloupe), Évelyne Trouillot's *Rosalie l'infâme/ The Infamous Rosalie* (2003, Haiti), Gisèle Pineau's *L'espérance-macadam/ Macadam Dreams* (1995, Guadeloupe), and Patrick Chamoiseau's *Texaco* (1990, Martinique). We watch powerful films such as *1.5 Stay Alive* in which music and culture are put on display alongside the damage being done to Caribbean islands which are more vulnerable to climate change than wealthier countries. The films were especially noteworthy for their steadfast inclusion of a call to action, demanding more from the audience. In each of these texts—both visual and literary, the environment comes to life as a character with the ability to influence the outcome of the plot. The novels in particular present land, sea, and all elements of nature as more than backdrop—they are history, character, content and form. In the context of the Caribbean, these relations become increasingly fraught. Based on what Kimberly Ruffin refers to as the "ecological burden and beauty," people of African descent in the Americas experience the land as both captivating and captivity due to the histories of colonialism and slavery (Ruffin, 2010, p. 2). Put differently, a study of the environment that intersects with the Black experience in the Americas introduces tensions that are difficult to resolve. At the same time, and as Erin Sharkey demonstrates with force in the collection *A Darker Wilderness: Black Nature Writing from Soil to Stars,* Black writers have embraced the environment through nature writing, activism, and advocacy work demonstrating a sense of belonging and rootedness in place despite the contradictions birthed from living in the Americas (Sharkey, 2023).

To complement our readings and provide a critical framework, we read widely from the fields of eco-criticism and eco-critical literary studies. We pore over essays by scholars like Evelyn O'Callaghan who argues that "Caribbean landscapes are texts in their own right, huge canvasses on which history has certainly painted very different pictures over time which can be read in different ways…" (O'Callaghan, 2015). We meditate on the presence of the land and all of its details—the impossibly verdant forests, lush mangroves, flourishing flora and fauna, hot tropical landscapes, pungent odors of the soil, flowing rivers and gurgling streams, the crashing waves of the Atlantic Ocean and the Caribbean Sea, and the broad expanse of the plateau—all make the natural world an exigent site of Caribbean cultural production both in literature and visual art. I take my students on a journey that appreciates the landscape, recognizes how it bears the weight of history, and invites them to consider how they might proactively work towards a better, more just future that advances racial justice and environmental justice. Because our focus is the Caribbean and I teach in the context of elite predominately white universities in the United States, the specter of tourism looms largely over our discussions. On several occasions I have had students whose only understanding of the Caribbean was informed by their experience as tourists. The class illuminated another aspect of these islands while simultaneously unveiling the environmental hazards of the tourism industry whose outsized negative effects on climate and sustainability they can no longer ignore.

While my class focused on Caribbean literature by French speaking authors, the context of African diasporic literatures in general is fertile ground upon which to observe how cultural workers take on overlapping commitments to racial and environmental justice. For example, "In [Gayl Jones] novel *Corregidora,* Black art makes for another imagining of relation, one that indicts anti-Black and anti-earth extraction," as Sarah Cervenak writes in *Black Gathering: Art, Ecology, Life*. Other scholars such as Camille Dungy in *Black Nature: Four Centuries of African American Poetry* and *Black on Earth: African American Ecoliterary Traditions* and Nathan Hale have also approached the environment keeping in mind the unique histories and lived experience of people of Africana descent. Scholars of Black Studies—in particular Black feminist scholars—have worked steadfastly to develop an ethic of relation to the land that is not extractive. "A small but growing group of researchers—including and especially environmental humanities scholars—have focused on the ways that gender, sexuality, citizenship, indigeneity, and nation shape the terrain of eco-logical inequalities, but those areas of scholarship remain in need of further development" (Pellow, 2016, p. 223). From my vantage point as a scholar of Haitian literature and Black feminism, I also turn to these contexts to understand representations of the environment. In Haitian culture, it is well known that spirit is nature and nature is spirit. That is to say, there is an additional spiritual connotation to and connection with every aspect of nature that we must also take into account. To that end, Maya Deren describes Vodou as "a religion of creation and life. It is the worship of the sun, the water and other natural forces" (Deren, 1983, p. 113). Similarly, the religion scholar Leslie Desmangles notes that "If Catholicism is identified with heaven, Vodou is associated with the earth. And if Haitian writers say that the *lwas* 'have to do with the earth' hence cannot be uprooted from Haitian life, it is because Vodou fulfills important functions in Haitian society" (Desmangles, 1992).

The trees offer one salient example of how elements of nature function as metaphor and method for impacting environmental justice that students can connect with. In Kettly Mars's novel *Kasalé* we notice the role that the trees play in the life of the protagonist for whom nature is deeply embedded in her life. As a practicing Vodouiyzan, Antoinette understands that nature is omnipresent and to be revered. She cherishes, protects, and communes with the land. In my book *Looking for Other Worlds: Black Feminism and Haitian Fiction,* I argue that there is a Black feminist ethic of care animating the 20th and 21st century works of Haitian women writers like Évelyne Trouillot, Kettly Mars, and Yanick Lahens. The book's final chapter, "Land and Sea: An Environmental Ethic," argues that the relationship between human activity and the natural environment communicates ethical concerns in the novels *Rosalie l'infâme/The Infamous Rosalie* (2003), *Kasalé* (2007), and *Bain de Lune/Moonbath* (2014), which highlight Black women's intimate relationships to nature according to a Vodou ethic. My arguments in the book were informed by the years I spent teaching "Francophone Caribbean Literature and the Environment" as well as my research on Black feminist ecocriticism and the role of Vodou in Haitian culture. In the chapter I also examine how the environment is represented in visual art by Haitian feminist artist Mafalda Mondestin. I share two images from a series entitled, "When the Island Holds All of the Secrets." The first is an image of a woman walking through the land, a crown perched on her head (see figure 2). In my analysis I wonder if she might be Madan Ayizan, the Haitian mother of the land. The second image features a set of ghostly dangling legs hovering ominously in a fuchsia sky. In the chapter's introduction I wonder where those legs come from. "The legs hover above what resembles tall stalks of sugar cane. The hanging legs in the context of cane fields conjure histories of violence, from the eighteenth-century sugar plantations that led to France's financial dominance in the world to the twenty-first-century *bateyes* (sugar mill settlements) in the Dominican Republic where human rights abuses against migrant workers of Haitian descent tragically occur on a daily basis" (Jean-Charles, 2023).

Figure 2. When the Island Keeps All the Secrets, Part I (Mafalda Mondestin, 2015)

Some of the most important thinkers in the field of Black Studies were also tireless advocates for environmental justice. This is especially the case for Black feminists like Audre Lorde and bell hooks. Ruminating on the legacy of Black feminist lesbian poet Audre Lorde, Alexis Gumbs writes, "There is no time that I can find in Audre Lorde's life where she was not thinking about our relationship with the planet." Similarly, bell hooks develops and environmental ethic of place and care in how she writes about her native Kentucky. Still, and as Chelsea Micheala Frazier has forcefully argued, although "Black feminist voices have often been seemingly absent from mainstream environmentalism and the intellectual movement that sprang forth from it in the early 1990s [...] Black Feminist Ecological Thought has been present and continues to evolve alongside an ecocriticism that often fails to recognize its existence and its intellectual and creative authority" (Frazier, 2020). In point of fact, "scholars in Black Studies...along with twentieth and twenty-first Black women writers and poets, have long engaged the intersection of anti-Black and anti-earth violence towards new models for social and ecological relation" (Cerenvak, 2023).

When we think carefully about where Black Studies and the environmental humanities intersect, we notice that scholars, artists, writers, intellectuals and activists have been sounding the trumpet of conscience as it is related to the environment. According to Fred Moten, Black Studies has an earth-saving mandate. He writes in "Black Topographical Existence": "We are a people is what we have been saying for generations to colonizers, invaders, and imperialists hell-bent on destroying us. And now, more than ever, Mother Nature too. The blessings of our islands is also their curse. Our geography gives us the year-round sun and beautiful beaches, but increasingly, in the age of extreme weather and climate change, we are on the front line of destruction.

CARIBBEAN LITERATURE AND THE ENVIRONMENT

In the canonical Caribbean novel, *Masters of the Dew,* Jacques Roumain's protagonist Manuel Bienaimé declares emphatically: *"Je suis ça, je suis cette terre là et je l'ai dans le sang"*/ [I am this land, and I have it in my blood]". Written by the founder of Haiti's socialist party, the novel explores themes such as community and collective work in order to more successfully care for the land. On the one hand Manuel was speaking to his particular positionality as a Haitian man who returns to his native land after having travelled to Cuba and who encounters a blighted and desiccated land suffering from drought. On the other hand, he might be speaking of the entire Caribbean and Black world in general. Our connection to the land and profound understanding of how and where we figure within it is a vaunted ideal for Black artists and authors.

Reading Caribbean literature that conveys environmental concerns, I have often been overwhelmed by the gorgeous and haunting power of the land. The lushness of the mangroves, the flora and fauna of tropical landscapes, the intricacy of the rhizome, the flow of great rivers, the crashing waves of the Atlantic—in Caribbean literature and art the natural world bursts with meaning. In recent years filmmakers have also joined in to create art that not only depicts the beautiful Caribbean landscapes, but also advocates for its maintenance and reveals how it is being affronted in the 21st century. Arnold Antonin's film *Sa lanmè di/Ainsi parla la mer* [So Spoke the Sea] (2021) draws from an ur-text of Haitian literature, Jean-Price Mars's *Ainsi Parla l'oncle,* a study urging Haitians to return to their African roots and to champion folklore, Vodou, and Kreyòl language. Antonin's film offers the wisdom of the sea, whispering in dulcet tones to the audience about how it has been afflicted and mistreated by human activity. The film describes myriad environmental crises enveloping Haiti, including plastic pollution, soil erosion, and vulnerability to both drought and flooding through so-called natural disasters. In a review of the film for *Journal of Haitian Studies*, Lois Wilcken writes, "Lanmè's central message that both [the sea] and her children depend on each other for their well-being plays out in thematic sections, each one articulated in the thoughts, experiences, and expressions of Haitian artists, activists, scientists, spiritual specialists, and workers" (Wilcken, 2023). Throughout the film it is this idea of relationship that is repeatedly emphasized. How humans interact with nature is the main lens through which the film conveys its message: the sea is in danger because of what humans have done to it both intentionally and through neglect. The film concludes with one of the many environmental experts featured advocating for the revival of mangroves and coral reefs for protection against the storms that have become increasingly destructive because of climate change. The film also ends with the sea offering her own conclusions for a better environmental future. "Renmen m pou m pwoteje w [Love me so that I protect you]." In other words, the relationship between humans and the environment can and should be one of love, care, and reciprocity. Films like *Ainsi parla la mer* manifest the intersections between Africana Studies and environmental justice, in this case situating the example of a particularly vulnerable island nation within a broader context of conquest, corruption, and neglect.

When I asked students to share their own reflections about how they experienced the class, I received a range of equally powerful responses. There was the student who lived in urban public housing and noted how infrequently he was able to immerse himself in nature while at home, unlike when he was on the verdant suburban campus of the university. Another graduate student, Laura, who went on to pursue a doctorate in comparative literature wrote the following to describe the experience. "The course occurred in the Spring of 2020, so things changed in March when we moved to Zoom. What could have been a traumatic change of dynamic was instead the consolidation of an already strong feeling of collaboration and collectivity. We had spent so many hours already discussing the vulnerabilities of the environment and our biology that turning to each other in class during the pandemic felt natural. I remember how we started each class sharing pictures of our promenades: flowers, lakes, trees. It made the topics of the class so tangible and close to our reality. We began sharing resources of different types (websites, apps, books, films, series, podcasts) on environmental justice, and the novels and readings we engaged with became the bridge to building this community, to share our feelings of biological vulnerability and our climate anxiety" (Zuleta, 2024). Laura's reference to "feelings off biological vulnerability and climate anxiety," speaks to the atmosphere in our classroom activated by an engaged pedagogical praxis that allows students to reflect and share in ways that build community and lead to action.

AFRICAN DIASPORIC ECO-LITERATURES

In the field of African diasporic literary studies alone, there is ample evidence of texts that lend themselves to an environmental justice and/or ecocritical lens in African American, Caribbean, and African traditions. Toni Morrison's ur-text about the experience of enslaved people in the United States, *Beloved* (1987) uses the tree as a lasting symbol of the violence of slavery. Dionne Brand's novel *At the Full and Change of the Moon* (2000) begins in Trinidad with a mass suicide of enslaved people in 1824 and then travels all over the world, each time calling the reader's attention to the role of the environment in African diasporic futures. One of the novel's opening passages brings the enmeshed relationships between humans, the environment, and history into great relief. "In this early morning when morning wasn't morning and nothing was anything yet; this morning when trees and grass and rock had not yet gathered themselves into their shapes, when life was not even life itself, when anyone could change into what they might be, this morning like any morning in the world for Marie Ursule was not a sign of anything certain. Hadn't life ceased to be certain long ago, hasn't every turning stood still, hadn't every stillness turned to motion long long ago? And what was memory when she felt it loop and repeat, when what she was about to do she had imagined done already, like a memory" (Brand, 2000). Beginning with the similar history of slavery, but in the different context of colonial Saint-Domingue (present-day Haiti), *Rosalie l'infâme/ The Infamous Rosalie* (2003) unflinchingly portrays how enslaved people and maroons use the land to their advantage even while it feels as though the same land conspires against them. Yanick Lahens's *Bain de lune/ Moonbath* (2014), offers a rousing representation about how the land matters to a rural village people across several generations. *Traversée de la mangrove/ Crossing the Mangrove* (1989) by Guadeloupe's Maryse Condé whose focus on the depth and breadth of the mangrove consumes the reader. In these works of fiction, the intimate relationship between people and nature dissolves the boundaries between them. They operate according to the same logic espoused by Afro-Puerto Rican writer and professor Mayra Santos-Febres who has said that "there is no separation between nature and humans," in fact, nature is a way of being rather than a cause of object to be saved. By *way of being* Santos-Febres means that for these individuals, "to be like a river to be like a mountain to be like a tree to be like the sea [in order to] reach good character."

To that end, I really *do want* my students to be like the land. M. Jacqui Alexander has written convincingly that, "the land holds memory." I hope that the students who took my Caribbean Literature and Environment class look back not only on their experience in the class but also on those images they created of the time they spent in the land. I hope that when they walk through their neighborhoods and experience the environment, it is never again in the same way. I hope that they think deeply about how their actions in the world affect the non-human environment and make different choices as a result. I want students who take classes in Africana Studies and in the environmental humanities to know that the two fields are interdependent and inextricably linked. "Given that Africana studies programs (those focusing on U.S. issues) primarily study the black experience per se and in the context of politico-economic dynamics, and given that environmental studies programs study, among other issues, the politico-economic factors affecting environmental quality and sustainability, it is crucial that the two work together" (Patterson, 2015, p. 10). When students understand how the two fields work together, and when they are activated by empowering pedagogy, they can imagine new, more sustainable futures and also work to create them.

CONCLUSION: RACIAL AND CLIMATE RECKONINGS

A few days after my seniors and second year master's students in "Caribbean Literature in the Environment," graduated in the most unceremonious of ways, the United States dove headfirst into another crisis. On May 25th, 2020, George Floyd was suffocated to death by a white police officer in Minneapolis, Minnesota. Darnella Frazier, an intrepid Black teenaged girl, recorded the killing on her cell phone and shared it with the world, irrevocably altering the course of history and Floyd's memory. While he was brutally pinned down by three police officers, Floyd's final words would become a haunting refrain: "I can't breathe." For eight minutes and forty-six seconds the police officer kept his knee on Floyd's neck even after he was unconscious.

Thinking about this moment now, and in the context of this essay on the relationship between Africana Studies and Environmental Studies, or the study of Blackness in relation to space and place, I cannot help but connect the question of breath and breathing to the air. As Jennifer Nash writes in *How We Write Now: Living with Black Feminist Theory,* "The words [I can't breathe] became a way of honoring Floyd, a strategy for marking how anti-Blackness operates by stamping out Black breath, and a way of naming a wish for Black life" (Nash, 2024, p. 33). The wish for Black life to be able to breathe life also relates to the air that we breathe especially when that air is riddled with racial and environmental injustice. Or, in the words of poet Cheryl Finney: "And the land remembers, even if we do not. For as the earth breathes, so do we. Because this is our home" (Finney, 2023). The Minneapolis air that George Floyd was breathing at the time of the arrest was no longer available to him. He could no longer breathe, the earth would no longer be his home, and that home had been hostile to him. His murder unleashed a wave of Black Lives Matter protests against racist policing, police brutality, racism, and the scourge of white supremacy thrusting us into what became known as the "Summer of Reckoning" throughout the United States and around the world. After the exhausting, demoralizing, death-laden, and grief-filled months of spring 2020, summer 2020 arrived with more death and righteous rage. As the summer continued, hundreds of thousands of people marched with masks on, protested while social distancing, demanding justice for George Floyd, as well as for Breonna Taylor and Ahmaud Arbery who had been killed earlier that year. Their activism was in the names of these, and more Black people killed unjustly, victims of white supremacy. Although class was over, the juxtaposition of the pandemic and the protests ignited more fires in my students who reached out to me regularly throughout the summer. Struggling to make sense of what they were living through once again; some were distraught while others demanded answers and others spurred to action. Many of them noted the injustice of a world that devalues Black lives, denies the humanity of Black people, and refuses to steward the life of the planet.

Taken together, the alarmingly rapid rate of climate change acceleration alongside the shockingly consistent incidence of racial violence tells us that there is much work to be done to eliminate racial and environmental injustice. While the future looks grim, I am encouraged by the words of Bryan Stevenson who reminds us that "hopelessness is the enemy of justice." I am not hopeless because I am inspired by a generation of students like those, I had the privilege of teaching in Spring 2020 who are committed to building a better world. These students and embodying Freire's model of praxis which couples' reflection and action in order to change the world. Our task as college and university educators is to educate and inspire them; in my view, offering more classes at the intersection of Africana Studies and Environmental Studies that embrace a pedagogy of experiential learning can help to further empower their unrelenting advocacy.

REFERENCES

Brand, D. (2000). *At the full and change of the moon: A novel*. Grove Press.

Cerenvak, S. (2023). *Black gathering*. Duke University Press.

Condé, M. (1989). *Crossing the mangrove*. Penguin Random House.

Deren, M. (1983). *Divine horsemen*. McPherson.

Desmangles, L. (1992). *Faces of the gods: Vodou and Roman Catholicism in Haiti*. University of North Carolina Press.

Finney, C. (2023). Memory divine. In E. Sharkey (Ed.), *A darker wilderness: Black nature writing from soil to stars* (pp. [insert page range]). Milkweed Editions.

Florence, N. (1998). *bell hooks' engaged pedagogy: A transgressive education for critical consciousness*. Bergin & Garvey.

Frazier, C. M. (2020, October 1). Black feminist ecological thought: A manifesto. *Atmos*. https://atmos.earth/black-feminist-ecological-thought-essay

Freire, P. (1970). *Pedagogy of the oppressed* (Ramos, M. B., Trans.). Herder and Herder.

Hooks, B. (1994). *Teaching to transgress: Education as the practice of freedom*. Routledge.

Jean Charles, R. M. (2023). *Looking for other worlds: Black feminism and Haitian fiction*. University of Virginia Press.

Nash, J. C. (2024). *How we write now: Living with Black feminist theory*. Duke University Press.

Natural Resources Defense Council. (n.d.). The environmental justice movement. https://www.nrdc.org/stories/environmental-justice-movement

O'Callaghan, E. (2015). Eco-criticism and Elma Napier's sense of place. In Brown, J. D., & Rosenberg, L. R. (Eds.), *Beyond Windrush: Rethinking postwar Anglophone Caribbean literature* (pp. 113–126). University of Mississippi Press. DOI: 10.14325/mississippi/9781628464757.003.0007

Paravisini-Gebert, L. (2011). Deforestation and the yearning for lost landscapes in Caribbean Literatures. In DeLoughrey, E., Gosson, R. K., & Handley, G. B. (Eds.), *Post-colonial ecologies: Literatures of the environment* (pp. 99–116). DOI: 10.1093/acprof:osobl/9780195394429.003.0005

Patterson, R. (2015). *Greening Africana studies: Linking environmental studies with transforming Black experiences*. Temple University Press. DOI: 10.2307/j.ctt14bsx8d

Pellow, D. N. (2016). Toward a critical environmental justice studies: Black Lives Matter as an environmental justice challenge. *Du Bois Review, 13*(2), 221–236. DOI: 10.1017/S1742058X1600014X

Roane, J. T. (2023). *Dark agoras: Insurgent Black life*. New York University Press.

Ruffin, K. (2010). *Black on earth: African American ecoliterary traditions*. University of Georgia Press. DOI: 10.1353/book11452

Sharkey, E. (2023). *A darker wilderness: Black nature writing from soil to stars*. Milkweed Editions.

Steinmetz, K. (2020, February 20). She coined the term intersectionality over 30 years ago. Here's what it means to her today. *Time Magazine*. https://time.com/5786710/kimberle-crenshaw-intersectionality/

Tack, S. (2023). Teaching for liberation: The manifesto assignment as an example of bell hooks' engaged pedagogy. *Australian Feminist Studies, 38*(109), 141–156. DOI: 10.1080/08164649.2023.2255931

Wilcken, L. (2023). Ecological wisdom in the marvelous realism of Vodou (by way of Jacques Alexis). *Journal of Haitian Studies, 29*(1), 90–96. DOI: 10.1353/jhs.2023.a922859

Zuleta, L. V. (2024, October). Personal communication.

ADDITIONAL READING

Benitez-Rojo. (1996). *Antonio. The Repeating Island: The Caribbean from a Postmodern Perspective*. Duke University Press.

Brulle, R., & Pellow, D. (2010). *Power*. Justice, and the Environment.

Buell, L. (2005). *The Future of Environmental Criticism: Environmental Crisis and Literary Imagination*. Wiley-Blackwell.

Campbell, C., & Sommerville, E. (2007). *What is Earthly Paradise? Ecocritical Responses to the Caribbean*. Cambridge Scholars Publishing.

Carrigan, A. (2012). *Postcolonial Tourism: Literature, Culture and the Environment*. Routledge.

Cervenak, S. J. (2021). *Black Gathering: Art, Ecology, ungiven Life*. Duke University Press.

Clark, T. (2011). *The Cambridge Introduction to Literature and the Environment*. Cambridge University Press. DOI: 10.1017/CBO9780511976261

Cristina, F. M., Peter, H., Owen, R., & Lesley, W. (Eds.). (2013). *Surveying the American Tropics: Literary Geography from New York to Rio*. Liverpool UP. DOI: 10.5949/liverpool/9781846318900.001.0001

DeLougherey, E. (Ed.). (2005). *Caribbean Literature and the Environment: Between Nature and Culture*. University of Virginia Press.

DeLougherey, E., & Handley, G. (Eds.). (2011). *Postcolonial Ecologies: Literatures of the Environment*. Oxford University Press. DOI: 10.1093/acprof:osobl/9780195394429.001.0001

Glotfelty, C. (1996). *The Ecocriticism Reader: Landmarks in Literary Ecology*. University of Georgia Press.

Hiltner, K. (2014). *Ecocriticism: The Essential Reader*. Routledge.

Lionnet, F. (2010). Critical Conventions, Literary Landscapes, and Postcolonial Ecocriticisms. In McDonald, C., & Suleiman, S. R. (Eds.), *French Global: A New Approach to Literary History* (pp. 127–144). Columbia UP.

Manigault-Bryant, J., & Constantine, J. (2024, September 30). Black Studies and the Production of Environmental Justice Knowledge. *Environmental Justice*, env.2024.0040. Advance online publication. DOI: 10.1089/env.2024.0040

McKibben. (1989). *Bill. The End of Nature*. Random House.

Monique, A. (2013). *Ariel's Ecology: Plantations, Personhood, and Colonialism in the American Tropics*. University of Minnesota Press.

Patterson, R. (2015). *Greening Africana Studies: Linking Environmental Studies with Transforming Black Experiences*. Temple University Press. DOI: 10.2307/j.ctt14bsx8d

Ruffin, K. (2010). *Black on Earth: African American Ecoliterary Traditions*. University of Georgia Press. DOI: 10.1353/book11452

Sharkey, E. (Ed.). (2023). *A Darker Wilderness: Black Nature Writing from Soil to Stars*. Milkweed Editions.

Tiffin, H., & Huggins, G. (Eds.). (2006). *Postcolonial Ecocriticism: Literature, Animals, the Environment*. Routledge.

Tinsley, Omise'ke Natasha. Thiefing Sugar: Eroticism between Women in Caribbean Literature. (2010). Duke University Press.

Chapter 7
Practicality Plus Creativity:
The Public Liberal Arts College

Janet Schrunk Ericksen
https://orcid.org/0000-0003-1375-260X
University of Minnesota, Morris, USA

ABSTRACT

This chapter examines the intersection of liberal arts education and sustainability initiatives at the University of Minnesota Morris (UMN Morris), a public liberal arts college. Through student voices, institutional history, and concrete examples, it demonstrates how the college has become a leader in sustainability while maintaining its core liberal arts mission. The article argues that the college's rural location, historical context (including the campus site as a Native American boarding school), and commitment to student engagement have created an environment where sustainability serves as both an educational framework and a lived practice. By examining specific initiatives and partnerships, including community collaborations and student-led projects, the piece illustrates how UMN Morris has effectively combined the traditional liberal arts emphasis on broad, interdisciplinary thinking with practical applications in sustainability, creating a model for environmental leadership in higher education.

DOI: 10.4018/979-8-3693-5454-4.ch007

PRACTICALITY PLUS CREATIVITY: SUSTAINABILITY AND THE PUBLIC LIBERAL ARTS COLLEGE

Student voices have always carried substantial weight at the University of Minnesota Morris, a place where sustainability firsts are catalyzed by a combination of student ambition, internal and external partnerships, and persistence. Any account of the sustainability efforts at this public liberal arts college within the University of Minnesota system, then, appropriately begins with a student voice:

> *When I started college at UMM, I thought the turbines were cool and composting was neat, but [sustainability] wasn't something I really focused on or cared about that much. But living there, seeing it in action, I began to appreciate it more and more. I took a class called Making Environmental Public Policy with Roger Rose, and that opened up my appreciation of sustainability as a whole and of policy as a specific. The high number of Native students and seeing how they thought about sustainability played a significant role, too, in the development of my thinking. I ended up working in the Sustainability Office, and I took Arne Kildegaard's course on the Green power revolution. I found the idea of renewable energy such an interesting topic—it was new, innovative, incorporated so many different areas (health, science, politics, communication). So by the time I graduated, sustainability had become a framework through which I thought about things—is it socially sustainable, economically, environmentally? (Rosemark, 2024)*

Such a transition from casual appreciation to well-informed, foundational, and multi-disciplinary advocacy captures beautifully one of the aims of liberal arts education. That it is tied here so explicitly to sustainability reflects the distinctive public liberal arts college experience at the University of Minnesota Morris. This particular student graduated in 2021 with majors in political science and economics and a minor in environmental leadership and stewardship. As an undergraduate, the student secured school board support and grant funding that brought two fully electric school buses to the local K-12 schools. The proposal succeeded because of its practical framing from the standpoint of saving costs and meeting needs. The school board's main concerns were range and cold weather, so the student contacted the bus company and arranged an online meeting between company representatives and the board (and the bus company later even dangled a job offer to the student). The initial idea was for one bus, but the school board "was so sold that they said, why not do two? The selling point was getting two new buses at a lower cost than their inefficient, older diesel buses" (Rosemark, 2024). This experience of increasing community sustainability is distinct to this student and at the same time reflects an intentional and extensive part of the public liberal arts college experience at UMN Morris.

A small campus in a small town, three hours from a major airport, UMN Morris has achieved a national profile and even some international notice for its sustainability efforts. As early as 2012, the college was listed on the Environmental Protection Agency's Top 30 On-site Generation list of the nation's largest green power users. The campus was a pilot participant in 2010 in the Association for the Advancement of Sustainability (AASHE) Sustainability Tracking Assessment and Rating System (STARS) and achieved STARS Gold certifications in 2012, 2015, 2018 and 2022, with another in process. The AASHE certification process, now coupled with attention to the United Nations' Sustainability Development Goals, continues to help the campus measure progress and guide next steps. Other formal recognitions have accrued, particularly in relation to a community partnership called the Morris Model, and while certainly sustainability has been an administrative priority of the college for decades, student interest—student pushes, really—have ensured attention to continuing improvement. Fueled by the blend of practical problem-solving and creative thinking that are core to a liberal arts education, partnerships have made possible projects that would otherwise be beyond the college's own reach. The case study that follows explains how one public liberal arts college has leveraged its rural location and student interests to advance sustainability both for itself and others.

Practicality and the Liberal Arts

The word *practical* is often distanced from liberal arts education. Several state legislatures have recently proposed or implemented funding models that give preference to STEM and vocational programs, and a fall 2023 article in *The New York Times* captures well an example of the discourse that characterizes such moves. The auditor for the state of Mississippi expressed the view that "state appropriations should focus more on engineering and business programs" rather than on "liberal arts majors" because of the need to graduate students "with some sort of practical skill that the world actually needs" (Hartocollis, 2023). Just as liberal arts education includes science and math, *practical* need not necessarily prioritize technical training over exploration, certainty over ambiguity, established methods over experimental approaches, or short-term applications over long-term adaptability. Problems rarely present themselves in neat, linear packages, and a narrow understanding of *practical*, one focused on immediate *practice* in opposition to larger thinking or longer-term applications, might actually be less useful in problem-solving, in preparing students for complex, evolving challenges than is a multidisciplinary, critical thinking-focused liberal arts education.

In 1960, founding UMN Morris was both a creative and highly practical maneuver: it created a more affordable public competitor to the state's private liberal arts colleges while also utilizing a campus site already owned by the state and extending access to the state's flagship university, which legislators were asked by the people to do. The oldest and largest campus of the University of Minnesota is the research-focused institution in Minneapolis and St. Paul founded in 1851. The five-campus UMN system now includes three on the eastern, more populous side of the state, and two on the western side. Those two, at Crookston and Morris, are located on the sites of former agricultural high schools, which themselves were established out of a desire to increase access to education. The Morris campus carries the additional history of being located on the site of a Native American boarding school that operated from 1887 to 1909, which remains an influence on the college today.

When the University of Minnesota Morris was founded, the U.S. Agricultural Census had just stopped asking if farms had electricity. The rural electrification process that began in 1936 had achieved its aims, and the vast majority of farms were utilizing electricity to improve the economy and their own standard of living (Cameron, n.d.). Those living in west central Minnesota recognized education as a key component of continuing improvement. The minutes of the 1957 meeting of the West Central Educational Development Association refer to the "tremendous need for higher education facilities in western Minnesota" and a year later the same group was advocating to the state legislature for "the Morris proposal" that would establish there "a complete four-year liberal arts" branch of the University of Minnesota (University of Minnesota Morris, 1958). A trifold informational brochure from the college's early years devotes a full column to explaining the "quality liberal arts education" at the University of Minnesota Morris, including the point that "a liberal arts education is not only aimed at training the students to make a living as an end, but also to become interested in living itself" (University of Minnesota Morris, 1960).

For anyone familiar with Latin, the *liber* in liberal arts, meaning "free," makes perfect and non-partisan sense as foundational to a curriculum that seeks to provide the education necessary for free citizens to function well in life. While the aims of liberal arts education at UMN Morris have remained largely consistent since 1960—focused on breadth, depth, and close working relationships between faculty and students—the parameters have, of course, changed as students and the world have changed. Will Katerberg in a 2016 blog post for Calvin University's Historical Studies Department noted that both critics and defenders of liberal arts education often do not acknowledge either the foundations of this form of education or the fact that it is not "a timeless ideal" but instead is "a messy, contingent, and evolving set of human practices." Even as it has changed, though, liberal arts education has held onto learning that is practical in a more complex way than is a sharp linearity between learning and specific jobs or tasks. Katerberg, citing scholar Bruce Kimball's work on the history of education, provides an overview of the development of liberal arts education from "two distinct streams" in ancient Greece and Rome: "The older stream, [Kimball] argues, emphasized rhetoric, grammar, and composing and delivering speeches in the democratic public life of city-states in Greece and in the Roman republic. A practical stream. The other stream, associated with Plato and Aristotle, emphasized reason and rational methods of pursuing truth and knowledge. A more purist stream" (Katerberg, 2016). Much changes as these two streams flow over the centuries, intermingling and shifting in their relationship and the subjects of focus, but fundamentally, both persist. Both are evident in the current liberal arts emphasis on engaging with multiple perspectives, on crossing traditional disciplinary lines, on small class sizes that foster engaged learning, and in the emphasis on "transferable intellectual and practical skills such as communication, analytical and problem-solving skills," as phrased in the American Association of Colleges and Universities' current definition of liberal education (American Association of Colleges & Universities, 2024). To more narrowly apply Cathy Davidson's (2017) description of what education needs to focus on now: liberal arts colleges play an important role in "equip[ing] students to thrive in this murky and often polluted new atmosphere that we now all breathe" (p. 6).

Sustainability and the Liberal Arts

UMN Morris continues to serve students in its region and well beyond it, and it continues to offer a high-quality liberal arts education. Its curriculum has shifted—sustainability leadership and green chemistry were not academic options in the twentieth century—but the college still centers both "distinct streams," the practical and the purist, within the strong community of a residential liberal arts college experience with emphasis on sustainability. Katerberg's review of the state of the liberal arts and their future includes the same questions that now shape key aspects of UMN Morris:

> *We need good science to understand how and why our climate is rapidly changing and to find practical technologies that can help slow (if not reverse) such changes and deal with their consequences. But the most complex problems associated with climate change are human factors. How do we live? Is consumer capital ism sustainable? What do we value? How do we envision the "good life"? Is it about toys and tools like the iPad I'm writing on? Is it about preserving wild forests and rivers and farmland fisheries? How do we decide, as individuals,*

and as nations and a global species? What price are we willing to pay to address climate change, or what ecology debts are we willing to dump on our progeny? There's more than enough need here for contemplation and activity, classical forms of training, modern disciplines and professions, STEM, STREAM, and STEAM, digital arts and traditional ones. (Katerberg, 2016)

In a 2006 article in the *Chronicle of Higher Education*, Frank Rhodes, former president of Cornell University, made the same point, writing about "Sustainability: The Ultimate Liberal Art." Rhodes considers the liberal arts at "both the smaller, traditional, independent undergraduate institutions [like UMN Morris] and the colleges of liberal arts and sciences within universities." He argues that the "concept of sustainability could provide a new foundation for the liberal arts and sciences," with sustainability defined as "the effort to frame social and economic policy so as to preserve with minimum disturbance earth's bounty—its resources, inhabitants, and environments—for the benefit of both present and future generations" (Rhodes, 2006). UMN Morris was already taking up this charge in 2006 and has continued to do so, blending the practical, useful, creative, and reflective into a model for living and learning, one in which students are encouraged to ask questions about and to understand sustainability as the student with whom this essay began learned to do, as "a framework through which [to think] about things—is it socially sustainable, economically, environmentally?"

Practicality and sustainability can appear too as factors of location in liberal arts education—or baccalaureate arts and sciences, to use the Carnegie classification that is perhaps marginally less obscure to those unfamiliar with the history of "liberal arts." An intentionally small, public, undergraduate institution is likely to fare better if it pays attention to its physical place and works to leverage whatever connections it can to that place and its people. That is true when the land on which the college operates once housed a Native American boarding school, and it is true when the surrounding economy has been and still is largely based in agriculture. It is true, too, when the climate includes winters with thermometer readings of thirty below zero (Fahrenheit), and when the wind blows with character and persistence across the open prairie that extends for hundreds of miles north and west of the campus. At a structural level at UMN Morris, creative thinking and practical connections—those that implement change effectively and efficiently and with consideration of available resources and contexts—have yielded local partnerships on facilities, including in a community-campus fitness center, a high school-university shared football stadium, and a university-community softball complex. And they take the form of sustainability projects, perhaps most notably a collaboration among the university, the school district, the town, and the county and in affiliation with a town in North

Rhine-Westphalia, Germany, about the same size as Morris, Minnesota, focused on advancing sustainable living by modeling it in small towns.

The University of Minnesota Morris boasts an impressive array of green energy projects, especially in partnership: a wind turbine, a biomass plant, a solar pasture, a residence hall that is the second LEED-Gold certified building on campus and is constructed with high thermal mass insulated concrete forms to prevent winter heat loss and summer heat gain. It has systems that allow students to study energy consumption and has on-site orchard, edible landscapes, and gardens. In 2005, UMN Morris received the Green Power Leadership Award from the U.S. Department of Energy and the U.S. Environmental Protection Agency for its accomplishments in utilizing renewable energy, and such encouragement helps keep the campus moving forward. The campus is a place where embracing the unfamiliar and challenging the status quo are valued, whether that means adapting to the quirks of the local coffee shop or testing new models of integrating renewable energy and agriculture. It is a place where the campus dining service includes local foods and where the plates, cutlery, and cups at events are all compostable—and actually composted. Students would never allow it to be otherwise.

Because of its rural location and the people who live and collaborate there—and because of the lived mission of the public liberal arts college that operates in this place—about 60% of the electricity for UMN Morris is powered by a pair of now iconic wind turbines that sit on a hill overlooking the campus. The University of Minnesota West Central Research and Outreach Center, which is adjacent to UMN Morris and partners with it on multiple projects, erected the first of these. In 2005, it was the first commercial-scale research wind turbine on public university land in the United States. It continues to operate alongside its companion turbine erected in 2011 and operated by UMN Morris. But the turbines are only the most visible evidence of the investments in sustainability in Morris. Visitors to campus are likely to start at the Welcome Center, built in 1915 and renovated in 2009 to become one of the most energy-efficient structures in Minnesota. That building was the first in the state to achieve LEED-Gold certification. It was also the first listed on the National Register of Historic Places to use chilled-beam technology, which uses water, not air, for cooling and heating, a far more efficient—and therefore financially practical—approach to staying warm in the cold winters of Minnesota (KFI Engineers, n.d.; Design Curial, 2012; University Relations, 2014). The campus's Green Tour might leave the Welcome Center for another warmth-producing innovation, the biomass gasification facility that serves as both a research site and a campus heating facility. Each of these projects is the result of partnership, risk-taking, and commitment. Each benefitted from a location that juxtaposed educational drive with rural location, undergraduate students with top notch researchers, and public goals with individual business needs, all tied together by a combination of practicality and creativity.

Solar

West central Minnesota has an abundance of sun, wind, and productive soils. Finding ways to utilize local resources effectively and efficiently means that four solar fields so far appear on the landscape at the University of Minnesota Morris. That solar energy development has been kept affordable by funding from grants, rebates, and a small loan (already repaid). After implementation costs, the solar arrays contribute to fiscal health by amplifying the power that the campus produces for itself and reducing the need to buy it from a utility company. One, consisting of locally produced panels, heats a swimming pool in the campus-community partnership fitness center. Another, behind the building that houses most of the science classrooms on campus, is a research-ready mix of fixed and sun-tracking panels. A third sits not far from the Green Prairie Community residence hall and contributes about 20% of that building's energy needs. The fourth and newest is an agrivoltaics array, and this one elegantly demonstrates both the intersection of agriculture and renewable energy and the aims of liberal arts learning.

On an early September sunny day in 2024, a group of people gathered in a field alongside a herd of grazing cattle and just beyond the Native American garden at the edge of the UMN Morris campus. The occasion was the dedication of an expansion to a solar array there, and those present included the state senator and a state representative for the region (both Republicans), the director of the University of Minnesota West Central Research and Outreach Center (part of a network of such centers throughout the state) and a dairy researcher who works there and was one of several UMN Morris alumni present, a local power company executive, an undergraduate student majoring in Spanish and psychology, and a couple of college administrators.

Agrivoltaics is the use of land for both agriculture and solar-photovoltaic energy generation. The word *agrivoltaics* is new enough that its spelling is not yet nationally settled (agro? agri?). That newness is intentional to a small campus with room to stretch, figuratively and literally. Larger than a laboratory and smaller than a full-scale implementation, the residential liberal arts college campus in rural Minnesota offers both the willingness to try out new ideas and space to do so on a site larger than a laboratory but contained enough to serve as a test location before scaling upward. And partnerships make so much more possible (including financially) than any one entity out on this windswept prairie could or would accomplish alone. The agrivoltaics array was constructed by the University of Minnesota Morris on a pasture owned by the West Central Research and Outreach Center (WCROC).

That center also owns the cows that graze under the solar panels, and WCROC researchers engage in research on multiple aspects of this mixed land use. The 500kW array is at least for now the largest dairy agrivoltaic field in the Midwest, and while it represents a much smaller amount of clean energy production compared to the wind turbines, it contributes especially to air-conditioning the campus. Wind and solar are complementary—often wind dies down when solar is strong, and Morris is windier in winter than summer. Ultimately the agrivoltaics may be most helpful as one component among other thermal decarbonization options, but this marriage of solar power and dairy farming reflects the broader institutional ethos of UMN Morris. As a campus ranked #1 in the United States in 2019 for renewable electricity generated per student, the agrivoltaics installation builds on existing sustainability infrastructure while pushing boundaries in both agricultural and energy innovation (Environment America, 2019). The project's success stems from its ability to serve multiple purposes: advancing sustainable energy practices, supporting agricultural research, and providing hands-on learning opportunities for students, all while maintaining productive farmland.

The project was publicly funded through a proposal to the Legislative-Citizen Commission on Minnesota Resources and sits on land owned by the state's land-grant university. It extends the long partnership between two units of that university: the University of Minnesota Morris and the West Central Research and Outreach Center. The two work well together despite their distinctly different focuses. The center operates as part of a statewide network of University of Minnesota Research and Outreach Centers and includes Extension. The center's administrative home is more than 150 miles east, on the Saint Paul campus of the University of Minnesota Twin Cities and within the College of Food, Agriculture, and Natural Resource Sciences (CFANS). While the faculty and staff at the center work with a small cohort of graduate students, their focus is on basic and applied research, and that aligns well with the undergraduate research emphasis at UMN Morris as well as the campus infrastructure commitment to sustainability. One of the best offsets of this has been the job opportunities and internships for students. One undergraduate student recently spent a summer internship working on a solar-powered robotic weeder, for instance, and one of the center's researchers who now focuses on electric and thermal energy usage in dairy farms was a UMN Morris undergraduate who worked at the center's dairy during her junior year.

Solar projects succeed at UMN Morris because people are willing to explore and to test new aspects of installation, and because multiple entities and interests work together. The agrivoltaics project provides 500kW of electricity to the Morris campus and more shade for the WCROC organic dairy herd. The land and cattle are owned by WCROC; the installation itself belongs to UMN Morris. Both were willing to try the new design, even when others were not:

Morris Vice Chancellor for Finance and Facilities Bryan Herrmann said contractors told them that installing solar panels 8 feet off the ground wouldn't work, but it had to work to graze cattle under them. So they found Hudson, Wisconsin, solar contractor Energy Concepts, which drove the metal posts 8 feet into the ground in order to support the panels, with the leading edge 8 feet in the air without the use of concrete footings. That choice was pleasing to those attending, as it meant when the solar panels someday reach the end of their life, they could choose to pull the metal supports and easily reclaim the ground. (Johnson, 2024)

The campus objective was primarily to reduce energy costs and test a model that could then be applied elsewhere, on a larger scale; the research center's objectives certainly overlapped those but were more focused on agricultural benefits. And both were willing to take a risk, to model being the creative problem-solvers identified as one of the priorities in a UMN Morris education. So that while the "'challenges of being 8 feet in the air certainly increased the difficulty of the project,'" as the head of the installation firm, Russel Gilberg, noted, the potential benefits and the shared understanding of educational mission drove it forward (Johnson, 2024).

Among what has already been learned from this agrivoltaics project is the necessity of solid and situation-specific engineering. Morris has wind, abundantly and sometimes fiercely, with little to diminish it as it roars across the prairie. The array has already beautifully survived 80-90 mph winds that caused significant damage on campus and in the surrounding area. An article in the Fargo, North Dakota-based agribusiness journal *AgWeek* detailed aspects of installation as well as agricultural value: "The posts are rated for the extreme conditions expected at the site. The panels are set at an angle to best take advantage of solar production. Shade is a byproduct of that. The panels are spaced so there is no shade hitting another panel during winter's lower angle of sun." Moreover, the new panels in this array "are bifacial, meaning they can absorb light reflected off the ground from the underside of the panels as well as the top of the panel. That should be especially productive when snow covers the ground" (Johnson, 2024). Creative innovation meets the practicality of testing on such a scale—will it work? Yes, it really does work. And more works, too.

The same *AgWeek* article explains the role of agrivoltaics in demonstrating "how energy and agriculture can work together." Cows have shown no hesitation in moving under the raised panels, and researchers are exploring what grows best in the shade of the panels. Brad Heins, a UMN-CFANS associate professor of organic dairy management who works at West Central Research and Outreach Center, points to such practical questions as does milk production vary between raised solar and open fields? The answer so far is no. Heins also cites research concerning the breathing rates of cattle, laid out in considerable detail for *AgWeek* readers. The research points to the possibility "that the solar array may decrease heat stress for cattle utilizing the shade of the array: the difference "can have pretty huge ramifications in milk production. So if we lower heat stress we can increase milk production, we can increase growth of heifers and have less stress on the animals as well'" (Johnson, 2024). If the benefits of a sustainability project are clear to multiple audiences, even if they are not the same benefits, the project has a much greater possibility of success and of seeding more success.

UMN Morris is working to become fully carbon neutral and to develop its reputation for innovation. The campus encourages its students, staff, and faculty to partner in wide array of projects that demonstrate both an intellectual openness and a practical approach to addressing environmental challenges. The rural location acts as an asset in this, not only but especially in relation to agriculture. The college offers no directly identified agriculture major, but graduates can and do go into ag-adjacent areas, such as law focused on agriculture and agribusiness, and they earn degrees in biology, chemistry, computer science, economics, business management, environmental sciences, and environmental studies. Any major can also be accompanied by a minor in sustainability leadership and participation in research, internships, and student organizations focused on sustainability. The college has long emphasized and excelled at student engagement in high impact practices, and the general education program requires experiential learning, such as undergraduate research, an internship, or study abroad (with donor funds supporting participation). Both farmers and liberal arts college students have strong interest in finding solutions to difficult problems. In the case of agrivoltaics, farmers may be driven by profit motives, and college is motivated by financial benefits, too. Energy is expensive, especially for a campus in a cold climate and with a core of old buildings on the National Historic Register. Interest in the future crosses educational and economic interests. Michael Reese, the director of the West Central Research and Outreach Center, underscores that: "This is a great challenge for us to be able to combine sustainable agriculture, sustainable energy and look not only next year down the road, but look five, 10, 20, 50 years down the road at what kind of impact we've had" (Johnson, 2024). That's also true for the college, of course, and that shared interest means there is more in Morris than solar.

Wind

UMN Morris was one of the first public colleges to generate on-site renewable power. Before doing so, the college engaged in a common financial analysis (a Net Present Value analysis) that made clear that investing in a wind turbine would be a good economic decision: in comparison to just buying power from a utility company, the wind turbine was clearly the better fiscal choice. Two University of

Minnesota-owned 1.65-megawatt wind turbines that together produce over 10 million kilowatts of power each year have basically allowed the campus to do budget planning with fixed costs for 60% of our electric bill for more than twenty years (Marquart, 2021). The campus now uses about half of this wind-generated electricity, and the other half supplies the local power grid.

The campus is distinctive in that a power line goes right to the campus from a wind turbine/solar installation, and the campus has an agreement to sell power to the local utility company, Otter Tail Power. The financial accounting of clean energy is somewhat complicated in this arrangement, but, in sum, the campus purchases renewable energy credits (RECs), not from some project far away but essentially buying back the clean energy attributes of the power that their own wind turbine makes. When the power company purchases the power, they purchase the electrons and the green attributes associated with it (the RECs). The RECs, though, are then purchased back by UMN Morris, which then retires them and, in so doing, moves closer to carbon neutrality. The turbines also play a role in campus culture, figuring in marketing and tours, social media and more generally in campus pride. A trail accessible from campus winds along the river and below the two turbines, facilitating a connection for students to their environment—and of course the trail includes educational signage, about prairie plants, animals, landscapes. The turbines are now aging, and the college has already had to find funding for repairs, but wind energy is so woven into the fabric of the place that replacement seems a necessity rather than an option. A challenge with an increasingly constrained budget will be securing the funding for that replacement, and the college may again have to do this through a loan, which is how the campus turbine was initially funded for the 2011 installation. Yet in just fifteen years, wind turbines have tripled in size, from 1.65MW to 5MW, so the campus has more to gain from reinvesting.

Biomass

The wind turbines, despite occasional problems, have on the whole functioned beautifully. The various solar arrays on and near campus have been almost problem free so far. Innovation, though, involves risk, and the UMN Morris biomass gasification plant attests to that. The plant was conceived as a model community-based energy system, displacing fossil fuels with locally obtained and grown biomass, like corn cobs, to provide a portion of campus heating and cooling needs. As planned, the gasifier would consume about 9,000 tons of corn residue and prairie grass biofuels per year harvested from farms within a 60-mile radius of Morris, and it would displace 8,000 tons of CO_2 emissions per year (University of Minnesota Morris, n.d.). Especially when something is first of its kind, however, things do not always proceed as expected. The biomass gasification facility was expected to begin functioning in the fall of 2008. In its first test run that fall, however, "plant operators found that a mistake had been made in the design of the gasifier, the $700,000 principal component used in converting corn stover and other biofuels into steam" (Downs, 2010). The gasifier had to be rebuilt. More than fifteen years later, after periods of inoperability, the facility has one 315kW biomass-fueled back-pressure steam-turbine generator that discharges steam to the campus heating loop or an absorption chiller—not what was initially planned, and not with the impact on either heating or the local economy that was anticipated (Kostka, 2020).

As Lowell Rasmussen, who served as Vice Chancellor for Finance and Facilities in the early 2000s and was key in defining the campus's sustainability path, explained: "'Part of it is we underestimated how variable the fuel stock was going to be in terms of its moisture and thermal properties. I think the gasification industry probably overestimated how easily they could make a transition from the fuel that they're very familiar with, which is wood, to [agriculture-based] fuel stock.'" The newness of the technology undoubtedly played a role in problems, and Rasmussen knew that was going to be a risk—being first at things carries greater risk. As the UMN Twin Cities campus newspaper reported at the time, "Rasmussen repeatedly stressed that fault cannot be specifically placed on any of the parties involved. 'It's not a real cut and dried, "Gee, we know exactly who's wrong here" [situation]. It's hard to hold someone responsible for something there are no previous data sets on,' Rasmussen said." Legal challenges between the contractor and subcontractor ensued, but ultimately an agreement was reached to repair the gasifier based on recalculated fuel density. The challenges of this creative attempt to serve practical opportunities taught everyone a great deal, and the pace of change in renewable energy technology often means jumping in to try new things. In this case, though, figuring all that out took so long that renewable energy technology has now largely moved past this biomass facility.

So much since the project began has changed in the world of biomass. The harvesting equipment farmers use has changed, so that corn cobs now come out in bits. This means they disintegrate in the field more quickly, and so farmers have less reason to bring them in for use as fuel. A persuasive incentive to bring in the cobs now would cost more than makes sense for the university. Waste wood—from various local sources—is the new the focus for local biomass, to cover 50% of the campus heating load, a move that will save money. But burning wood is still burning carbon, so the campus is exploring other paths toward full carbon neutrality and reduced heating costs, including thermal energy batteries, or solar steam, and geothermal options. New technologies lie ahead.

Compost

The town of Morris is surrounded by miles of corn and soybean fields, and the university's biomass plant began with that recognition and with the desire to make something useful from something that was waste. Something similar motivated the campus's cold-weather composting, and UMN Morris may be one of the only campuses that not only held a celebration for composting—on the occasion of diverting an estimated one million pounds of organic material from the landfill—but did so in February, when in Minnesota composting would seem to be hampered by the environment. Motivated by students requesting the option and developed by facilities staff willing to figure out how to make things work in a place with long winters and temperatures well below zero, composting on campus began small, in 2012, "when waste was collected in tub containers in Turtle Mountain Cafe (TMC). Efforts to collect compostable material expanded from TMC to the dining hall, then to the Student Center, residence halls, and, finally, to most campus buildings" (Supply Chain Council of European Union, 2022). In the fall of 2019 and in partnership with the campus's food service provider, Sodexo, the dining hall purchased a tub grinder that condenses food waste and removes excess water from it before it goes to be composted, making it lighter and more easily managed. Like so many other sustainability projects on the campus, the effort is a cooperative one, and that is a huge factor in the campus's successes. In a press release about the grinder, Vance Gullickson, who coordinates recycling and composting for the campus, characterized the investment of staff in such initiatives: "'the compost process still requires some amount of moisture to operate properly, so we are always learning and trying new things here at Morris.'"

So successful was—and is—the cold weather composting that what began on campus has now expanded to the county. UMN Morris graduate Sydney Bauer, who as an undergraduate was a standout in sustainability efforts, went on to develop and coordinate the Stevens County Organics Recycling Program:

> *"When I was a student and Office of Sustainability intern," said Bauer, "I focused on campus energy systems. As a graduate, I am working to implement our county-wide organics program. The campus plays an important role in the county because they were an early adopter of organics recycling in Stevens County—and because they have been able to test and showcase important tools that will improve future work. The grinder is a good example of UMN Morris working to improve the efficiency of this process. Removing excess water from food waste is beneficial because waste is processed by ton, and these tools allow for the excess water weight to be removed from the hauling and disposal costs of organic materials. Institutions with large quantities of food waste would ben-*

efit from learning about these tools and considering the benefits for their own facilities." (Sugimura, 2021)

In every campus building on campus, composting bins sit in the halls alongside recycling and trash bins. And so normalized are both composting and recycling that we tend to stand blankly in front of the solitary waste disposal options that we encounter when we travel to other campuses, unsure of how to proceed. It just feels wrong to throw away so much into "trash." Similarly, when UMN Morris hosts visitors for events that include a meal, members of the campus community tend to stand by the bins and be directive: "yes, that plate is compostable. Yes, your utensils also go into the compost. And, yes, even that plastic cup goes into compost—it's a compostable bioplastic." The utensils are made from potatoes; the cups are made from corn. Given local crops, this reflects a pairing of the practical and the creative at this liberal arts college. And the campus also saves energy and water (and in turn money) by not even using trays in the dining hall. Small choices contribute to the bigger effort.

Because of composting and recycling, the remaining campus waste is drier and more easily burned in a nearby incinerator. Less easily reduced waste from other local entities is transported much farther, to a larger incinerator operation in North Dakota. And for now, at least, the campus pays nothing for composting—from student big idea to a significant financial gain.

Educational Integration

In looking back over his sustainability involvement at UMN Morris, the former vice chancellor for finance and facilities emphasized the extent to which sustainability had already become an integral part of the campus culture by 2010:

A lot of what Morris really is doing well is integrating the whole package of sustainability. It isn't just one segment of our curriculum or one aspect of our campus operations that we can typify and be a model or a leader in sustainability for. It's understanding [that] it's an integral approach, it's local foods, it's recycling, it's conservation. It's academic understanding and diligence and understanding the issues of carbon. And then it's also our physical facilities that become research labs in a sense, for promoting the issue of sustainability." (Rasmussen, 2010)

Fifteen years later, that continues to describe the campus well. At the core of sustainability at UMN Morris, from the start and still, are the students who choose to pursue their bachelor's degree in a place where their voices are so significant.

One of the most frequent explanations voiced by prospective students in answer to the question, "why Morris?" is its sustainability reputation, along with their desire to *do* something related to environmental change or green energy or more sustainable living right from the start of their college education. Driven in part by the Office of Sustainability, UMN Morris has offered curriculum development funds to incentivize the inclusion of sustainability in the curriculum, whether in courses focused on it somehow or in courses that include the subject within the framework of another focus. The college offers courses such as Chemistry for the Curious Citizen: The Role of Chemistry in the Environment and Everyday Life, Environmental Science and Place-Based Education, The Environmental Imagination: Reading and Writing about the Natural World, Environmental Problems and Policy, Bottomdwellers in an Ocean of Air, and Environmental Ethics, Communication and the Environment: Analysis and Criticism, Culture, Climate, and Collapse in Medieval Iceland, among others. Course offerings have grown into such things as a green chemistry track within the chemistry major and a minor in sustainability leadership, both in addition to majors offered in environmental studies and environmental science. The sustainability leadership minor nearly, though, did not make it into existence, and its development provides an example of why good ideas sometimes stall out even when aligned with institutional mission and supported by external funding.

The practical creativity that fuels so many of the sustainability efforts at UMN Morris was recognized by a generous external grant in support of sustainability efforts. One outcome was the Sustainability Leaders for the Future program, replicable across other college campuses. In addition to encouraging the development of courses and course-content, the program led to fellowships for faculty, staff, and students that gathered like-minded people to advance various projects and supported visiting speakers and related events. The program also initiated a Sustainability Literacy Assessment - Morris (SLA-M) for first-year and graduating students to

- Examine students' attitudes regarding sustainability
- Determine whether and how students are learning about sustainability at Morris
- Identify how much students know about problems relevant to regional, national, and global sustainability (University of Minnesota Morris n.d.)

All good efforts, with good outcomes. But a key part of the grant proposal required the implementation of a new academic major or minor, and discussions about what should be part of it and what could and could not actually be taught or learned in a classroom almost derailed that effort. What is "sustainability"? And can "leadership" really be taught in classes? These are just the sort of questions one would expect from university faculty, with highly honed creative and critical thinking skills, and the flexibility of both words as well as their popular use only amplified the problems. Eventually, though, the recognizability of the words was deemed an asset by enough people to approve their use, and the new sustainability leadership minor at UMN Morris became an official option. Sustainability is indeed an expansive term, and it is perhaps overused, but particularly in the context of a campus with so much contextual support for paying attention to environmental impact, not having such an obvious academic option seemed far more problematic than the potential challenges of a "sustainability leadership" minor.

Student engagement with sustainability extends well beyond any one academic program, as is evident at the campus's annual Undergraduate Research Symposium, where student presentation topics attest to the range and prevalence of sustainability awareness. Over the last few years, these topics have included, to name just a quick sample that highlights creative intellectual engagement from multiple disciplines: "The Hidden Carbon Footprint of AI Models: GPU-aware Carbon Modeling with Dynamic Lifetime" (advised by a computer science professor), "Optimization of Mangosteen Dye-Sensitized Solar Cells" (advised by a chemistry professor), "Feminism as an Approach to Fighting Anthropogenic Climate Change" (advised by a studio art professor), "Renewable Energy Consumption, CO2 Emissions and Economic Growth: Evidence from Developing Countries" (advised by an economics professor), "Characterization of the Pomme de Terre River Macroinvertebrate Community Prior to Restoration of Natural Flow Conditions" (advised by a biology professor), and "Exploring Native Peoples' Attitudes Toward Native Food Sovereignty Within Their Home Communities" (advised by a psychology professor).

Environmental awareness on the campus is inflected by the fact that 25-30% of the student body is Native American or Indigenous, and the campus is federally designated as a Native American Serving Non Tribal Institution, one of only four campuses so designated in the upper Midwest. The campus student government's annual Sustainability Forum brings in a keynote speaker and showcases multiple projects, and it focused in 2022 on Indigenous perspectives and projects. The invited speakers included representatives from 8th Fire Solar, an "Anishinaabe-run nonprofit based on the White Earth Nation Reservation" (Hazzard, 2022). They build solar thermal panels as "a viable and economically feasible energy source" that boosts other systems and makes a significant difference in costs and in warmth. A solar thermal panel system cannot provide all the heat a home in Minnesota needs in the winter, but the sun shines a lot in the cold winter months in Minnesota, and solar thermal is inexpensive and easy to install. The manufacturers do not "think of solar thermal as a silver bullet, but more like a part of your golden buckshot" (8th Fire Solar, n.d.).

Such inspirational sustainability work has a responsive audience at UMN Morris, for all those interested in connecting sustainability to their education as well as their communities and cultures. Paisley Sierra provides another such example. Sierra graduated in 2024 with degrees in environmental studies and Native American and Indigenous studies, and while still a student served as the coordinator for the UMN Morris Intercultural Sustainability Leaders program. A descendant of the Oglala Lakota Nation, she also worked on finding ways to connect Indigenous languages, ideas, and sustainability. Out of that grew a project with a Dakota artist:

> *"This project aims to promote indigeneity (Indigenous identity), sustainability, and community. My initial vision was to create signage around campus that 1) prompts reflection on the land, the seasons, and the cyclical processes of nature, and 2) raises awareness and provides representation for the people and languages that have long been part of this region. In an era where many of us work primarily in offices or on computers, students often overlook the natural world beyond our academic environment. We tend to measure time in terms of 'semesters,' 'months,' and 'school years,' rather than through the lens of the natural world's cycles. This project's purpose is to foster a dialogue within our campus community about the environment, as well as Lakota, Dakota, and Ojibwe languages and cultures. It aims to enhance our campus environment and create a supportive space for Indigenous students." (Goodnough, 2024)*

On a campus that once housed a boarding school that worked to eradicate Native American languages and cultures, such work gains even greater importance, and it functionally connects academics and sustainability, both of which to be most effective need to include multiple perspectives and experiences.

Since student-faculty working relationships form a core element of an undergraduate liberal arts college experience, the campus encourages faculty to engage not just in advising undergraduate research projects but in their own sustainability-related research. And hiring with this in mind helps, too: one area of research (and multiple publications) by a recently hired assistant professor in computer sciences concerns energy efficiency in computing; an economics and business management faculty member just published an article on "The Heterogeneous Impact of Climate Change on House Prices." A music professor and the head of keyboard studies, Ann DuHamel, widely performs excerpts from "Prayers for a Feverish Planet," a project she conceived that drew more than 170 submissions of music composed on the topic of climate change, from musicians in 35 countries. And continuing the theater program's history of engagement with sustainability, which dates back to at least 2010 and a production of Shakespeare's *As You Like It* in which the set and costumes were made entirely from recycled materials, a faculty member, Lucas

Granholm, was recently recognized with the Kennedy Center American College Theatre Festival (KCACTF) and the Association of Theatre in Higher Education (ATHE) Innovative Teaching Award (only eight national awards given) for his work incorporating sustainable design practices into the classroom. With faculty and students paying attention to sustainability, the creative possibilities just keep multiplying, and repeatedly, they save resources, both materials and money, for the college and beyond.

One of the highest impact choices the university has made for advancing sustainability was to create an Office of Sustainability and to hire a full-time director of it. The Sustainability Director was the first in the University of Minnesota system, in 2006, and the director regularly meets with the chancellor. UMN Morris had the good fortune to hire a remarkably dedicated and enthusiastic Director of Sustainability, Troy Goodnough, who still occupies the role. Although an office of one, as is common for many offices and roles on a small campus, Goodnough excels at seeing and motivating combinations of creativity and practicality for sustainability projects within the framework of the liberal arts college experience. Regular campus updates on sustainability initiatives and progress, campus events, engagement opportunities for faculty, staff, and students, news stories, grant writing—all this and more come out of the office. Both the Sustainability Office and the UMN Morris Center for Small Towns (located next to each other) also organize internships related to sustainability, in both the narrowest and broadest senses of the word, and both help coordinate faculty and student research. Internships with the neighboring research center often focus on renewable energy projects, such as a recent summer's work with a solar-powered robotic weeder, and students also hold sustainability-related work study and other jobs on and off campus.

Morris Model

A key component of so much of the sustainability work at UMN Morris is partnership: students and faculty/staff; staff/faculty and larger community; students and community; campus and research center, campus and local school district, campus, city, and county. National news has taken notice of one of these partnerships, the Morris Model. *The New York Times* in July of 2022, "As Federal Climate-Fighting Tools Are Taken Away, Cities and States Step Up," cites Morris, Minnesota as one example, noting that "By removing partisan politics from community discussions about climate policy, it's somctimes possible to reach a consensus that's been difficult to achieve on a national level." As the article explains, the University of Minnesota Morris campus leans left politically, while surrounding farming communities lean right. But both communities broadly support — and have helped to shape — the "Morris Model," which calls for reducing energy consumption 30 percent by 2030, producing 80 percent of the county's electricity locally by 2030 (thus guaranteeing it comes from renewable sources) and eliminating landfill waste by 2025. "We've never focused on climate as being the thing to talk about, because you don't have to," said Blaine Hill, the city manager, noting the benefits of lower energy bills and more local economic activity from the locally produced power. "You can go around that and just start working on stuff." (Astor, 2022)

The Morris Model began in the early 2010s with five initial pillars: energy conservation, clean energy, community resilience, cultural exchange, and celebration. One motivator for UMN Morris to participate in a city-campus partnership stemmed from its Second Nature commitment, which included campus-community resilience planning, and in 2014 the campus participated in a Morris Area Climate Dialogue with the city, the county's Institute for Agriculture & Trade Policy, and other partners. In 2015, the campus was recognized with Environmental Initiative's award for Community Action, framed around the Morris Model. A Morris Model 2018 planning session determined three ambitious major goals, none of which directly name climate change:

1. Produce 80% of the energy consumed in the county by 2030
2. Reduce energy consumption 30% by 2030
3. No land-filling of waste generated within the county by 2025 (Morris Model, n.d.).

The cultural exchange piece of the Morris Model has been amplified as part of the Climate Smart Municipalities project sponsored by the University of Minnesota's Institute on the Environment, and since the fall of 2015, the Morris group has been working with the Saerbeck community to exchange and advance sustainable living (Climate Smart Minnesota, n.d.). Exchanges between the two cities have connected farmers and university faculty, mayors and students. The UMN Institute on the Environment supports the partnership, too, by including both the UMN Morris sustainability director and students on a winter break study abroad course to Düsseldorf, Saerbeck, and Münster, Germany, called "Germany: Leading the Renewables Revolution." Students return even more energized for environmental action, with real-world experience of how things can be done differently to improve sustainability—and their participation has been enthusiastically supported with donor funding. In turn, UMN Morris and its neighboring research center host sustainability interns from Germany every summer.

So successful has been the Morris Model that it went on to attract attention from author and journalist Edward Humes. Among the books Humes has written is *Garbology: Our Dirty Love Affair with Trash*, which he describes as "a story about how wasteful Americans are without knowing it," and more recently, *Total*

Garbage: How We Can Fix Our Waste and Heal Our World (Humes, n.d.). Humes has visited Morris and met several UMN Morris students and graduates, including Sydney Bauer, who started the West Central Minnesota Climate Network and led a Climate Strike event locally; Bauer is among the dedicatees at the start of Humes' recent book, *Total Garbage*. In July of 2023, Humes published a piece in the *Wall Street Journal* that elevates the work of the Morris Model and stresses its foundation in the alliance of creativity and practicality. Referring to Morris as an "environmental Nirvana," Humes notes that "It's fair to say that environmental and climate concerns have never been front of mind when it comes to votes and policies in Morris. But residents will talk all day long about rural self-sufficiency, high energy and fuel costs, saving tax dollars and eliminating costly inefficiency and waste" (Humes, 2024). Those interests were something that UMN Morris and the West Central Research and Outreach Center could work with, and they found allies:

> *[UMN Morris Sustainability Director] Goodnough's bet was that the common-sense, cost-saving goals the farmers prized could lead them to choices that also happened to be good for the environment. But could it really be as simple as changing the terms of that conversation? Yes, says Blaine Hill, the recently retired city manager who helped make it happen. "We never made it about climate. We just did it because it makes sense. And the more we did, the more we wanted to do."*

While the town of Morris has only a little more than 5,000 residents, the impact of the Morris Model extends well beyond that. The former city manager for Morris had plenty of visitors interested in how their own small towns could do what Morris has done. But as Humes explains,

> *"What aspect of the Morris Model most helped Blaine Hill sell the idea to mayors and other town officials? It was the solar-powered Morris liquor store. "Morris broke the ice," says Fergus Falls mayor Benjamin Schierer. "Now we're working with towns as small as 150 people who happen to have a municipal liquor store, and we can show them how to put a 40-kilowatt solar system on it and that the community can benefit. The big idea we can grab from the Morris Model is: Once you open that door, the possibilities are endless." (Humes, 2024)*

That point drives not just the Morris Model but also the campus-specific projects at UMN Morris, and together, the two entities accomplish more. A student who graduated from UMN Morris in 2022 with majors in environmental studies and sustainability studies and who has continued to work in sustainability recently described one of the relationships that undergirds the Morris Model, that between the campus sustainability director and the West Central Research and Outreach Center director: “It’s a powerful pair, Troy [Goodnough] and Mike Reese—they represent two sides of the same coin in the sustainability fight. They definitely don’t agree on a lot of things, but they do agree that sustainability makes sense economically, for our communities, for our prosperity—we’re doing this in a town of 5000 in a county of 10,000” (Peck, 2024).

CONCLUSION

When I started writing this account of sustainability at a public liberal arts college, the task was daunting, and as I wrote, I kept learning how much I would have to leave out. The fact that this essay is not a comprehensive account of sustainability at the college, however, contributes to an important point: the more I looked for it, the more I found, and sustainability has truly become such an integral part of the college’s existence and its student experience that separating out examples to sketch the framework is the only workable option. Sustainability at UMN Morris is a habit, a frame of mind, an embedded aspect of life—and one that fits well with our fiscal limitations. In leadership, staff, and students, the campus has acted on a long commitment to sustainability in both the broad and the more environmentally specific senses of the word. Maybe that was driven from the start by the challenges and opportunities inherent in convincing the state to open a liberal arts college in a rural location 150 miles west-northwest of Minneapolis. In many ways, and in sustainability initiatives in particular, being a little far away is actually one of the campus’s significant benefits. The small campus provides the same degree as its much larger and urban research-focused sibling—students at all five campuses in the system earn the University of Minnesota degree—and UMN Morris offers a genuine community with easy access to and support for an abundance of high-impact practices, in a setting that encourages exploration. There is intellectual and physical space for students to try out new things and to investigate for ideas for themselves, and the campus culture encourages that, as this case study shows.

What lies ahead for the living laboratory that is UMN Morris? In 2022, the West Central Research and Outreach Center and UMN Morris together launched the Center for Renewable Energy Storage Technology (CREST) with the goal of using research and demonstration projects to advance energy storage options. UMN Morris already exports half of the power from the wind turbines to the local utility, but of course it is not always windy and not always sunny. Advancing technologies that can improve the usage of intermittent renewable power, then, is a need. The West Central Research and Outreach Center is a leader in wind-to-hydrogen-to-ammonia research in the county and already has a one-of-a-kind ammonia production skid.

The center is currently scaling up their ammonia production and they are part of the Midwest Hydrogen Hub initiative. UMN Morris wants to demonstrate technologies that allow the campus to store electrons in Minnesota's challenging climate. If the university can leverage corporate interests to help with funding, innovative projects such as this become even more practical, benefitting both the company that wants to try out a new technology and the campus, for teaching and research purposes as well as cost savings. CREST has initiated a partnership with a local power company (Otter Tail Power) and a microgrid company in Minnesota to install a 1-megawatt battery on the UMN Morris campus; the project goal is to install a non-lithium-based battery in 2025. The rural landscape already includes storage containment, from large grain elevators to ammonia storage, and CREST envisions a future where energy storage is just a new normal part of the landscape, much like wind turbines and solar installations have become.

The implementation framework that works at UMN Morris starts with conservation, builds partnerships, integrates with the curriculum, and engages community both on and off campus. The public liberal arts college plays a distinct role in advancing sustainability, as a recent graduate explains it:

> *In general the liberal arts education breaks your worldview open. I experienced that in a couple of different settings. I had two years at a community college before Morris, and I can contrast my sister's experiences with my own—she went to a private liberal arts college. Lots fewer people attend UMN Morris because their parents wrote a great big check. At UMN Morris, I met people from every background of life, with a wide range of experience. . . . Another aspect that's more UMN Morris-specific that has factored heavily in my work is learning about Tribal sovereignty and Indigenous knowledge, and that's a huge factor in environmental work, in sustainability. The public liberal arts vision helped support and expand how we approach Indigenous studies on the campus. I didn't take a specific Native American and Indigenous Studies class, but it filtered in anyway. And it is practically helpful in my current work to be familiar with this greater range of knowing. The liberal arts lens also helped me adjust to and open up to the rural perspective.* (Peck, 2024)

Students who choose the practical and creative path of a liberal arts education continue to be at the center of sustainability for UMN Morris, where practical application and theoretical understanding reinforce each other, preparing students to address complex environmental challenges through multiple disciplinary lenses. The UMN Morris case demonstrates the power of sustainability in higher education when it emerges from the core values of the institution rather than being treated as an add-on initiative, and the partnerships between the university, local community, and regional organizations illustrate how one liberal arts college has leveraged rural location as an asset rather than a limitation. These collaborations not only enhance educational opportunities but also create sustainable models that can be adapted by other small communities and institutions, showing how liberal arts education can be both locally relevant and globally significant.

REFERENCES

8th Fire Solar. (n.d.). Why solar thermal. Retrieved December 2, 2024, from https://www.8thfiresolar.org/why-8th-fire

American Association of Colleges & Universities. (2024). Trending topic: What is liberal education. Retrieved December 2, 2024, from https://www.aacu.org/trending-topics/what-is-liberal-education

Astor, M. (2022, July 1). As federal climate-fighting tools are taken away, cities and states step up. *The New York Times*. https://www.nytimes.com/2022/07/01/climate/climate-policies-cities-states-local.html

Cameron, L. A. (n.d.). Rural electrification administration in Minnesota. MNopedia, Minnesota Historical Society. Retrieved December 2, 2024, from https://www.mnopedia.org/thing/rural-electrification-administration-minnesota

Climate Smart Minnesota. (n.d.). About. Retrieved December 2, 2024, from https://www.climatesmart-mn.org/about

Condon, S. (2015, November 10). "GOP debate fact check: Was Marco Rubio right about welders vs. philosophers?" CBS News. https://www.cbsnews.com/news/republican-debate-fact-check-was-marco-rubio-right-about-welders-versus-philosophers/

Davidson, C. N. (2017). *The New Education: How to Revolutionize the University to Prepare Students for a World in Flux*. Basic Books.

Design Curial. (2012, February 20). University of Minnesota Morris' Welcome Centre wins LEED Gold. Progressive Media International Limited.

Downs, R. (2010, January 28). Morris biomass plant sees setback. *The Minnesota Daily*. https://mndaily.com/226520/uncategorized/morris-biomass-plant-sees-setback/

Engineers, K. F. I. (n.d.). University of Minnesota Morris Welcome Center. Retrieved December 2, 2024, from https://www.kfi-eng.com/project/welcome-center-u-of-m-morris/

Environment America. (2019, April 4). New rankings compare colleges and universities on progress toward clean energy. https://environmentamerica.org/media-center/new-rankings-compare-colleges-and-universities-on-progress-toward-clean-energy/

Goodnough, T. (2024, October 25). Expanding sustainability perspectives with poster art and Dakota language. University of Minnesota Morris. https://morris.umn.edu/news-events/news/expanding-sustainability-perspectives-poster-art-and-dakota-language

Hartocollis, A. (2023, November 3). Can humanities survive the budget cuts? *The New York Times*. https://www.nytimes.com/2023/11/03/us/liberal-arts-college-degree-humanities.html

Hazzard, A. (2022, November 3). Native-run solar firm aims to lower heating emissions and costs. *Sahan Journal*. https://sahanjournal.com/climate-environment/solar-heating-8th-fire/

Humes, E. (2024, July 3). A conservative farm town went green without really trying. *The Wall Street Journal*.

Humes, E. (n.d.). Edward Humes: Pulitzer Prize winning author. Retrieved December 2, 2024, from https://www.edwardhumes.com/

Johnson, M. (2024, September 23). UMN Morris mixes cows and solar on Midwest's largest agrivoltaic pasture. *Agweek* (ND).

Katerberg, W. (2016, February 12). The liberal arts—Present, past, and future. *Historical Horizons*. https://historicalhorizons.org/2016/02/12/the-liberal-arts-present-past-and-future/

Kostka, K. (2020, October 2). University of Minnesota Morris carbon neutrality report. Ever-Green Energy. https://www.ever-greenenergy.com/posts/university-of-minnesota-morris-carbon-neutrality-report/

Marquart, C. (2021, July). Wind over Morris. West Central Research and Outreach Center, University of Minnesota. https://wcroc.cfans.umn.edu/research/renewable-energy/wind-over-morris

Morris Model. (n.d.). Planning. Retrieved December 2, 2024, from https://morrismodel.org/planning/

Peck, G. (2024, October 11). Personal communication.

Rasmussen, L. (2010). Lowell Rasmussen interview. University of Minnesota Morris Stories, 32. https://digitalcommons.morris.umn.edu/stories/32

Rhodes, F. H. T. (2006). Sustainability: The ultimate liberal art. *The Chronicle of Higher Education*, *53*(9), 71.

Rosemark, S. (2024, October 11). Personal communication.

Sugimura, L. (2021, November 23). UMN Morris improves composting with grinder in dining hall. University of Minnesota Morris. https://morris.umn.edu/news-events/news/umn-morris-improves-composting-program-grinder-dining-hall

Supply Chain Council of European Union. (2022, February 19). UMN Morris achieves a million pounds of compost. https://scceu.org/umn-morris-achieves-a-million-pounds-of-compost/

University of Minnesota Morris. (1958). Executive Committee Meeting Minutes, May 1958. UMN Morris Founding Documents collection, Briggs Library. https://digitalcommons.morris.umn.edu/founding_docs/21

University of Minnesota Morris. (1960). A Liberal Arts Education Pamphlet [Brochure].

University of Minnesota Morris. (n.d.). Sustainability Goals & Initiatives. Retrieved December 2, 2024, from https://morris.umn.edu/sustainability-umn-morris/sustainability-goals-initiatives

University of Minnesota Morris. (n.d.). Sustainable Academics & Coursework. Retrieved December 2, 2024, from https://morris.umn.edu/sustainability-umn-morris/sustainability-goals-initiatives/sustainable-academics-coursework

University Relations. (2014). Green Prairie Community Awarded Prestigious LEED Green Building Certification. Campus News Archive, 105. https://digitalcommons.morris.umn.edu/urel_news/105

Chapter 8
Leading to Green

Elsa M. Núñez
https://orcid.org/0009-0006-2337-3119
Eastern Connecticut State University, USA

ABSTRACT

The integration of sustainability into higher education is increasingly recognized as a critical pathway for addressing global environmental, social, and economic challenges. This chapter explores the evolving role of Eastern Connecticut State University in promoting sustainable development through curriculum innovation, campus operations, community engagement, and long term initiatives. By embedding sustainability principles across disciplines, higher education institutions not only prepare students to become responsible global citizens but also serve as living laboratories for sustainable practices. This chapter highlights best practices, institutional frameworks, and policy recommendations that support the transition toward more sustainable academic environments. It also examines the barriers to implementation, including institutional inertia, funding limitations, and the need for interdisciplinary collaboration. Ultimately, the chapter argues that higher education has a unique and transformative potential to lead the sustainability agenda, fostering a culture of critical thinking, innovation, and ethical responsibility.

INTRODUCTION

"I'd put my money on the sun and solar energy. What a source of power!
I hope we don't wait until oil and coal run out before we tackle that."

Thomas Edison speaking to Henry Ford and Harvey Firestone in 1931

When I arrived at Eastern Connecticut State University as its sixth president in 2006, I was fortunate to have inherited a campus deeply committed to environmental stewardship. Led by a visionary leader in former President David G. Carter Sr.,

DOI: 10.4018/979-8-3693-5454-4.ch008

Eastern had been actively engaged in energy conservation and other sustainability efforts for many years, going back to the first Earth Day in 1970. As a residential liberal arts university, Eastern was ideally suited to be a testing ground for environmental initiatives — advancing green skills and technology, engaging in empirical research, and supporting policy development. Falling within the Masters College and University Carnegie Classification, Eastern is deeply connected to its local and regional communities, and most of its students graduate to stay in Connecticut. Creating a generation of climate-conscious citizens is central to the University's commitment to a sustainable future.

More than 20 years ago, Eastern's leaders understood the power of alternative energy sources such as solar and geothermal and were part of a growing movement that also understood the dangers that dependence on fossil fuels represented. In so many ways, Eastern was ahead of its time. In 2000, the Board of Trustees of the Connecticut State University System created a Center of Excellence in Sustainable Energy Studies at Eastern to lead the University's academic programming for sustainability and in 2001 the Board created the Institute for Sustainable Energy (ISE), an arm of the University committed to advocating for environmental solutions through policy change, K-12 curriculum, and technical assistance and outreach to schools, municipalities, colleges and universities, nonprofits, and other organizations. Initially the ISE was funded by $6 million in grants, using funds allocated by the Department of Public Utility Control and the Connecticut Energy Conservation Management Board, with membership including state utilities Connecticut Light and Power, United Illuminating, and Yankee Gas.

It was clear to me when I arrived in 2006 that my task was to support these existing programs while also finding new ways to engage in preserving our natural resources, knowing that the academic and scientific communities were constantly unveiling innovations to ensure a sustainable future. I also understood that my role as president was not to be the main advocate of a cause as all-encompassing as sustainability, but rather to foster a culture where everyone on campus had a stake and a role to play.

It was important to nurture responsibility for a shared sustainable future at multiple levels, with institution-wide programs, office-level initiatives, and each campus member's personal commitment to environmental stewardship. Everyone can recycle, turn off lights when not in use, buy food locally, and more.

Facilitating and supporting a sustainability culture meant it was my job to empower many champions on campus — mitigating climate change cannot fall on the shoulders of any single individual or team. For a culture embracing sustainability to be effective, we also needed to continue to institutionalize our vision in the University's daily operations. Over time, we have built our success on (1) Commitment (committed to carbon neutrality and other sustainability goals, manifested by Eastern

President David G. Carter's signing of the Talloires Declarations in 2004 and my signing the American College and University Presidents' Climate Commitment in 2007)[1]; (2) Planning (the long-term 2009 and 2020 Climate Action Plans and shorter, annualized planning); and (3) Execution (effective strategies implemented broadly across the campus).

To provide a framework to execute Eastern's environmental vision, and to serve as a mnemonic to inspire personal participation, the University describes itself as a house built on three sustainability "pillars" — combining Academics (the Center for Sustainable Energy Studies), Administration (the Green Campus Committee and its programs), and Outreach (the Institute for Sustainability).

With the United Nation's adoption of 17 Sustainable Development Goals in 2015, Eastern's sustainability framework has evolved to more broadly address the equity and social justice themes found in the UN's goals. Those goals include traditional, environmental concerns — clean energy, climate action, clear water — as well as issues of social justice and equity, including hunger, poverty, education, and gender equality.

As a university, we are deeply committed to the United Nations Brundtland Commission's definition of sustainability: "Meeting the needs of the present without compromising the ability of future generations to meet their own needs, (un.org).

This chapter is organized into three parts. In the first part, I explore each of Eastern's three "pillars of sustainability," detailing how they have come to pass, and the progress made in each area. Then, I share highlights of Eastern's 2020 Climate Action Plan, which guides our initiatives and our embrace of the UN' broad definition of a sustainability to reorient the Center for Sustainable Energy Studies to become the Center for Sustainability and the Institute for Sustainable Energy became the Institute for Sustainability. The chapter concludes with a brief discussion of how a college president can be directly involved in advocating for a sustainability culture on their campus and a sustainable future for our planet and its people.

THE THREE PILLARS OF SUSTAINABILITY AT EASTERN

Pillar 1: Academics: Helping to build a green economy.

The Center for Sustainable Energy Studies, created by the Connecticut State Universities Board of Trustees in 2000, is aligned with the Environmental Earth Science (EES) Department, where a Sustainable Energy Science concentration was introduced into the Environmental Earth Science major in 2006. A Sustainable Energy Science minor is also available. Over time, students have interned for public policy agencies; received a NASA fellowship to research alternative fuel cells; built a wind turbine for a local school in Jamaica; and installed solar collectors on campus to supply solar hot water, all in preparation for green careers.

In 2022, the name of the Center for Sustainable Energy Studies was changed to the Center for Sustainability Studies, to reflect the broader emphasis established in the UN's 17 Sustainable Development Goals. The center's mission focuses on engaging in research and providing curriculum that actively engages Eastern students with sustainability issues, including climate change and the disproportionate impact that environmental crises have on marginalized communities and minority populations.

Students in the Environmental Earth Science Department are supported by the Smith, Snider, Clebnik Founders of EES Endowed Fund, which was launched in 2009 to honor the founders of the Environmental Earth Science Department and the EES major — Professors Henry Snider, Sherman Clebnik and Raymond Smith. EES alumni are employed in private consulting firms, government agencies, nonprofits, and other organizations. With this fund they are able to support present and future EES students with their generous donations. Income from the endowment is used to pay expenses related to student research, and field experiences, including off-campus fieldwork and travel to professional conferences.

In addition to coursework leading to a concentration in Energy Science in the Environmental Earth Science Bachelor of Science program, the Environmental Club, Outdoor Adventure Club, and Habitat for Humanity Club provide students with many opportunities for co-curricular activities in the broad field of sustainability. The Environmental Club advocates for the importance of minimizing negative human impact on the environment by using sustainable practices on campus and throughout the community. The club works to educate its members, as well as the Eastern campus, through hands-on activism, community service, and discussions around the topics of environmental health and awareness. The Outdoor Adventure Club promotes healthy lifestyles through exploration of the natural environment, while Habitat for Humanity impacts homelessness by building homes in the community.

Hands-on Learning Informs Sustainable Energy Studies. Included with the Center for Sustainable Energy Studies designation were funds to create an endowed chair, and in 2004, Dr. Fred Loxsom was hired as the first Endowed Chair in Sustainable Energy Studies. During his eight years at Eastern (2004–12), Dr. Loxsom developed the concentration in Sustainable Energy Studies for the Environmental Earth Science major, worked closely with students preparing for green careers, and collaborated with the Institute for Sustainable Energy and the Office of Facilities Management and Planning to create a sustainability culture at Eastern. He was truly a champion of the environment and worked tirelessly to advance sustainability on campus, in Connecticut, and beyond.

Dr. Loxsom has since retired, but the University continues its outstanding coursework in the environmental sciences. Students have on- and off-campus internship opportunities, as well as the opportunity to work with faculty mentors on research on a variety of sustainability topics.

The Institute for Sustainability, for example, has been the training ground for more than 150 interns during the past 20 years. One innovative internship opportunity took place during the 2018–19 academic year when seven Eastern interns conducted a massive conversion to energy-efficient lighting at Three Rivers Community College (TRCC) in Norwich, Connecticut. The six-month project saw the students research lighting choices and replace almost 9,000 incandescent light bulbs with LED lighting, saving more than $55,000 in annual energy costs for the community college moving forward. Both Eastern and TRCC are members of the Connecticut State Colleges and Universities System. This project was not only an example of "helping one of our own" but also demonstrated how Eastern's sustainability programs are community-based and supportive of the broader eastern Connecticut region. As proud as I am of our sustainability efforts on campus, our outreach in the local region rings even louder, and speaks to our mission as a public university.

Blending classroom lecture with co-curricular clubs and exciting fieldwork gives EES students and students in other majors an immediate opportunity to impact climate change as they prepare for green economy careers. Biology students have studied arachnids in the southwest U.S. to advance antibiotic research, Health Sciences students have interned in public health facilities in Ghana, West Africa, and EES students have studied erosion on the southern shores of New England.

An emerging initiative on campus is to teach environmental literacy across the entire curriculum. To that end, the Center for Teaching, Learning and Assessment has sponsored several faculty development workshops to help faculty integrate the UN's 17 Sustainable Development Goals into their courses. At this point, more than 125 Eastern courses have some connection to sustainability. These range from Sustainable Cannabis and Energy and the Environment to Oceanography and Public Health Nutrition.

Climate Change Research. Given a broader definition of sustainability, it is critical that funding for faculty and student research also be broad-based. Much remains to be done in this area. In "The Future of Academic Research on Climate Solutions," Burrage, Carter and Goldstein (2022) found that the majority of federal grants in their sample went to the geo-sciences, engineering, biology, and computer science.

In their review of FY20 and FY21 Department of Energy and National Science Foundation grants, the authors found that 85 percent of the grants went to geo-sciences, engineering, biology, and computer science, and only 5 percent was allocated to the social sciences. "Funders should prioritize social science and encourage collaborations between social science, natural science and engineering fields," said Burrage, Carter and Goldstein (2022).

The researchers also found disparity in the equitable distribution of grants. While Hispanic-Serving Institutions accounted for 16 percent of enrolled students, they received only 8 percent of the grants in question. In addition, 95 percent of all grants went to R1/R2 institutions and only 5 percent to master's and baccalaureate-level colleges and universities. The authors also proposed greater collaboration between college researchers and industry partners, (Burrage, Carter & Goldstein, 2022).

Burrage, Carter and Goldstein also observed a lack of collaboration between university researchers and their local communities, and argued for creating partnerships to encourage "equity and climate justice." Such partnerships range from active community participation in research projects to ethnographic research to crowd-based data collection that documents local climate change. Such collaboration can "empower marginalized communities and encourage local entrepreneurship," (Burrage, Carter & Goldstein, 2022).

Eastern has responded to this concern for more community engagement proactively, with another example of how student advocates are taking the lead on sustainability. In this instance it has been our composting program. It turns out that food waste is a huge part of the world's solid waste problem. Connecticut's major waste-to-energy facility, located in Hartford, closed in July 2022. Connecticut joins states across the Northeast who are in competition for capacity to manage their waste, driving up the state's cost of sending excess waste elsewhere. The largest component of landfills is food waste. This is especially problematic as food waste produces methane gas when it is in landfills.

Eastern's student-run Garden Club has partnered with the CLICK (Commercially Licensed Cooperative Kitchen) in our hometown of Willimantic, to compost food waste. Assuming every resident student produces two pounds of food waste a month that was turned into compost, more than 25 tons of food waste could be diverted from landfills.

Composting allows organic materials to break down naturally, releasing carbon dioxide that can be absorbed by plants. Compost can also be used as soil fertilizer, reducing reliance on chemical fertilizers and pesticides. Nutrient-dense compost also increases soil filtration and helps combat flooding and erosion.

Undergraduate Research. Eastern's undergraduate researchers reflect the need to broadly address the multiplicity of sustainability issues identified by the United Nations. At Eastern's 24th annual CREATE Conference (Celebrating Research Excellence and Artistic Talent at Eastern), student research topics related to sustainability ranged from transportation systems such as high-speed rail to mental health among college students, healthcare inequities, improving pre-natal care for at-risk populations, and the impact of plastic waste on oceanic plant and animal species.

At many colleges and universities, faculty-mentored research is limited to graduate students. At Eastern, undergraduates often begin working with faculty as early as their first year. With climate change being a global issue, building sustainability experience as early as possible is a priority at Eastern. I am so proud of our student researchers for taking the lead in discovering new sustainability solutions across a range of academic disciplines.

For example, Michaela Belden '24 used her field experience on the big island of Hawaii in 2023 to inform her research on sustainable agriculture. An Anthropology major, Michaela interned at Big Island Farms (BIF), an organic vegetarian center in Honokaa, Hawaii. Gardens at BIF include a citrus orchard, an avocado orchard, and native hardwood and softwood trees.

Concepts Michaela learned during her internship included "permaculture" (developing a self-sustaining agricultural ecosystem); and "symbiosis" (mutually beneficial relationship between two or more species). Permaculture principles include local resourcing, biodiversity, waste reduction, harvesting and storing food during abundant periods, composting, and use of renewables.

While in Hawaii, Michaela learned of the attempted assimilation of Hawaiians into American monoculture, and the desire by indigenous people to sustain and protect their heritage, including agricultural practices.

In fall 2023, Computer Science students Ian Valeta and Diego Hernandez Ramirez championed sustainability by showcasing their skills to win a coding contest against teams from Yale, UCONN, Trinity College, and other schools. Their idea — developed in 24 hours of non-stop coding — was a mobile app that allows users to play a game to plant trees. As users cut down trees with a digital axe, they are connected to the nonprofit "Digital Humani," which links donors to reforestation projects. The app integrates the Cash App by Square to bring together existing consumer behavior (game playing) and existing services (Cash App and Digital Humani) to engage users in sustainable actions. As Hernandez Ramirez explained to us, "For example, a user purchases 100 virtual credits for $10. We can set 90% of the revenue to be sent to a reforestation service. This reforestation service (Digital Humani) will plant one tree for $1. So, for that user transaction, nine trees will be planted."

Whether field trips to study geology, undergraduate research into antibiotics, composting programs, or internships out in the community, Eastern students are using applied learning opportunities to ground their academic work. Beyond the sciences, overtime Eastern expects to build sustainability into every student's plan of study. Having a strong academic base for Eastern's sustainability goals gives meaning and context to the work we have done over the years to green our campus.

Pillar 2: Greening the Eastern Campus

Eastern exhibits best practices in mitigating climate change and demonstrating environmental stewardship on campus. Every person on our campus is a consumer of energy and natural resources. How the University "walks the talk" has a direct impact on the environment. It also motivates and informs our students, faculty, and staff, and provides students many opportunities as a living laboratory. As president, it has been my task and pleasure to encourage students, faculty, and staff at every opportunity to participate in the University's "green campus" initiatives.

To facilitate sustainable actions on campus, in 2007 Eastern created the Green Campus Committee, which continues and is coordinated jointly by the Institute for Sustainability, the Center for Sustainability, and the Facilities Management and Planning Office. The committee has organized Earth Day, hosted faculty lectures, led the campus's recycling program, inventoried greenhouse gases, and endorsed building structures on campus that conform to the U.S. Green Building Council's Leadership in Energy and Environmental Design (LEED) guidelines.

Additional initiatives include composting, the Thrift Warriors thrift shop, Shawn's Cupboard food pantry, our waste reduction partnership with Chartwells Food Services, energy conservation programs, the Science Building's 400KW Fuel Cell, the Garden Club, electric car charging, and more.

Converting to green energy. Eastern's approach to alternative energy sources reflects a broad-based, campuswide commitment to mitigating climate change. One green energy practice that is impressive in its scale and uniqueness at the time was the geothermal heating and cooling system in the University's High Rise Apartments, a residence hall for upper level students. In 2001, Eastern's Office of Facilities Management and Planning secured a grant from the U.S. Department of Energy to replace the building's heating and cooling system with Connecticut's largest geothermal installation. The system featured three 800-foot vertical wells that use the ground's natural temperatures to heat and cool the building. Prior to the installation, the building had no air conditioning. Then, the system for the 72,000-square foot building is 30 percent more efficient and saves Eastern $50,000-$70,000 in heating and cooling costs each year until its decommission in 2024.

In 2005, the University installed two 2.5kW solar photovoltaic (PV) systems on roofs of two residence halls to power perimeter and emergency lighting, and six smaller PV arrays to provide LED lighting for bus shelters around campus.

An additional example of pursuing alternative energy sources took place in 2006 when Eastern received a Supplemental Environmental Program Grant of $47,000 from the Connecticut Department of Environmental Protection to pilot an innovative renewable energy alternative, B20 — a mixture of 80 percent diesel fuel and 20 percent soybean oil — to heat our south campus. Finally, since 2018, all of Eastern's electricity has come from green, renewable sources.

Conservation and recycling. Eastern's recycling programs are especially noteworthy for their scale, their campuswide participation, and the statement of commitment that they make.

Whereas turning to alternative energy sources is chiefly an administrative decision, each member of the campus community can and does contribute to energy conservation and recycling programs at Eastern.

Recycling of plastics, glass, aluminum cans, and other consumables has been taking place for more than 60 years. Recycling in America goes back to 1690, when the Rittenhouse Mill in Philadelphia began recycling linen and cotton rags to produce paper for Bibles and newspapers, (New England Recycling Council, 2019).

At Eastern, in addition to the typical recycling of paper, plastics, aluminum cans, and glass, we also participate across the campus in recycling disposable containers and other materials. To add a bit of excitement to the mix, our residence halls participate in a competition to see who can recycle the most during the annual Warriors Cup, which also features academics, service activities, and athletic contests. Residential students interested in the environment also live together in our Green Theme housing accommodations.

Another example of a student-based recycling project is the "Warriors Don't Waste" program at the end of each spring semester. For several years — before and after the COVID pandemic — Eastern students living on campus have contributed to the University's overall recycling efforts by cleaning out their rooms. At the close of each spring semester, students have collected more than a ton of bedding, household goods, clothes, and other items in a clean-out of residence hall rooms to benefit local charities.

In 2023, the "Warriors Don't Waste" program was organized by senior Jacob Sweeney as part of his internship with Eastern's Institute for Sustainability, showcasing two pillars of Eastern's sustainability agenda. Students at six residence halls donated 2,253 pounds of items that were delivered to the Covenant Soup Kitchen, the No Freeze Shelter, and the Salvation Army in Willimantic. That included 605 pounds of clothes, 108 pounds of bedding, 909 pounds of food, 134 pounds of toiletries, and 497 pounds of "miscellaneous" items that ranged from pots and pans to fans, microwaves, and fridges.

Recycling a Building!

Recycling cardboard, paper, and plastic has become a weekly routine for most American households, as well as every office and residence hall on the campus of Eastern Connecticut State University. Eastern has taken sustainability a step further by recycling an entire building as part of its demolition!

In what is perhaps the most dramatic recycling project at Eastern was the May 2024 demolition of "Eastern Hall," a portable building that has been used for classrooms and office space for more than 20 years. Eastern Hall, built in 2000 as a temporary classroom and office building next to Eastern's library, has had its bits and pieces recovered, re-used, and recycled to make way for a new green space. Following completion of the demolition in late May 2024, the building's former footprint has been replaced with sod, shade trees, and native plantings.

The building was in use as late as the fall 2023 semester. After it was decommissioned, the utilities were cut off. Some of the older-style furniture that was no longer used on campus was donated to the National Guard. Furniture was also donated to Eastern's new "Thrift Warriors" thrift shop on campus, and 30–40 percent was repurposed on campus. Metal furnishings were recycled at Willimantic Waste.

As the building came down, construction waste materials were processed for their content at a recycling facility in East Windsor, Connecticut, under a contract with USA Waste and Recycling. About 65 percent of the sheetrock and carpeting was recovered for re-use as well. USA also handled general debris disposal for the project through its Hartford waste recycling center. Concrete and masonry were also hauled off and recycled, as well as mixed metals.

When everything was hauled away, planting began. Native highbush blueberry shrubs and deciduous sweetgum trees were planted, along with some non-native species. The energy savings from removing Eastern Hall from service is estimated at 166,500 kWh annually. An the green space left behind has become a favorite gathering spot for students to study, play, and bask in the spring sun.

LEED Construction. The U.S. Green Building Council promotes the construction of energy-efficient, high performance buildings through its Leadership in Energy and Environmental Design (LEED) certification program. Eastern's David G. Carter Sr. Science Building (dedicated in 2008), was granted Silver LEED status, while other University buildings i.e. the South Residential Village and the Fine Arts Instructional Center were constructed following LEED certification guidelines. These include the use of environmentally friendly materials, recycling rainwater, incorporating natural surroundings (the Science Building has a bioswale running along its west side), and reducing energy use by up to 30 percent. The 174,000 square foot Science Building is also equipped with daylighting controls, a grey-water system, and variable speed drives.

When I arrived at Eastern in 2006, "greening the campus" was taking place collaboratively across departments but without a plan or an oversight group. With our Climate Action Plan in place and with the leadership of the Green Campus Committee, we have made significant progress in showcasing our University as a sustainability laboratory for our students and a symbol of our commitment to carbon neutrality. I am very proud of the work that our students, faculty, and staff have done these past 18 years of my presidency to advance our reputation as a green campus.

Pillar 3: Outreach and Advocacy

Former President David G. Carter Sr. understood the importance of spreading the sustainability message across Connecticut and beyond. With funding from the state's utilities and the support of the State of Connecticut, the Institute for Sustainable Energy (ISE) was formed in 2001. Early ISE initiatives were ably led by William Leahy, executive director from 2002–14. Mr. Leahy forged relationships with Connecticut's utility companies, worked closely with municipal and educational officials, and collaborated with Eastern faculty and staff to advance sustainability on our campus and throughout Connecticut. He was one of the University's chief sustainability advocates and his legacy continues today.

The Institute for Sustainability is the University's sustainability outreach arm and has been a key player in Connecticut's environmental movement. Early on, the institute was funded through state grants, contracts and project clients. It was charged with four main areas: (1) conducting educational outreach, (2) developing energy solutions, (3) supporting the development of public policy, and (4) dispensing information resources.

At the outset, ("Living Green," 2003), the ISE published a set of recommendations for the "Greening of Eastern." Those included:

- Modify the University's purchasing policy to emphasize the use of recycled and sustainable resources. A great example of this is Eastern's use of thousands of reams of recycled photocopier paper each year.
- Support an active on-campus recycling effort on campus, with a goal of recycling as much as 50 percent of the campus waste stream annually. (From 2013-21, 42 percent of materials were diverted by recycling, composting, donations, or resale.)
- Create curriculum and other learning opportunities to ensure that all Eastern students gain environmental literacy.
- Reduce solid waste. (From 2013 to 2021, Eastern saw a 39 percent reduction in total waste generated).

- Conserve energy. (Our Encellium motion-sensing lighting system has saved the University hundreds of thousands of dollars by shutting off lights and related HVAC systems in unoccupied rooms.)
- Purchase green power. (100 percent of Eastern electricity was generated by green sources as of May 2018)
- Enhance water and wastewater policies and practices.
- Provide planned and safe disposal of hazardous materials.
- Sustainable transportation. (commuter carpooling, on-campus shuttle bus system)
- Implement sustainable food service. (local food purchasing, Chartwell's tray-less cafeteria)
- Improve land use. (20-acre arboretum)

Much of the work during the early years of the ISE was driven by the Talloires Declaration, the outcome of a four-day retreat of college presidents in Talloires, France, in 1990, (ulsf.org, 2015). The vision of those 22 presidents and chancellors was to inspire leadership among colleges and universities as role models, living laboratories, and educational centers. Dr. Jean Mayer, President of Tufts University, was the host for the sessions, which resulted in 10 key objectives:

- Increase awareness of environmental sustainable development
- Create an institutional culture of sustainability
- Educate environmentally responsible citizens
- Foster environmental literacy for all
- Practice institutional ecology
- Involve all stakeholders
- Collaborate for interdisciplinary approaches
- Enhance capacity of primary and secondary schools
- Broaden service and outreach nationally and internationally
- Maintain the movement

In 2004, Eastern President David G. Carter Sr. signed the Talloires Declaration, joining an alliance of more than 85 U.S. universities and more than 300 colleges and universities in 40 countries, adding Eastern to a global initiative to promote green energy and reduce the use of fossil fuels. At the time, Eastern was the only signatory among Connecticut's public colleges and universities, demonstrating early on that Eastern was an environmental leader in higher education and in the State of Connecticut.

Educational outreach. Among the ISE's focal points has been the development of K-12 curriculum and related teacher training to more than 2,000 Connecticut teachers. Similarly, the institute has provided training for building code officials under the authority of the Connecticut Department of Public Safety.

In 2004, the ISE hosted and coordinated the UN Children's Conference on the Environment involving more than 450 children from 50 nations coming to Connecticut to learn about biodiversity, energy conservation, and indigenous peoples. Through Eastern's leadership, 82 workshops, 31 field trips and 8 seminars were held over the five-day event. The institute has also coordinated a climate challenge contest for students in grades 4-12 called "Keep Connecticut Cool," with more than 700 students participating over the years.

The institute frequently works with government and community groups seeking environmental guidance. Early in its infancy, it was asked by the Connecticut legislature to evaluate power supply and conservation opportunities in southwest Connecticut. Later the institute was asked to do a similar study of Long Island Sound's natural resources. Providing research documentation and testimony to advance energy- and environment-related legislation has also been an element of the Institute's outreach efforts.

In 2009, Eastern developed a plan to "green" the Mystic Aquarium. That same year, an energy audit for The Hole in the Wall Gang Camps in Ashford, Connecticut, enabled the nonprofit camp for children with serious illnesses to develop their own sustainability initiative.

The Institute for Sustainability continues to work closely with local and regional communities to advocate for sustainable solutions. In early 2024, Eastern hosted a local panel discussion of community leaders from Windham/Willimantic. Windham Mayor Tom DeVivo said, "Connecticut has a solid waste problem," noting that the state currently ships 850,000 tons of waste out of state annually. DeVivo said new waste-to-energy facilities need to be built, and positively noted that Windham/Willimantic is one of the 63 townships currently certified by Sustainable CT, a nonprofit organization created to advance sustainability throughout Connecticut's 169 townships.

All these activities point to Eastern's commitment to sharing its expertise with local communities and state organizations. This was part of Dr. Carter's initial vision for the Institute for Sustainable Energy, and remains a central part of my own personal commitment to mitigating climate change on campus and in Connecticut.

Research and community engagement are natural partners. In one example, Bryan Oakley, professor of Environmental Earth Science and an expert in coastal geology partners with the Watch Hill Conservancy of Westerly, Rhode Island. Oakley and his students have been studying the impact of erosion on the shores of southern New England. For more than a decade, Oakley has been observing and recording the effects of climate change on the beaches of eastern Connecticut and Rhode Island. In combination with rising sea levels, the impact of storm surges caused by more frequent, intense storms have seen nearby Napatree Point move 300 feet in the past 85 years. Oakley and his students work closely with the Watch Hill Conservancy to counter the effects of sea level rise, although retreating from the coast may be the only sure solution over time.

Public Education. The Institute for Sustainability also engages in public education efforts, producing and distributing newsletters, news articles, and social media posts. The message of this information campaign is that individuals and families can contribute to sustainable communities at home.

The food we eat is a good place to start. Unused food constitutes a quarter of all landfill content. Not only is this wasteful, it generates dangerous methane gas. In addition, up to 25 percent of global greenhouse gas emissions (GGEs) are generated by food, agricultural processes, and associated land use.

Reducing food waste and shifting to a plant-based diet can significantly reduce our carbon footprint. For instance, if an individual person was to adopt a single meatless day per week, it would save 7,000 gallons of water and avoid 416 lbs. of carbon emissions a year. Buying locally produced food reduces shipping expenses and related fuel emissions; composting is another way to reduce solid waste in landfills.

Partnering with local schools. It has always been a pleasure for me to see Eastern faculty, students, and alumni making contributions to our local community of Willimantic, Connecticut. One exciting local partnership is the work of alumnus Nick Kukla '19 at the Charles H. Barrows STEM Academy in Willimantic. Kukla teaches eighth-grade science and was recently named a "sustainability advocate" by the UN's Global Schools Program. The program, which began in 2019, covers 1,560 schools in more than 100 countries. In addition, Barrows students complete a problem-solving 8th grade capstone research project that challenges them to understand and propose solutions to the issues outlines by the UN's 17 Sustainable Development Goals. Implemented by English Language Arts teacher, Christian Kollegger, these interdisciplinary project are then publicly presented at Eastern in partnership with the Institute for Sustainability.

This past year, the students addressed such problems as food insecurity in Willimantic, resulting in donations to local food banks and selling bracelets to raise attention to under-age marriages around the world.

Sustainable CT. Like Dr. Carter and the founders of the Institute for Sustainable Energy, I believe in supporting sustainable communities across Connecticut. Under the leadership of Lynn Stoddard, ISE director from 2014-22, Eastern and other partners created Sustainable CT in 2017. This independent organization has a vision to support municipalities across Connecticut that want to become more sustainable and resilient. With funding from the Emily Hall Tremaine Foundation,

Hampshire Foundation, and Common Sense Fund, the program spent five years building its capacity at Eastern before becoming a separate 501(c) 3 nonprofit organization in 2022.

Sustainable CT offers best practices, technical assistance, and funding support for towns and cities seeking to build sustainable economies, develop equitable solutions to mitigate climate change, reduce greenhouse gas emissions, and support thriving local art and culture.

To date, the organization has granted more than $4 million to support local efforts. In addition, the organization's Fellowship program supports college students who work on sustainability programming at the local and regional level. Sustainable CT looks at sustainability in its broadest scope — beyond the natural environment— to include community resilience, social justice, and public health.

As of April 20, 2024, 134 of Connecticut's 169 townships were registered with Sustainable CT. Of those 134 communities, 63 have been certified, including 37 at the Bronze level; 24 at the Silver level; and two at the Gold level. Certifiable actions include:

- Fostering an inclusive and equitable community
- Supporting a thriving local economy
- Stewardship of land and other natural resources
- Vibrant cultural ecosystem
- Resilience planning
- Clean and diverse transportation systems
- Renewable energy infrastructure and operations
- Inclusive engagement, communication, and education
- Materials management
- Health and wellness opportunities
- Healthy, efficient, and diverse housing
- Homelessness prevention

Creating a three-pillared approach to sustainability at Eastern — Academics, Greening the Campus, and Outreach to the Community — has been the University's formula for advancing our resilience and helping to mitigate climate change. This approach has engaged the entire campus community and local and statewide partners. It has become a major part of our culture and value system. Our students — as often as the administration and the faculty — develop and implement strategies that speak to our ongoing commitment to climate action. I could not be prouder of our campus community for their commitment and personal involvement in our collective efforts.

EASTERN'S 2020 CLIMATE ACTION PLAN

While the "three sustainability pillars" at Eastern (Academics, Administration/Green Campus, Outreach) has served as a useful tool to represent the University' environmental stewardship, the realities of budgets, staffing, and operations require a planned response to our sustainability goals. In 2009 and again in 2020, the University articulated its plans for a sustainable future in the form of our Climate Action Plans.

The University's latest 2020 Climate Action Plan, in addition to accelerating our carbon neutrality goal from 2050 to 2030, contains key strategies for the future — short-term and long-term — as well as indicators of achievement that have been realized since Eastern's first Climate Action Plan was released in 2009.

Key indicators. Over the next 10 years, indicators of the University's progress will include:

(1) **Reducing energy consumption and increasing energy conservation.** This will include supporting the movement from gasoline-powered to electric vehicles; continuing to use green electricity, fuel cells, geothermal and other alternative fuels; and utilizing Eastern's motion-sensing lighting and building control system to conserve electricity and natural gas.

(2) **Managing solid waste.** Reducing solid waste, continuing to work with Chartwells Food Service to reduce and divert food waste, and enhancing Eastern's recycling efforts are all part of the University's solid waste management plan.

(3) **Reducing greenhouse gas emissions.** The goal is to reduce emissions by 3,200 metric tons over a 10-year period. Eastern uses the Sustainability Indicator Management and Analysis Platform (SIMAP) to measure greenhouse emissions. Currently, 50 percent of Eastern's emissions are due to natural gas consumption; 86 percent of all emissions are due to natural gas and commuter travel.

(4) **Educating students.** In addition to preparing students for careers in a green economy, we want all Eastern graduates to practice active sustainability habits as consumers. Over time, courses in sustainability literacy will be added to the Liberal Arts Core curriculum.

(5) **Climate mitigation and resilience.** It will be important for Eastern and other communities to continue to reduce greenhouse gas emissions while pursuing climate resilience or adaptation practices. For instance, as shorelines recede, oceanside communities will need to consider relocating.

(6) **Align with Connecticut's planning.** Eastern's planning reflects the work of the State of Connecticut in producing its own climate action plan ("Building a Low Carbon Future for Connecticut: Achieving a 45 percent reduction in greenhouse gas emissions by 2030"; 2018)

Achievements. In addition to the many University accomplishments described in other parts of this chapter, the 2020 Climate Action Plan also noted other progress being made at Eastern.

- Since 2009, the University has reduced copier paper consumption by 50 percent.
- Automated building control systems: in addition to motion-activated management of lighting, Eastern's automated building control system is also linked to HVAC systems, further reducing energy consumption.
- In addition to retrofitting residence hall showers with low-flow shower heads, all campus lighting is being retro-fitted with LED bulbs.
- Since 2010, vegetarian options have been available in the Dining Hall.
- Since 2018, Eastern has used compostable dining ware for campus events.
- 20 percent of faculty/student researchers are engaged in studying sustainability topics.
- In 2017, Eastern worked with the Town of Windham to develop a Resilience Report for the town and the University.

Future Plans. The 2020 Climate Action Plan also lays out a series of planned strategies for the coming years. These include:

- Developing a more sustainable purchasing policy to take advantage of bulk purchases.
- Developing a pest management plan that will prioritize native and pollinator friendly plants while also mitigating the use of pesticides.
- Increasing the use of locally sourced foods in Dining Services, including campus grown food.
- Reducing the sale of plastic water bottles on campus while creating water bottle refill stations.
- Transitioning all University forms to online submissions.
- Exploring the installation of photo-voltaic cells for new construction and/or roof renovations.
- Converting buildings from fuel oil and propane heat to low carbon sources.
- New construction should include no fossil fuel obligations and/or net-zero energy design.
- Investing in electric buses for the shuttle fleet, transitioning the entire car fleet to EVs.
- Supporting sustainable transportation options, including a bike-friendly campus (bike racks, bike shares) and a carpooling program for faculty and staff.

- Integrating a sustainability literacy requirement into the Liberal Arts Core at Eastern.

RESPONDING TO THE UN'S SUSTAINABLE DEVELOPMENT GOALS

For more than 20 years, Eastern has been a central player in advancing sustainability causes on its campus and throughout Connecticut. Our "three-pillared" approach (academics, campus administration, and outreach) and our climate action planning have focused, in large part, on the physical world and the conservation of natural resources — electrical energy management, recycling, food waste reduction, and other traditional environmental initiatives.

In recent years we have turned to the United Nation's 2030 Agenda for Sustainable Development for guidance; it includes 17 Sustainable Development Goals that were published in 2015. These span an array of social conditions and advocate for an equitable approach to improving life on our planet. In addition to environmental and climate change goals, the UN's sustainability goals also reflect social justice and equity issues. They serve to guide nation states, non-government organizations, institutions like Eastern, and other players in an international effort to conserve resources and secure a stable, sustainable, and equitable future for the world's people and our fragile planet.

In examining the UN's sustainability goals, it is evident to me that (1) they appropriately reach much further than simply preserving our physical and natural environment; (2) they place an emphasis on achieving equity on a host of issues that have too often left people behind; and (3) in this broader context, Eastern can bring renewed focus and breadth to its sustainability initiatives.

A turning point in Eastern's sustainability evolution was when the University mapped its curriculum and other campus activities against the UN's 17 Sustainable Development Goals. Not only did it become clear that we had many more courses (125+) than we realized that were responsive to the UN's goals, it was also evident that courses focused on science and the environment remained critical to maintaining a sustainable campus and promoting sustainability beyond Eastern. Just as Environmental Earth Science courses spoke to a green economy and clean energy, courses in such programs as Social Work, Psychology, and Early Childhood Education also show our commitment to social and environmental equity. In operationalizing our own sustainability agenda, the Eastern community is impacting each of the UN's 17 sustainability goals, in the curriculum, operations, student activities, and community partnerships.

No Poverty

Eastern's Center for Community Engagement (CCE) coordinates volunteer and service learning programs in our local community. Among the many service opportunities are weekly programs with local soup kitchens and homeless shelters. More than 1,000 students contribute more than 23,000 hours totaling $573,000 in donated time to the local community supporting the mission of the Windham No Freeze Shelter, the Covenant Soup Kitchen, and the Windham Area Interfaith Ministries.

Zero Hunger

"Shawn's Cupboard" is Eastern's on-campus, student-operated food pantry. Created by and named after student Shawn Dousis '19, the pantry is stocked with produce and nonperishable food items donated by the local community. Student volunteers at Shawn's Cupboard collect food donations, stock shelves and provide customer service to student patrons. The annual "Plunge for Hunger," Organized by Eastern's men's rugby team, with participation from other athletic teams and community members, raises funds for the local Covenant Soup Kitchen. Finally, the annual "Day of Giving," is a Thanksgiving feast for the community. Begun in 2007 by student Jason Budahazy '09, the yearly event feeds more than 400 people each year — patrons of the local soup kitchen and food pantry, senior citizens, and other people who can use a well-cooked meal.

Good Health and Well-Being

Each year, for at least 12 years, the men's soccer team partners with the National Marrow Donor Program. Since 2012, Eastern has registered 2,328 volunteers for NMDP, 34 of whom have become donors. Eastern's new Bachelor of Science in nursing (BSN) program, established in 2024, is an example of how higher education, government, and the private sector can partner together for the public good.

Although 3,000 new nurses are needed each year in Connecticut, nursing schools in the state graduate only 1,900 new candidates, according to the Governor's Workforce Council. The University received a $1.2 million grant in 2023 from the Connecticut Health Horizons project, established by the state to address the nursing shortage in Connecticut. Funding supports instructional equipment and three new faculty positions. Hartford HealthCare is constructing a 6,700-square foot clinical simulation lab at Windham Hospital for use by Eastern students and faculty starting in fall 2025.

Inclusive and Equitable Education

Aligned with the institutional mission to engaged students from diverse backgrounds in a transformative liberal arts learning experience that provides knowledge and skills to lead enriching and purposeful life, Eastern is a host institution for Jumpstart, a national AmeriCorps program created to help preschool children from low-income neighborhoods develop their literacy skills. As a native of Puerto Rico whose home language is not English. Established in 1993, with programs in 15 states, Jumpstart aims to end the cycle of child poverty by offering language, literacy, socialization skills, and emotional development for preschool children from disadvantaged communities.

Gender Equality

Members of the Eastern campus community who take "Safe Zone" training serve as advocates for the LGBTQ+ community by learning language and practices to better understand and support gender/sexual identity. The University's Pride Center exists to "provide a space to safely, bravely and proudly explore and celebrate students' LGBTQ identity."

Clean Water and Sanitation

The bioswale filtration system located just outside of the David G. Carter Sr. Science Building collects rainwater runoff for irrigation and simulates a natural riverbed full of native vegetation. Three new water wells installed on the Eastern campus in September 2023 are being used by Environmental Earth Science (EES) students for hands-on learning in hydrology relevant to careers in environmental consulting and management.

Affordable and Clean Energy

Eastern is home to two fuel cells, a 400 kW phosphoric acid fuel cell installed in April 2012 behind the Science Building, provides electricity and onsite heating from the fuel cell's byproduct heat. The second cell was installed on campus in 2021 near the Admissions Building. It produces 1,150 kWh and was designed to produce half of the total campus electrical load during peak demand. As of November 2021, this fuel cell has helped avoid 545,320 lbs. of carbon dioxide and 27,833 gallons of gas not used, the equivalent of 53 cars removed from use for one full year. Leveraging a federal grant, we now have 11 electric vehicle charging stations available throughout campus and in Eastern's two parking garages. The stations are free and available to students, faculty, staff, and the general public.

Decent Work and Economic Growth

Accounting students have participated in the Volunteer Income Tax Assistance (VITA) program since 1999, helping to steer millions of dollars in tax refunds back into local economies.

Eastern's new Cannabis Cultivation and Chemistry minor is attracting students to the emerging field of cannabis for medical and other purposes. The minor began in spring 2022, with a focus on the hemp industry. Eastern is a licensed hemp producer with the Connecticut Department of Agriculture. The program is rooted in the field of plant biology and uses the University's greenhouse and laboratories. The "Pathways from Poverty" program sponsored by the U.S. Department of Labor and the American Recovery and Reinvestment Act (ARRA) was designed to create more jobs throughout the nation. The ISE actively supported Green Collar Job enterprises recommended by the State of Connecticut and by the ARRA. One of those initiatives was Green-Up-Bridgeport, a workforce development program funded in 2010 by a competitive, U.S. Department of Labor grant of $4.4 million awarded to The Workforce Inc.

Industry Innovation and Infrastructure

Eastern is home to five buildings constructed to standards set forth by the U.S. Green Building Council's Leadership in Energy and Environmental Design (LEED)'s certification program. These buildings are energy efficient and sustainably constructed. In 2015, Eastern was selected as one of seven higher education institutions nationally to participate in the Department of Homeland Security's Campus Resilience Pilot Program, which was created to develop emergency readiness plans for campuses in the face of man-made and climate-related challenges.

Reduced Inequalities

One program at Eastern I am particularly proud of is our association with TheDream.US, a national foundation for undocumented students. In 2016, Donald Graham, former publisher of the Washington Post, launched TheDream.US foundation, dedicated to providing financial support to undocumented college students from states that limit their ability to receive a public university education. Eastern became one of the program's two pilot universities and today is the largest TheDream.US partner, with more than 350 students having been enrolled as Opportunity

Scholars. With a 97 percent retention rate and an 89 percent four-year graduation rate, students in the program are demonstrating the program's success and their personal dedication. In 2020–21, all four Student Government Association officers were Opportunity Scholars, just one example of the on-campus leadership of these undocumented students.

Sustainable Cities and Communities

In 2005, ISE surveyed Connecticut schools and determined that Connecticut's schools were among the nation's most inefficient in their energy usage. The institute's response was a benchmarking program that helped more than 400 schools, 100+ municipalities, and 120 State of Connecticut buildings to develop energy efficiency programs. For instance, our local Windham Public Schools adopted the institute's recommendations for lighting improvements, water conservation efforts, and air quality enhancements, saving local residents more than $500,000 annually. Institute staff have also given legislative testimony, assisted in drafting public policy around the environment, written high school curriculum, trained more than 2,000 teachers, and partnered with other organizations to expand public awareness of climate change and the need for a range of sustainable practices.

Responsible Consumption and Production

The Thrift Warriors shop on campus is available to students to donate and take (free of charge) gently used clothing, school supplies, and housewares. Students enjoy "thrifting" and realize they are helping the environment in the process. As student Aidan Stewart told us, thrifting is a way to "help someone who might not be able to afford (new clothes) right off the shelf." Student Jannik Wille enjoys thrifting as a social activity, a way to enhance his wardrobe and, most importantly, a sustainable pastime. "It's a great way to help the environment." Chartwells, Eastern's dining services partners, are industry leaders in reducing food waste and supporting food security programs in the community. In 2020, for the second consecutive year, Eastern and Chartwells placed eighth nationally in the Food and Dining category of the Sustainability Tracking, Assessment & Rating System (STARS).

Climate Action

Climate action was at the center of Eastern's sustainability initiatives long before I arrived on campus in 2006. It has continued to be an emphasis during my 18-year tenure as Eastern's president. Three themes describe our climate action efforts: (1) renewable energy; (2) waste reduction and recycling; and (3) energy conservation.

Perhaps the most impactful example of energy conservation on our campus is Eastern's modern building controls system, which links our lighting and HVAC systems to reduce electrical consumption in unoccupied spaces.

Life Below Water

Biology students and faculty have traveled to the Caribbean for more than five decades to study tropical marine biology. Since 1968, Eastern biology professors have accompanied students on a Tropical Biology field course to the Caribbean basin. Destinations have changed over time, initially taking students to Bermuda. In 1984, the Biology Department introduced Jamaica as a second location, with Belize taking its place in 1986. Bermuda and Belize alternated each year until 2001, when San Salvador Island in the Bahamas replaced Bermuda. Costa Rica replaced Belize as a destination in 2008.

Life on Land

The Arboretum is Eastern's on-campus nature preserve, containing 20 acres of walking trails and more than 30 native tree species. The arboretum features a deciduous forest, meadows, streams, a pond, and surrounding wetlands. Students can study native plant-animal interactions in this biodiverse setting, and learn how to monitor stream flow, water quality, groundwater, soil-water, and pond levels. Students from the Environmental Earth Science and other majors use the arboretum as a living hydrology and botany laboratory. Art students have installed innovative art exhibits among the Arboretum's forest. Even the preschool children in the campus's Child and Family Development Resource Center enjoy the green space. "The Arboretum is really a gem for a semi-urban campus like Eastern," said Patricia Szczys. "It's an essential green space and also sequesters carbon dioxide as part of our Climate Action Plan."

Peace, Justice and Strong Institutions

The JEDIs (Justice, Equity, Diversity, Inclusion) are a group of faculty who are determined to create a campus-wide culture that supports Eastern's Black, Indigenous, and other People of Color (BIPOC) students. Founded in spring 2021, this informal faculty organization seeks to address racial injustice in the classroom and throughout campus. In summer 2022, representatives attended a Virtual Summer Institute on Racial Justice and Public Liberal Arts to discuss best practices taking place at institutional members of the Council of Public Liberal Arts Colleges (COPLAC), of which Eastern is a member.

Partnerships for the Goals

For more than 20 years, Eastern has worked closely with local, state, and national organizations to advance sustainability on our campus and beyond. In some cases, those partnerships have brought needed grant funds to campus to support our initiatives. Partners have ranged from state utilities — Connecticut Light and Power — to state agencies such as the Connecticut Clean Air Fund. The University will continue developing strategic partnerships at the local, state, regional, and national level to share expertise and best practices.

THE PRESIDENT'S ROLE

Since arriving at Eastern in 2006 as its sixth president, I have made it my task to continue the vision of Dr. David Carter and his team before me, taking an active role when appropriate, and advocating for and supporting a sustainability culture across the campus. Allocating resources where needed, championing initiatives both on- and off-campus, and leading by example are ways I have tried to personally advance Eastern's sustainability goals.

In 2007, I signed the American College and University Presidents' Climate Commitment (ACUPCC), pledging that Eastern become a model green campus and achieve carbon neutrality by 2050. I am proud to have been one of the first 50 charter signatories; today, 700 presidents representing all 50 states have made the commitment.[2] In 2015, I co-signed the Presidents' Climate Leadership Commitment, administered by Second Nature. In 2020, we moved up our timeline to achieve carbon neutrality by 2030 in our updated Climate Action Plan. Included in the plan are net zero energy buildings, climate literacy across our campus, the use of renewable energy resources, and much more. Eastern lies within the Eastern seaboard corridor that runs from Boston to Washington, D.C., one of the most densely populated regions in our nation. By accelerating our carbon neutrality goals, we can support the country's environmental objectives while leading by example.

As a member of the Board of Directors of Second Nature and that organization's Climate Leadership Steering Committee, it has been my pleasure to sign on Eastern as one of more than 700 members of the Second Nature Climate Leadership Network. This position has allowed me to advocate for sustainability and resiliency solutions that are equitable in their reach. Reducing carbon pollution and increasing communities' resilience to climate hazards can directly impact those most harmed by climate change. Policymakers across the nation have clear evidence that people of color, people living in low-income urban neighborhoods, and other marginalized populations are disproportionately harmed by climate change and environmental issues.

In addition to empowering sustainability champions across campus and supporting their work, I have been fortunate to have had opportunities to advocate personally and directly for a sustainable and equitable future for our planet's inhabitants. I have delivered several major speeches on sustainability at regional and national conferences, including the AASCU National Conference in Virginia in October 2008; and Green Jobs and Campus-wide Sustainability in Boston in October 2011. During the pandemic, I moderated a panel on sustainability with Rafael Reif, former MIT president, and former Senator and Secretary of State John Kerry, who most recently was special envoy to President Biden on climate. I have served on the board of our local Covenant Soup Kitchen for many years, and have also supported other networking opportunities with state and national organizations such as the Connecticut Alliance for Campus Sustainability.

These opportunities have not only allowed me to share my Eastern experiences with other universities and organizations, but they have also expanded my knowledge base for the benefit of our University.

CONCLUSION

We are proud of our progress. Eastern has received numerous recognitions for its commitment to sustainability. In addition to being named a Green Campus for the past 14 years by the Princeton Review, Eastern received Gold status in the 2023-24 STARS (Sustainability Tracking, Assessment & Rating System). Eastern was one of only three gold schools in Connecticut and one of 138 colleges and universities nationwide to be Gold-rated by the Association for the Advancement of Sustainability in Higher Education (AASHE). AASHE's rating system has more than 900 participants in 40 countries.

In previous years, the University has received the Energy Star Leadership in Energy Efficiency Award from the U.S. Department of Energy, the EPA's Environmental Merit Award, and been named a "Cool School" by the Sierra Club. In addition, the David G. Carter Sr. Science Building has silver certification from the U.S. Green Building Council's Leadership in Energy and Environmental Design program.

As Eastern's president for the past 18 years, it has been my job to serve as cheerleader, facilitator, advocate, and resource provider for our campus community's sustainability initiatives. It has been my distinct pleasure to watch our sustainability culture grow and mature, as students, faculty, and staff have found ways to make a difference in protecting our world and its people — collectively and individually.

REFERENCES

Burrage, A., Carter, T., & Goldstein, A. (2022, March). *The future of academic research on climate solutions* (p. 3). Second Nature. https://secondnature.org

New England Recycling Council. (2019, November 19). *[Website]*. https://nerc.org

University Leaders for a Sustainable Future. (2015). *The Talloires Declaration*. https://ulsf.org

Chapter 9

University Climate Planning:
Building Community Partnerships to Drive Action

Leah Bamberger
Northeastern University, USA

ABSTRACT

Northeastern University has adopted an innovative approach to centering justice and equity in its Climate Justice Action Plan (CJAP). The plan was co-created with a leading environmental justice organization in Boston called Alternatives for Community and Environment. This partnership has helped strengthen relationships between the university and city and community partners. It also creates opportunities for experiential learning and applied research. By working with community partners who are on the frontlines of the climate crisis, dealing with intersecting crises like housing and economic inequality, Northeastern's bold plan seeks to address climate and sustainability at the systems-level. It raises the bar for campus sustainability, making connections to issues like student housing, supplier diversity, and inclusive hiring practices. This work demonstrate how higher education institutions sustainability offices can be a contributing force towards a regenerative, equitable, and healthy campus, neighborhood, and city.

DOI: 10.4018/979-8-3693-5454-4.ch009

SECTION 1: INTRODUCTION

Solving the climate crisis requires changes that will reshape every aspect of human civilization including our economy, culture, environment, governance structures, and educational systems. Through teaching, research, and their own operations, higher education has an important role to play in this transition.

Many universities have become more engaged with their local communities, striving to serve as drivers of social, environmental, and economic progress in their surrounding communities. As educational institutions, they are centers for learning, research, and innovation. Yet, universities have the potential to extend their influence far beyond campus boundaries by working collaboratively with local governments and community members. This paper explores how universities can amplify their societal impact by forming strategic partnerships that address local challenges in alignment with global imperatives, including climate justice and sustainability. Central to this discussion is a case study of Northeastern University's Climate Justice Action Plan—a model that exemplifies how a university can work in close coordination with its surrounding community and local government to create a more resilient, just, and sustainable future.

This chapter is organized into five sections, each of which contributes to understanding the broader impact of university, community and government collaborations. Following this introduction, the second section provides an overview of the traditional and evolving roles of universities in society. The heart of the chapter is the case study of Northeastern University's Climate Justice Action Plan, covered in Section 3. This section provides a comprehensive analysis of how Northeastern has engaged both the City of Boston and local community groups to create a climate justice-focused initiative. Following the case study, Section 4 examines the benefits and strategies of engaging local communities for mutual benefit while the final section address future trends and practical challenges. In the conclusion, the paper calls for universities to deepen and formalize their community partnerships, using Northeastern University's Climate Justice Action Plan as a model for transformative, sustainable change.

Through these sections, the chapter presents a comprehensive view of how universities, local governments, and communities can collaborate to tackle pressing societal challenges. By examining both the broader context and a specific case study, this paper seeks to provide actionable insights for university leaders, policymakers, and community organizers looking to create lasting positive impact.

SECTION 2: THE ROLE OF UNIVERSITIES IN ADDRESSING SUSTAINABILITY

Societal Contributions of Higher Education

Historically, universities have held a central role in advancing society by fostering intellectual development, promoting knowledge creation, and supporting social and cultural growth. They have served as hubs for academic excellence, offering a platform for individuals to deepen their understanding and engage in critical thinking across diverse fields. In medieval Europe, for example, universities like Bologna, Paris, and Oxford became the earliest institutions dedicated to learning, where scholars gathered to discuss philosophy, theology, and the sciences (Haskins, 1957). These institutions laid the foundation for what universities are today—centers of higher education that not only teach and disseminate knowledge but also uphold scholarly traditions and ethics. Universities have long been viewed as custodians of culture, ensuring that knowledge is preserved, critiqued, and passed down to future generations, thereby maintaining societal continuity and cultural heritage (Altbach, 2016).

In addition to their academic role, universities have traditionally contributed to economic and social development. They provide the workforce with skilled graduates and contribute to research and innovation that drive technological and economic advancement (Brennan, 2008). As societal needs have evolved, so too has the role of universities, extending beyond the classroom to include active participation in addressing social issues and contributing to public policy. Research universities play a key role in fostering innovations that address complex global challenges like climate change and public health crises, positioning themselves as essential drivers of positive change in society (Etzkowitz & Leydesdorff, 2000). Climate change, or global warming, was first documented in peer-review publications in the 1800s. Scientists like Joseph Fourier, John Tyndall and Svante Arrhenius laid the groundwork for our current understanding of climate change and the greenhouse effect. (IPCC, 2007) All three of these had strong ties to academic institutions.

Taking the Lead on Climate Action

In this way, universities have transitioned from traditional centers of learning to dynamic institutions that shape public opinion, influence policy, and contribute to the social and economic well-being of communities and nations alike. As such, higher education has played a leading role in addressing the climate crisis through. First and foremost, through research, as described above, and through teaching. By the early 2000s, universities were also taking the lead with campus-wide sustainability efforts.

Colleges and universities were quick to organize in the early 2000s to demonstrate their commitment to the cause. The American College & University Presidents' Climate Commitment (ACUPCC) was launched in 2007 to mobilize higher education institutions in addressing climate change. Prompted by growing concerns over climate impacts and the role of educational institutions in leading societal change, the commitment aimed to drive actionable climate strategies across campuses. Institutions that signed the ACUPCC committed to achieving carbon neutrality as soon as possible, integrating sustainability into curriculum and campus culture, and developing and implementing action plans with regular public progress reports.

This initiative was primarily led by Second Nature, a nonprofit organization, in partnership with the Association for the Advancement of Sustainability in Higher Education (AASHE). Leaders such as Anthony Cortese, co-founder of Second Nature, and Michael Crow, president of Arizona State University, played key roles in establishing the ACUPCC framework (Second Nature, n.d). Since its inception, the ACUPCC has evolved into the Climate Leadership Network, continuing its work in helping institutions set and meet ambitious climate goals (Second Nature, n.d).

The Rise of Sustainability Offices in Colleges and Universities

University sustainability offices began to emerge prominently in the mid-2000s, driven by increasing environmental awareness on campuses and the need for a coordinated approach to sustainability efforts. Early pioneers established these offices to centralize and expand on-campus initiatives in areas like energy efficiency, waste reduction, and community engagement. By providing a strategic framework, these offices have enabled universities to implement sustainable practices across operations, curriculum, and partnerships with the surrounding community.

At present, approximately 65% of U.S. colleges and universities have dedicated sustainability offices or staff responsible for coordinating sustainability initiatives across campus. This trend reflects the increased prioritization of environmental goals within higher education, such as carbon neutrality and sustainable campus operations. AASHE, through its Sustainable Campus Index and Sustainability Tracking, Assessment & Rating System (STARS), has documented the growth and impact of these offices, showing how sustainability roles have become integral in university operations (AASHE, 2020).

These offices have helped accelerate the adoption of sustainability practices. Take green building practices for example. Higher education institutions were among the first to adopt the LEED (Leadership in Energy and Environmental Design) certification as standard for their building, showcasing its benefits in reducing energy use, minimizing environmental impact, and promoting healthier spaces. Recent estimates put the number of LEED-certified and registered projects in higher education at just over 9,286, spanning about 1.1 billion square feet of real estate on U.S. campuses. This translates to approximately 20% of all LEED-certified projects when considering the number of universities and their commitment to sustainability (U.S. Green Building Council, 2024).

The Lewis Center at Oberlin College was one of the first LEED-certified buildings at any college or university. It's known for features like solar panels, composting toilets, and a "living machine" that treats wastewater on-site, embodying the principles of regenerative design. Universities like Yale, Stanford, and the University of California system were also early adopters of LEED, incorporating sustainability not only in new buildings but also in renovation projects. By doing so, they demonstrated the feasibility and cost benefits of sustainable design to a wide audience, including students, staff, and local communities. As a result, these institutions set a visible precedent, leading other sectors to embrace LEED standards as a way to meet environmental and operational goals (U.S. Green Building Council, 2021).

Colleges and universities have increasingly integrated LEED as part of their sustainability strategy, emphasizing its value not only for environmental impact but also for enhancing campus appeal to students, staff, and donors. This adoption is driven by students' environmental awareness, institutional goals, and benefits like reduced operating costs and improved building performance (Vinson, 2024).

Furthermore, the visibility and influence of university-led LEED projects contributed to changes in local policies and codes, as municipalities observed the economic and environmental benefits of green construction. This ripple effect has contributed to the rapid expansion of LEED certifications across the U.S., reinforcing the role of higher education as a catalyst for broader sustainability practices (AASHE, 2020; Second Nature, 2020).

Adapting to the equity-focused sustainability field

The field of sustainability has evolved significantly since the early 2000s, shaped by an increasing awareness of the connections between environmental issues and broader social challenges. Events such as the COVID-19 pandemic and the widespread protests following the murder of George Floyd highlighted links between climate change, inequality, and public health, prompting many organizations to reevaluate their approaches to sustainability. In response, mainstream environmental organizations began to incorporate principles of environmental and climate justice, recognizing the importance of addressing social inequities in environmental policies. This shift has led to a greater examination of the predominantly white leadership in the sustainability sector and increased scrutiny regarding the historical exclusion of low-income communities and communities of color from environmental decision-making. Consequently, equity is now acknowledged as a foundational component of sustainability, emphasizing the need for inclusive practices that address the social dimensions of environmental challenges. (Phillips et al., 2020).

With this evolving focus, there is a growing recognition that addressing climate change and building a sustainable society requires systemic change. The Just Transition Framework illustrates the interconnectedness of natural resource extraction and labor exploitation, showing how certain communities are disproportionately affected by these processes (Jenkins et al., 2020). Many disenfranchised communities, from rural Appalachia to parts of Louisiana, have endured a disproportionate burden from industrial pollution and environmental degradation. These areas, often termed “sacrifice zones,” face heightened exposure to health risks without adequate support or participation in decision-making processes (Lerner, 2010). This pattern has led to calls for an inclusive approach to sustainability that prioritizes both environmental and social justice.

As society adopts a renewed focus on environmental justice, higher education institutions face the challenge of integrating climate and environmental justice principles into their academic programs, research, and campus operations. Meaningful engagement with communities most affected by environmental and social crises is essential for effective climate justice. Data show that low-income and minority communities are most impacted by environmental hazards. For instance, African American communities in the United States are exposed to 56% more air pollution than they generate, while white communities experience 17% less pollution than they generate (Mikati et al., 2018). Addressing such disparities requires a shift toward inclusive decision-making processes, ensuring all communities have a voice in policies that impact their health and environment (Bullard et al., 2008).

Many universities, however, face obstacles managing “town and gown” relationships due to the significant impacts they can have on surrounding neighborhoods. Like other large institutions, universities often occupy extensive land, drive up housing demand, and contribute to traffic congestion, which can strain relationships with local communities (Gavazzi & Fox, 2015). Academic research, particularly in low-income areas, is sometimes viewed as extractive, where residents become subjects of studies without tangible improvements in their community conditions (Israel et al., 2005). While public or community affairs offices often attempt to address these issues by establishing scholarship funds or investing in local infrastructure, these actions may not fully address deeper issues of equity and empowerment (Maurrasse, 2001).

To more effectively serve as community partners, universities have an opportunity to align their missions with local priorities and concerns, particularly through campus sustainability initiatives. By extending sustainability efforts beyond campus boundaries, institutions can foster mutually beneficial relationships that build trust and encourage collaboration. For example, community members may gain access to academic resources, research, and decision-making processes, while universities benefit from real-world problem-solving opportunities that enhance their teaching and research (Fitzgerald & Primavera, 2013). Such partnerships are vital for universities aspiring to become active agents of climate and environmental justice. By aligning campus sustainability with community needs, higher education institutions can significantly elevate their societal impact and reinforce the practical relevance of their research and academic programs (McNall et al., 2015).

SECTION 3: CASE STUDY – NORTHEASTERN UNIVERSITY'S CLIMATE JUSTICE ACTION PLAN

Background

Northeastern University has long been a leader in experiential learning, research, and impact. Much of this work has focused on climate action and sustainability. In 2007, University President Joseph Aoun was a founding member of the American College & University Presidents Climate Commitment. In 2010, the facilities division released the University's first sustainability plan, which made a commitment to reducing greenhouse gas emissions 80 percent by 2050. The university has made great progress since then, reducing emissions by 42 percent[1]. However, it has become clear that tackling the root cause climate change requires more than a simple emissions reduction strategy.

Northeastern was established in 1898 by the Boston Young Men's Christian Association as an all-male institute before being incorporated as Northeastern College in 1916, gaining university status in 1922. Through much of its early history, it served working professionals who were striving to advance their education while they continued their careers. It is in these roots that the now famous co-op model was born. Students were encouraged to take time off from learning to focus on their professional careers and help them pay for their education. Most of its students were hyper local, so there was a clear benefit to the surrounding community.

In the late 1980s and early 1990s, the University began to grow. Over the course of a short period of time, the university was transformed from a commuter school to a highly competitive research university. Today, Northeastern draws students from around the world who are attracted to their strong reputation, focus on experiential learning and real-world impact, and global presence.

With this growth, however, community tensions have increased. Rising cost of living, growing inequality, and a housing crisis have made institutes of higher education targets in communities across the country. Local community members and politicians have challenged the universities plans for growth, even then those plans include additional housing for its students. Furthermore, the recently-elected Mayor Michelle Wu had run on a strong Green New Deal platform. Both housing and climate were top priorities for the City. As a result, Northeastern began to see its climate and sustainability work as more than just a way to meet its campus commitments and satisfy students and faculty. It was going to be an important strategy for managing city and community relations.

Getting Started

Northeastern University launched its Sustainability group in 2022 as part of its Planning, Real Estate and Facilities division. It was an expansion of the sustainability team that was previously housed in Facilities and was created in response to a culmination of factors. One, students and faculty were calling for action on climate justice specifically. Two, Mayor Michelle Wu was elected, bringing to Boston a focus on climate justice and sustainability. Three, key university development projects were being challenged and held up by opposition from neighbors and city planning processes who were challenging the impact of this development on their communities.

Northeastern's Sustainability team had early conversations with stakeholders to help shape the vision of the Climate Justice Action Plan. Informed by talking to climate justice leaders on campus, in the community, and in City Hall, Northeastern established four goals that persisted throughout the planning process. They are:

- Make our campuses models of a sustainable, healthy, interconnected, and just community.
- Contribute to building a just and regenerative economy, both locally and globally.
- Enact just research and academic practices that advance equitable solutions to the climate and intersecting crises.
- Build equitable relationships with local communities to model place-based justice and sustainability in action.

The last goal was particularly important to start working on as soon as the planning process began. This plan had to be developed in partnership with communities that are most impacted by our campus. The process needed to be collaborative, help us build relationships, and not be extractive of community member's time or energy. This meant that community partners were not just stakeholders in the process, they would be co-creating the plan with us. The City and Community Engagement team at Northeastern was a key partner throughout the process.

Through a competitive-bid process, a consultant team that included One Square World (1SW) and Alternatives for Community and Environment (ACE) was selected. 1SW, a non-profit dedicated to creating liberatory systems for racial and environmental justice, facilitated the process and ACE, a leading environmental justice organization in Roxbury, was part of our core project team and helped ensure community interests were represented throughout the process.

Designing for Accountability

One of the first things the team did was co-create a steering committee to serve as an accountability mechanism throughout the process. The committee was made up of community members, students, staff, and faculty. It was important that half the committee be community members so the process would remain centered on community priorities. The project core team, which was made up of Northeastern and ACE staff created the following purpose statement for the steering committee:

To: gather a group of stakeholders who will collaborate with their communities, contribute to, and advise the direction of the climate justice action plan

In a way that: supports participants to bring their strengths and unique roles towards a common goal; is accountable to those most impacted, to the work we are doing now and to future accountability; honors individual expertise and autonomy; explicitly honors lived experience as expertise; is regenerative and not extractive; and participation is mutually beneficial.

So that: ensure that the process is aligned with the priorities and needs of local frontline communities with whom NU is in relationship with and the NU community (in Boston).

The core team also came up with criteria (Table 1) for identifying steering committee members. It was important that all members had an orientation towards justice and shared values around community engagement. Finding the right balance between participants that would be available for the process, have some ability to affect change at Northeastern, and has an orientation towards the values of the process helped us ensure that participants would be able to contribute effectively.

Table 1. Steering committee selection criteria

External Community Members	Northeastern Members
Is a member of and is actively accountable to populations most affected by climate injustice (Roxbury, Mission Hill, as well as citywide) and affiliated with BIPOC-led and base building organizations.	Is a member of and is actively accountable to populations most affected by climate injustice.
Demonstratively justice-oriented. I.e. understands systems of oppression and has values that center community.	Brings a justice lens to their work at Northeastern.
Leadership skills/experience/potential/interest with systems and organizations	Power to implement or influence change
Available for the process	Available for the process

Sustainability staff were responsible for providing recommendations for Northeastern participants and ACE was responsible for community recommendations. Through a consensus process, 16 committee members were identified and agreed to participate.

Table 2.

Angel Howell	Associate Director and Senior Organizing Team Lead, New England Community Project
Armani White	Executive Director, Reclaim Roxbury
Mari Gashaw	Community Engagement Manager, Boston Ujima Project
Ryan "Rei" Fielder	Director of Coalitions, Policy, and Grassroots Organizing, Boston Ujima Project
Hakim Sutherland	Director of Youth Organizing, ACE
Emmanuelle De Barros	Director of Development and Community Engagement, ACE
Sophia Owen	Senior Attorney and Director of Environmental Justice Legal Services, ACE
Beya Jimenez	Director of Belonging and Engagement, NU PREF
Leah Bamberger	Executive Director of Sustainability, Northeastern
Rebecca Riccio	NU Juffali Family Director, Social Impact Lab
Becca Berkey	Director of Community-Engaged Teaching and Research and the John Martinson Honors Program
Chimel Idiokitas	NU Assistant VP for Community Relations
Broov Lacet	Owner and Operations Manager at Thomas S Carlson Corp
Lisa Susser	NU Manager for Office of Belonging
Meher Kaur Khanna	NU Student
Mikayla von Ehrenk	NU Student

Building a Shared Understanding of Systems

One of the first things the steering committee was tasked with was to attend environmental justice trainings that were designed to give all participants a shared understanding of systemic oppression, racism, and the connections to environmental issues. These trainings were developed and facilitated by 1SW and were conducted over the course of three sessions that took 13 hours in total. ACE was an active participate and trainer, helping connect the concepts to local issues.

The training had almost 50 participants from the neighborhood and Northeastern. By extending the training to participants beyond the steering committee, the learning experience and the plan was extended to additional stakeholders. This base would help with engagement efforts throughout the process.

Understand community priorities and concerns through lived experience

Following the trainings, the core team planned a series of community conversations to better understand community priorities and concerns. Through these interactive meetings, participants were ushered through an online survey while also connecting in groups. Rather than ask participants how Northeastern could reduce its carbon footprint or better address climate justice, the conversation and associated survey focused on capturing the expertise in the room. This meant understanding community members experience as Boston residents and neighbors of Northeastern and students experience on and off campus. The conversations and surveys were structured around the four elements, earth, air, water, and fire, to focus feedback on environmental issues. For example, questions included, "What has your experience with air been like in Boston? Do you or your loved ones struggle with respiratory conditions like asthma? Do you experience poor air quality in Boston? Do you smell gas leaks?" For fire, questions were asked about energy burden and how people heat and cool their homes. Finally, participates were asked to describe what their ideal day in Boston would look like, what their current relationship to Northeastern is, and what they would like it to be in the future.

The community conversations and surveys were open to community members, students, faculty, and staff. Boston residents who were not affiliated with Northeastern were eligible for a stipend if they attended a community conversation, completed the survey, and agreed to be contacted in the future for any follow up questions. Compensating community members for sharing their expertise was a critical part of the planning process, helping ensure the relationship was not extractive.

Using Art to Connect and Learn

Art was an important tool throughout the process. It was a prominent feature of the trainings, helping convey the humanity in the work and the impact that environmental justice issues have on people's lives. For example, 1SW and ACE used the video, 35sqft and Rising to talk about the importance and impact of air quality issues in Boston. This compelling story, told through poetry and animation, shares the narrative of a parent whose child suffers from asthma and is combating indoor and outdoor air quality issues that trigger the asthma.

The survey results were also shared creatively through art. Three animations were made to share the ideal day of an NU student, worker, and community member. The stories were a compilation of what was heard from various survey responses. Lastly, the project team contracted with the artist D.S Kinsel, who collaborated with Boston-based artist EYEVAN to create signs based on quotes from the CJAP surveys. The project was an extension of Kinsel's *Sign O The Times: Protest Sign and Archive Reproduction* which aims to reimagine and recreate protest signs from the #blacklivesmatter movement and civil actions that have taken place worldwide since the beginning of the 2020 pandemic. In addition to the quantitative analysis of the survey feedback, these art pieces were used to help share survey results back to the Steering Committee. They were also featured in the two final community conversations in March 2024, where the draft plan was presented.

These final meetings showcased the draft strategies and actions and again asked participants to complete an online survey following a gallery walk of the art and draft plan. The meeting included local artists who shared their reflections on the plan and the themes that can up through the process through spoken word, poetry, and hip hop. Through these performances, along with the animations and signs, we were able to center community voice in the meeting and limit the amount of time Northeastern was holding the space. Our role at this time was to listen to feedback and answer questions.

Over 150 people attended the two meetings, and we received 100 survey responses. Again, community members not affiliated with Northeastern received a stipend for attending the meeting and filling out a survey. Over half of the attendees were community members, which was a resounding success.

Finalizing the plan with an eye towards accountability

In the spring of 2024, the draft plan was finalized based on feedback from these final community conversations and it was released online a few months later. Prior to publication, the final conversations at the Steering Committee focused on accountability. Community members needed assurance that the plan was not going to sit on the shelf and that it would be implemented with continued input and collaboration. To that end, the steering committee agreed to continue to meet three times a year to help oversee implementation. The plan's website also built in transparency but showing what department at Northeastern is responsible for implementing each action, what the status of the action is, who the partners are, and who they are accountable to. With this established, the Steering Committee endorsed the final plan and, as of the fall of 2024, is continue to work closely with the university on implementation.

Results and Discussion

The Climate Justice Action Plan will shape the University's forthcoming Institutional Master Plan and related investments in community that will enable Northeastern to better address the climate and sustainability priorities of our neighbors, while moving away from the transactional nature of traditional community benefits. The plan identified many existing NU programs, initiatives, and research areas that are contributing positively to the surrounding community. This includes things like efforts to reduce air pollution from buildings and commuting, establishing career pathways from the local vocational high school to Northeastern's trades team, community engaged teaching and research practices, and procurement policies that support local, minority-owned businesses. The plan seeks to build upon and expand these efforts in ways that increase their mutually beneficial nature. The plan also touches upon more challenging long-term issues like the impact of students on housing, gentrification, and access to campus for community members. These issues may not offer quick-fix solutions, but it is critical to identify them in the plan as longer term goals.

Throughout the process it was important to be honest with partners about expectations and what progress could be made in the short term versus long term. This helped build trust and focus the work. Relationship and trust-building was a key part of the planning process from start to finish. The relationships are ultimately what hold people accountable to each other and to the work. These relationships and the accountability that comes with them, are in many cases, more important than the plan document itself. Northeastern intends to continue to work with these new partners to make progress toward our shared goals and implement the strategies and actions in the Climate Justice Action Plan.

SECTION 4: SUSTAINABILITY AS A PLATFORM FOR PARTNERSHIPS

Universities and Local Communities Addressing Climate Challenges Together

Northeastern University's Climate Justice Action Plan illustrates how universities and local communities can collaborate to advance climate resilience and sustainability. By co-creating a plan with community partners like ACE (Alternatives for Community and Environment), Northeastern developed an approach that recognizes and elevates community priorities. This partnership-based model strengthens trust and fosters shared responsibility, a vital step for universities that aim to be true partners in climate resilience. Such collaborations allow universities to connect research, teaching, and campus operations to local and regional climate priorities, bridging the "town and gown" divide.

As universities engage with local communities, they become better positioned to integrate knowledge from diverse stakeholders into their sustainability efforts. This can be advantageous for campus planning and operations as well as academics and research. Students, faculty and researchers can benefit from these partnerships as they gain access to real-world challenges, research partners, and hands-on learning experiences.

Initiatives rooted in shared learning and community engagement not only yield practical solutions but also advance climate justice. For example, Northeastern's approach to listening to the lived experiences of neighboring residents helped ensure that their Climate Justice Action Plan addressed specific, urgent concerns, from air quality to energy burdens, that disproportionately affect these neighborhoods.

Creating Policy Alignment and Overcoming Barriers: Aligning University, Community, and Government Policy Goals

Northeastern's Climate Justice Action Plan demonstrates the power of policy alignment across university, community, and government goals. By centering community voices and goals, Northeastern's plan aligns closely with Boston's broader climate policies under Mayor Wu's administration. This alignment fosters greater impact, as policy coherence across institutions accelerates sustainable outcomes, especially for frontline communities. Universities and cities can benefit immensely from synchronizing their sustainability objectives to support resilience and climate justice for the entire urban ecosystem.

An example of this in action is Northeastern's engagement with community heat resilience, which started with an ask from the Mayor for institutions in Boston to do more to help communities respond to extreme heat. Northeastern leveraged its community partnerships to hold educational events on heat resilience throughout the summer and now hosts pop-up cooling stations with drinking water and misters during heat waves in the summer. These are strategically positioned as close to the edges of our campus in order to serve the greater neighborhood. Northeastern also has taken an ambitious approach to its campus landscape program, achieving level II certification as an Arboretum and planting and maintain trees in the surrounding community. All of these efforts help further build relationships and address community challenges on and off campus.

However, aligning policies across institutions is not without challenges. Universities must overcome institutional inertia, competing priorities, and the complexities of working across diverse political and regulatory frameworks. Successful alignment requires universities to build strong relationships with local governments and community organizations, creating joint initiatives that foster collective action.

Key to Northeastern's success was allowing for a lengthy planning horizon of a full year, hiring community-based facilitators that had strong relationships with key community partners, and ensuring that community participants were hired and compensated as consultants to the process. This helps address power dynamics and ensure partners are seen as valued contributors to the process. Additionally, Northeastern's process of ensuring continuous community engagement—through mechanisms like the Steering Committee and regular feedback loops—was essential to building a strong relationship and overcoming barriers of trust. The university's commitment to transparency and shared decision-making, including its alignment with Boston's climate goals, offers a model for how academic institutions can play an influential role in shaping policy, particularly when those policies serve the needs of the most vulnerable communities.

In overcoming common barriers to such alignment, universities can establish joint initiatives, create transparent decision-making processes, and maintain an accountability framework, all strategies Northeastern employed. By making climate justice and sustainability goals part of its Institutional Master Plan, Northeastern underscores the importance of such alignment and provides a model for other universities to use sustainability and climate action as a framework for broader impact.

Measuring Success and Accountability

Effectiveness of climate action must be measured through both quantitative and qualitative metrics. Key performance indicators like reductions in greenhouse gas emissions, improvements in local air quality, and increases in community investment serve as benchmarks of success. Northeastern's Climate Justice Action Plan employs diverse metrics, tracking not only emissions and energy use but also improvements in community well-being, access to resources, and equitable economic opportunities.

Metrics are only one part of the accountability equation because numbers alone cannot tell the full story. Ongoing relationships between the university and community are critical to ensuring the plan is being implemented in a way that is responsive to community priorities and concerns. In Northeastern's case, this means continuing the Steering Committee meetings and regular report-outs from the university to the members of the Committee. Leveraging both quantitative metrics and direct feedback through partnerships provides a clearer picture of the plan's impact, ensuring it remains accountable to both the university and its community stakeholders.

SECTION 5: CONCLUSION, FUTURE TRENDS, AND OPPORTUNITIES

The Future of University-Government-Community Collaboration

As the climate crisis deepens, the role of university-government-community partnerships will grow in significance. Future collaborations will likely focus on adaptive resilience and innovative funding models that enable universities to share resources more equitably with their surrounding communities. Emerging opportunities include leveraging technology for real-time environmental monitoring, supporting workforce development in green jobs, and implementing community-owned renewable energy projects. Such initiatives can build on existing partnerships while addressing gaps in current climate and social justice work.

As climate impacts worsen, so will the need for resilient, adaptable partnerships that address both immediate and long-term challenges. The increasing frequency of extreme weather events, urban heat islands, and environmental health disparities underscores the importance of university-community collaborations that can rapidly respond to these evolving conditions. The permanence of higher educational institutions make them an important player in planning for and adapting to climate change impacts at the local level.

Restating the Importance of Partnerships

Effective partnerships are the foundation of transformative climate action. Northeastern's Climate Justice Action Plan illustrates how a university's relationship with its surrounding community can drive meaningful progress toward climate justice and resilience. These partnerships help universities fulfill their mission while addressing local challenges in a comprehensive and sustainable way.

Call to Action

To maximize their impact, universities must deepen their relationships with local communities and governments. Key actions include creating advisory boards with significant community representation, establishing transparent decision-making structures, and co-developing climate justice policies that prioritize local concerns. By embedding these principles into their core operations, universities can move beyond transactional benefits toward genuine collaboration and trust-building.

Closing Thoughts

As hubs of knowledge and innovation, universities are uniquely positioned to lead transformative change for a more just and sustainable future. By prioritizing equitable partnerships and community engagement, universities can not only enhance their campus sustainability efforts but also contribute to the broader societal shift toward a regenerative and inclusive economy. Northeastern's Climate Justice Action Plan serves as a powerful example, inspiring universities across the country to engage meaningfully with their communities and pave the way for a resilient, equitable future.

REFERENCES

Altbach, P. G. (2016). *Global Perspectives on Higher Education*. JHU Press. DOI: 10.56021/9781421419268

Association for the Advancement of Sustainability in Higher Education. (2020). Sustainable Campus Index. Retrieved from https://www.aashe.org/resources/sustainable-campus-index/

Brennan, J. (2008). Higher Education and Social Change. *Higher Education Quarterly*, *62*(2), 384–386.

Bullard, R. D., Mohai, P., Saha, R., & Wright, B. (2007). *Toxic Wastes and Race at Twenty: 1987–2007. A Report Prepared for the United Church of Christ Justice and Witness Ministries*. United Church of Christ.

Etzkowitz, H., & Leydesdorff, L. (2000). The dynamics of innovation: From National Systems and "Mode 2" to a Triple Helix of university–industry–government relations. *Research Policy*, *29*(2), 109–123. DOI: 10.1016/S0048-7333(99)00055-4

Fitzgerald, H. E., & Primavera, J. (2013). *Community Engagement and Higher Education: International and Local Perspectives*. MSU Press.

Gavazzi, S. M., & Fox, M. (2015). *For Town and Gown: Building a Sustainable Relationship between a College and Its Host Community*. Myers Education Press.

Green Building Council. (2021). LEED in higher education: A case study on the growth of green buildings on campuses. Retrieved from https://www.usgbc.org/resources/higher-education-leed-case-studies

Haskins, C. H. (1957). The Rise of Universities.

Intergovernmental Panel on Climate Change. (2007). The Physical Science Basis: Contribution of Working Group I to the Fourth Assessment Report of the Intergovernmental Panel on Climate Change. Cambridge University Press. https://www.ipcc.ch/site/assets/uploads/2018/03/ar4-wg1-chapter1.pdf

Israel, B. A., Schulz, A. J., Parker, E. A., & Becker, A. B. (2005). Community-based participatory research: Policy recommendations for promoting a partnership approach in health research. *Education for Health*, *14*(2), 182–197. PMID: 14742017

Jenkins, K., Sovacool, B. K., & McCauley, D. (2020). Humanizing sociotechnical transitions through energy justice: An analytical framework. *Energy Policy*, •••, 117.

Lerner, S. (2010). *Sacrifice Zones: The Front Lines of Toxic Chemical Exposure in the United States*. MIT Press. DOI: 10.7551/mitpress/8157.001.0001

Maurrasse, D. J. (2001). *Beyond the Campus: How Colleges and Universities Form Partnerships with Their Communities*. Routledge.

Mikati, I., Benson, A. F., Luben, T. J., Sacks, J. D., & Richmond-Bryant, J. (2018). Disparities in distribution of particulate matter emission sources by race and poverty status. *American Journal of Public Health, 108*(4), 480–485. DOI: 10.2105/AJPH.2017.304297 PMID: 29470121

Phillips, C. A., Caldas, A., Cleetus, R., Dahl, K. A., Declet-Barreto, J., Licker, R., & Merner, L. D. (2020). Compound climate risks in the COVID-19 pandemic: Intersectional perspectives on environmental justice and climate change. *Environmental Justice*, *13*(3), 97–108.

Second Nature. (n.d.). 15-year celebration. Second Nature. Retrieved 10/15/24 from https://secondnature.org/15-year-celebration/

U.S. Green Building Council. (2024, September). Applying LEED to higher education projects. U.S. Green Building Council. https://support.usgbc.org/hc/en-us/articles/12091439417875-Applying-LEED-to-higher-education-projects

Vinson, D. (2024, February). The growing influence of LEED certification in higher education. Private University Products and News Magazine. Retrieved from https://pupnmag.com/article/the-growing-influence-of-leed-certification-in-higher-education/

ENDNOTE

1 As of 2021 and relative to a 2005 baseline year and square feet of real estate

Chapter 10
Transforming Education for a Sustainable Future: Climate Action in Higher Education

D. Horner
The American College of Greece, Greece

Pavlos Vlachos
https://orcid.org/0000-0002-0885-7143
The American College of Greece, Greece

Stella Apostolaki
https://orcid.org/0000-0002-5010-5287
The American College of Greece, Greece

ABSTRACT

In the Times Higher Education Impact Ranking 2023, The American College of Greece (ACG) was rated the 1st most sustainable private higher education institution in Greece and the 4th among the 18 public and private universities/institutions rated in the country. 2,031 universities from 121 countries/regions applied in the Impact Ranking 2023 and were assessed and ranked for their contributions to the United Nations' 17 Sustainable Development Goals (SDGs). Among the 2,031 institutions globally, ACG scored in the top 7% for 'Responsible Consumption and Production', top 12% for 'Climate Action' and top 17% for 'Partnerships for the Goals'. In Greece, ACG attained the top scores for all three of these SDGs.

DOI: 10.4018/979-8-3693-5454-4.ch010

1 INTRODUCTION

In the *Times Higher Education* Impact Ranking 2023, The American College of Greece (ACG) was rated the 1st most sustainable private higher education institution in Greece and the 4th among the 18 public and private universities/institutions rated in the country. 2,031 universities from 121 countries/regions applied in the Impact Ranking 2023 and were assessed and ranked for their contributions to the United Nations' 17 Sustainable Development Goals (SDGs). Among the 2,031 institutions globally, ACG scored in the top 7% for 'Responsible Consumption and Production', top 12% for 'Climate Action' and top 17% for 'Partnerships for the Goals'. In Greece, ACG attained the top scores for all three of these SDGs.

These most recent ratings are consistent with a broader pattern of distinction and recognition of ACG as a sustainability leader in Greece. For example, ACG has achieved four STARS ratings (Bronze – 2016, Silver – 2018, Gold – 2020 & 2023) by the Association for the Advancement of Sustainability in Higher Education for its sustainability practices in the areas of Academics, Research, Engagement, Operations, Planning & Administration as well Innovation & Leadership. In 2023, the *Princeton Review* Guide to Green Colleges listed ACG as one of the Greenest Colleges for the eighth consecutive year.

1.1 Sustainability at The American College of Greece: A Strategic Approach

ACG's comprehensive sustainability initiative is situated in the context of the ACG Strategic Plan, which was adopted initially in 2009 following the appointment of current ACG president, David G. Horner, in July 2008 and features a distinctive "one-page" format (see https://agb.org/trusteeship-article/a-one-page-strategic-plan/) with four institution-wide goals:

Goal One. Achieve high standards of performance across all educational programs and leverage education for individual, economic, cultural and social impact by: strategically expanding, diversifying and enhancing educational programs; strategically growing and diversifying enrollment

Goal Two. Cultivate a high performance, sustainable organizational culture based on transparency, mutual support, collegiality, and continuous improvement by: enhancing collegial governance rooted in a highly functioning board, faculty and staff: assuring externally competitive and internally equitable compensation

Goal Three. Establish ACG's educational leadership position in Greece, broaden awareness in the Greek diaspora and build a culture or constituency engagement and philanthropy by: investing in targeted marketing and communications, advancement (e.g., alumni, fundraising), and public affairs / public events initiatives

Goal Four. Assure a comprehensive, sustainable financial and infrastructure model by: diversifying revenue sources: developing and physical and technological resources; strengthening financial management; implementing a comprehensive sustainability initiative

To support these four major goals, annual priorities are established, which then guide the operating plans of the various operating units of the College. Overall progress is tracked and projected (e.g., ten-year retrospective and prospective) in a Strategic Plan Dashboard with approximately eight targets per major goal (e.g., STARS Rating under Goal Four). As part of the implementation of the ACG Strategic Plan, the Center of Excellence in Sustainability (CoES) was established in 2011.

While connected most directly to two ACG undergraduate degree programs, Environmental Studies and Biomedical Sciences, CoES promotes the inclusion of sustainability elements in numerous courses across the ACG curriculum and acts as the central clearing-house and organizing body for information, communication and collaboration for sustainability related research, grant activity, and student activities. CoES, in collaboration with its Advisory Boards and the ACG Office of Public Affairs – Office of Sustainability, the CoES, is also responsible for formulating, reviewing, updating, and submitting for approval the annual ACG sustainability strategy.

The ACG Office of Public Affairs – Office of Sustainability, led by its Sustainability Director, is responsible for ACG's external (e.g., STARS, United Nations Academic Impact, United Nations Global Compact, National Alliance for the Elimination of Food Waste, Times Higher Impact) and internal reporting (e.g., President's Annual Report); organizing community-engaging activities (e.g., events, awareness campaigns, community contribution/volunteer activities); recommending and promoting new ACG sustainability policies, and working with ACG Technical Services & Operations for implementation of campus sustainability best practices.

2 SUSTAINABILITY AND CLIMATE CHANGE EDUCATION AT ACG

The importance of sustainability and climate change education in fostering environmental awareness, protection and conservation is undeniable. Highlighting the need for sustainability and climate education in university curricula prepares students for the challenges of climate change and towards a sustainable future.

The latest findings on climate change according to the 6th Assessment Report of the IPCC (2023) indicate an increase in the global surface temperature reaching 1.1°C above the pre-industrial reference year of 1850. Global greenhouse gas emissions have continued to increase, in unequal manners due to unsustainable energy use, land use and land-use change, lifestyles and patterns of consumption and production. The measured atmospheric CO^2 concentrations in 2019 were 410 parts per million, higher than at any time in at least 2 million years. Similarly, the concentrations of methane (1866 parts per billion) and nitrous oxide (332 parts per billion) were higher than at any time in at least 800,000 years. And, according to recent research promoting the 'planetary boundaries framework', six of nine planetary boundaries are transgressed (i.e., biosphere integrity, climate change, novel entities, land system change, freshwater change, biogeochemical flows).

Human-induced climate change has caused widespread adverse and rapid impacts on the atmosphere, ocean, cryosphere, nature, biodiversity, and people in an unequal distributed manner across systems, regions and sectors. Approximately 3.3 to 3.6 billion people live in contexts of high vulnerability to climate change. Millions of people are exposed to acute food insecurity and reduced water security due to climate change, with the largest vulnerability located in less developed regions of Africa, Asia, Central and South America, Small Islands and the Arctic. In addition, human mortality or risk for life from climate related extreme events, such as floods, droughts and storms, were 15 times higher in highly vulnerable regions for the period 2010 - 2020, compared to regions with very low vulnerability.

Further, climate change has affected the livelihoods of individuals through the destruction of homes and infrastructure, loss of property and income, human health and food security, with adverse effects on gender and social equity. Severe economic damages from climate change have occurred in several sectors, such as agriculture, forestry, fisheries, energy, and tourism.

The Mediterranean basin and Greece are considered hot spots of climate change with an average temperature increase exceeding the global average. The Copernicus Service of the European Commission (2023) identifies year 2023 as the warmest calendar year on record for the region. 2023 reached a record number of days with 'extreme heat stress', of more than 46°C. Summer also saw the largest area of Europe affected by at least 'strong heat stress' of any day on record, with 13% of the continent, and 41% of southern Europe, experiencing 'strong', 'very strong' or 'extreme heat stress' in July. Furthermore, the eastern Mediterranean has experienced a record ocean water temperature increase of up to 5°C.

Climate change impacts on the productive and economic sectors and introduces new challenges in the job market primarily affecting young people. There is currently an increasing need for 'green jobs' or professions with acquired 'green' skills. Sectors in pressing need for 'green' expertise include: ecosystems and biodiversity conservation; protection of natural resources; jobs in the productive sector – agriculture, animal breeding, aquaculture, fishing, in which knowledge on new practices and techniques, often digitized, is required; the energy sector with emphasis on renewable and/or alternative sources; the tourist sector; circular economy and management of ecosystems; 'green' entrepreneurship and 'green' marketing; sustainability reporting (ESG framework). Education plays a pivotal dual role: on the one hand, education is necessary to acquire the 'green' skills to address contemporary challenges; on the other hand, jobs in education related to climate education at all levels (pre-K to graduate education) are on the rise. Consequently, this new era introduces new challenges but also new opportunities for universities.

2.1 Undergraduate teaching on climate, climate change & sustainability

ACG, given its awareness of and concern over contemporary global challenges, is placing emphasis on environmental and sustainability-oriented education, primarily through its Environmental Studies program but also through incorporating relevant concepts in teaching material and the experiences of all taught disciplines.

ACG was the first independent educational institution in Greece to introduce an Environmental Studies program. Launched in fall 2010, the program aims to develop students' understanding of the root causes and multiple dimensions of environmental problems (ecological, social, economic, political) and provide them with the necessary training to effectively address these problems in an increasingly complex world.

The program provides students with theoretical knowledge and skills to:

- enter the professional world and pursue careers related to environmental issues,
- become environmentally informed and socially responsible citizens, and
- undertake graduate studies in environmental fields.

The Bachelor of Science in Environmental Studies (ES)[1] is an interdisciplinary program that provides the students with a sound background and skills, necessary to address global environmental problems. Students explore current environmental conditions and global concerns, e.g., climate change, biodiversity loss, population growth, degradation of natural resources, pollution, use of energy resources, and the way these issues affect the planet, the species and human societies. In an increasingly complex world, there is an urgent need for sustainable solutions and for mitigation and adaptation options to address global problems at multiple levels: environmental, social, economic, political. Students of this program obtain skills that make them competitive in the pursuit of 'green' and sustainability-related careers and are equipped to undertake graduate studies in environment-related fields.

The ES program combines knowledge from diverse disciplines in order to provide students with a better understanding of the complex nature of environmental problems by examining their varied dimensions. Courses in environmental science, biology, ecology, chemistry, and geology offer students a broad understanding of the structure and functioning of natural systems as well as the new challenges imposed by the dual global problems of climate change and biodiversity loss. Students examine the various dimensions of sustainability – environmental, social, economic, political – and acquire knowledge of environmental policy and principles of environmental management that are essential components of problem solving.

During their course of studies, by participating in activities and initiatives of the CoES and ES faculty, ES students have the opportunity to: participate in climate and sustainability related research projects; gain relevant field work experience; be exposed to significant cases of climate change mitigation and adaptation, and contribute towards innovation for climate adaptation and sustainable development. The following list includes examples of sustainability and climate focused courses:

Introductory ES program courses serving also the undergraduate Liberal (General) Education requirement and thus available for all ACG students include: *Ecosystems and Biodiversity; Energy Resources and Pollution; Environmental Geology; Environmental Ecology.*

Intermediate level courses in the ES program with a focus on or a relation to climate change include: *Responses to Climate Change; Principles of Environmental Management; Environmental Policy and Legislation; The Economy and the Environment; Integrated Methods in Environmental Analysis; Geographic Information Systems*; *Environmental Health; Environmental Chemistry.*

Advanced ES program courses with high relevance to climate change include: *Air Quality and Global Atmospheric Changes*; *Sustainable Food Production: Soil and Environment*; *Conservation of Wildlife and Mediterranean Ecosystems*; *Sustainable Use of Resources and Waste Management*; *Water Resources: Threats and Sustainable Management*; *Sustainable Cities*; *Education for the Environment and Sustainability*; *Special Topics in Environmental Studies*; *Environmental Governance in the European Union.*

Outside the ES program, an International Honors Program (IHP) course, *Greening the Campus, focuses* on environmental challenges and responses, especially related to climate change, on the ACG main campus. Students engage in sustainability-related projects, become involved in sustainability-related research, and help to establish and enhance a sustainability culture on campus. The *Greening the Campus* course typically integrates implementation of selected SDGs based on current UN, international, and national priorities, and is formulated around SDG 13 on Climate Action.

The CoES, through its Academic Advisory Board (AAB), has recently developed a Concept Note that acts primarily as a guide aiming to define ways on how to enhance sustainability in existing courses across the ACG curriculum, in new relevant academic offerings and faculty research. This guide also sets criteria according to which a course can be considered to address sustainable development.

Further, ACG offers experiential learning opportunities on climate education through field trips and field courses in which students are exposed to climate related problems but also to innovations for climate change mitigation, adaptation and/or building resilience. The ACG main campus is often utilized as a living lab with student driven research and use of the campus premises for sustainability and climate related educational experiences, e.g., monitoring of campus biodiversity, energy & water audits, photovoltaic panel installations, a meteorological station.

2.2 Graduate teaching on climate, climate change & sustainability

At Alba Graduate Business School (AGBS), ACG offers high quality graduate level education on sustainability challenges for business leaders and decisions-makers. AGBS has been a member and signatory institution of the United Nations Principles for Responsible Management Education (PRME) since 2009.

The private business sector is often identified as the main culprit contributing to transgressing planetary tipping points (e.g., global cycles of phosphorus (P) and nitrogen (N) have been significantly disrupted by agricultural and industrial activities). For this reason, a core sustainability and ethics course has been added in the AGBS MBA and MSc curricula, and academic directors and professors across all degree programs have a mandate to incorporate sustainability topics in their course outlines. This mandate is important because it not only fosters interdisciplinary collaboration (e.g., sustainable finance, socially responsible human resource management, corporate due diligence in the supply chain), but also because it conceptualizes sustainability not as a separate subject but rather as a core element within every management discipline. AGBS faculty are encouraged to use real world social responsibility and environmental sustainability cases studies (e.g., the role of pharmaceutical companies' sales strategies in driving the opioid and heroin crisis in the U.S.; VW's dieselgate scandal) as well as 'serious games' such as:

- The Net Zero Game: Developed by the *Financial Times* in partnership with Infosys, players adopt climate policy decisions with the goal of achieving net zero emissions by 2050.
- The *Financial Times* Purpose vs. Profit Trade-off Game: Players assume the role of the CEO of a recently publicly traded tech company (Flexbird) and

must make business decisions highlighting the difficult trade-offs or tensions business leaders face when trying to balance profit with purpose.

2.3 Research

ACG faculty are becoming increasingly involved in preparing research proposals for grants and implementing relevant projects, many of which are related to climate change. An emblematic project is the SDG4U ERASMUS+ EU funded project 'Engaging SDGs for Transformative Education and Enhanced Sustainability in Universities'. This project focuses on identifying and strengthening innovative educational practices to fulfil the commitments to the 17 UN SDGs in universities.

Faculty research on climate change also addresses monitoring climate and atmospheric parameters, assessment of ozone and solar radiation, assessment of energy resources, and alternative energy sources. Research efforts also focus on mapping the impacts of climate change on biodiversity and ecosystems, on water resources, on the oceans, society, human health and well-being. Planning to increase resilience to climate change and participatory processes to promote climate justice are also prioritized in faculty research. Faculty-led research projects often actively engage students in an effort to further promote climate education.

The establishment of the Theodore Papalexopoulos Chair in Sustainability at AGBS in 2021 deepened ACG's commitment to and resources to support sustainability-related research, enabling the Chair holder to produce and organize research outcomes and events including: publications in the *Financial Times* and listed academic journals, a book on 'Stakeholder Capitalism', hosting sustainability paper development workshops by top management journals (*Journal of Management Studies*).

A novel sustainability research project, which is now ongoing in partnership with the CoES and the ACG Center of Excellence in Food, Tourism & Leisure (CoEFTL), is the *Hellenic ESG Barometer*. The *Barometer* was developed with two objectives in mind: (1.) to record and compare the perceptions of three distinct populations (employees, top management teams, general public) with regard to corporate ESG (environmental, social and governance) strategies and actions, and (2.) to provide organizations with credible data on which they can base future ESG related dialogue with employees. Below, are selected findings of the Hellenic ESG Barometer (data collected, December 2022 – April 2023, using psychometrically validated measurement instruments):

- A perceived ESG performance gap exists between external and internal stakeholders: the general public is more critical of ESG performance compared to internal stakeholders (top management team and employees).

- A majority of the general public agrees that ESG affects their consumption and their employer seeking behaviors; however, they are not ready to sacrifice salary for higher ESG organizational commitment.
- The general public attributes ESG activity to profits and competition rather than moral obligation, whereas top management attributes ESG more to moral obligation.
- Supply chain is the most challenging ESG area for all three stakeholder groups.
- The general public does not trust institutions, including big firms and NGOs; this drives their attitudes towards ESG as well. Small businesses and scientists enjoy higher levels of trust.

2.4 Community Engagement & Partnerships

ACG actively participates in public dialogue on climate change as well as on climate networks/initiatives at national and international levels through collaborations on climate related topics with organizations, including the World Meteorological Organization (WMO), the CLIMPACT National Network of Greece, the National Observatory of Greece, the ATHENA Research Center, the National Center for Scientific Research – Demokritos, as well as with other Greek and international universities.

In addition, ACG hosts public lectures, conferences and events, film screenings, panel discussions on climate and sustainability related issues. It also collaborates with local, national and international environmental entities and organizations on awareness raising activities, climate-related Ideathons & Hacathons, gaming, and engagement in the public dialogue. Several campaigns and events open to the ACG community, to the neighboring Municipalities, and to regional and national stakeholders are also organized. In more detail, students involved in Ideathons & Hacathons co-organised by the ACG and the Municipality of Aghia Paraskevi have shaped and proposed sustainable interventions/solutions to issues that concern the Municipality, such as harvesting and handling of rainwater, urban planning and enhancement of green open public spaces with amenity, recreation and social cohesion in mind, biodiversity conservation etc. ACG students also have the opportunity to graduate with the title of Student Scholars of the Center of Excellence in Sustainability in recognition of their active participation and successful completion of sustainability related projects, often related to supporting community groups, education on sustainability related matters through art or in Open Days and Events. Within this context, groups of students, primarily members of the Environmental Studies, the Biomedical Sciences, Sociology and International Honors Program Societies, have been actively supporting and educating refugee kids while they have designed and implemented numerous educational games on the UN 17 Sustainable Development Goals (SDGs) and on selected SDGs, for school children and the wider public. Student Creative Projects that are characterized by multi-disciplinarity and inter-disciplinarity are prepared as part of the curriculum of the 'Greening the Campus' ACG International Honors Program and are displayed to the ACG community in an effort to raise awareness on sustainability matters; many creative projects combine arts, science and humanities. Furthermore, student work on sustainability topics that often form part of Capstone projects has been presented in conferences of relevant thematic.

Student feedback received through the ACG course evaluation system, at the end of every academic term, is proof of the student appreciation to the sustainability related teaching and efforts within the ACG. Students also have the opportunity to evaluate the importance of selected courses to the SDGs through a Mapping Tool that has been developed for this purpose within the framework of the SDG4U E.U. Erasmus+ project, in which the ACG is a partner through the CoES. Those valuations demonstrate high awareness level on the SDGs among students and their ability to associate the SDGs with the course they are taught.

2.5 Green Campus Initiatives

ACG as a 'green' campus, introduces and implements 'green' initiatives such as: environmental monitoring and management, policy for sustainable practices, conservation of campus ecosystem and biodiversity, recycling programs, water management, energy audits, and green building certifications. The ACG energy strategy aims to minimize the carbon footprint through collaboration with an electricity provider that produces 100% "green energy" from renewable sources of energy (solar and wind) and provides the corresponding official state and corporate certificates. This initiative is significantly reducing ACG's carbon footprint in equivalent CO2 emissions. In summer 2024, solar panels were installed on numerous ACG roofs. Real time data on the energy generated by solar panels and the tree equivalent of CO_2 emissions savings is at permanent display at the Deree Central Building for awareness raising reasons.

3 LOOKING AHEAD: ACG SUSTAINABILITY STRATEGY

Going forward, the ACG Sustainability Strategy (ACG SS) aims to provide a paradigm of promoting the principles of sustainability, collaboration, integration, inclusiveness and, innovation throughout ACG and in alignment with the 17 SDGs – in particular with Goal 4.7: *'by 2030 ensure all learners acquire knowledge and skills needed to promote sustainable development, including among others through education for sustainable development and sustainable lifestyles, human rights, gender equality, promotion of a culture of peace and non-violence, global citizenship, and appreciation of cultural diversity and of culture's contribution to sustainable development'*. ACG SS is intended to support ACG to deliver education and research that effectively promotes sustainable practices and communities and to follow exemplary practices in the College's internal operations and governance.

ACG SS was introduced in 2024 by the CoES in collaboration with the ACG Office of Public Affairs – Office of Sustainability; is planned to take effect in January 2025 with an initial Annual Action Plan, and will be assessed utilizing a set of Key Performance Indicators (KPIs). ACG SS draws on the first Sustainability Strategy issued by the Office of Public Affairs in collaboration with the CoES in 2016, STARS results, and earlier documents developed by the Green Campus Committee (2009 – 2011), and is aligned with the more recent ACG Strategy on Research. ACG SS is built on five pillars of SDG 4.7: Education; Research & Innovation; Campus (Facilities & Operations); Community; Governance.

3.1 Education

ACG provides education from Pre-Kindergarten through graduate studies. ACG SS projects a consistent emphasis on issues of sustainability across this broad educational spectrum by: updating curricula, including increased emphasis on interdisciplinarity and multidisciplinarity, to expose students to contemporary sustainability challenges at local and global levels; faculty training on incorporating sustainability topics into their teaching; encouragement of student internships and leadership programs focused on sustainability; introducing new programs of study with a sustainability focus at the undergraduate and graduate levels; expanding co-curricular and extra-curricular activities (e.g., workshops, certifications). Key outcomes for sustainability-focused curricula include students being able to define sustainability, understand ecosystem limitations, and apply sustainable development concepts to global challenges. In sustainability-supportive courses, students learn about systems thinking, social responsibility, and the implications of their actions on a global scale.

3.2 Research

ACG SS prioritizes research on sustainability related and climate change topics through funded and non-funded research, especially research based on multidisciplinary and interdisciplinary approaches that address all four components of sustainability: environment, society, economy, and governance.

3.3 Campuses – Facilities & Operations

ACG has numerous buildings in several locations/campuses, including the main campus in Aghia Paraskevi, situated in a unique, protected area covering part of the Mt. Hymettus Forest. ACG SS seeks to minimize the negative impacts of ACG's current and projected physical infrastructure by reducing carbon emissions and resource depletion while also lowering operating costs. ACG SS foresees upgrading existing buildings and constructing new buildings in accordance with sustainable/ green construction principles (e.g., LEED certification) to minimize energy and water consumption as well as the carbon footprint while: maximizing the positive impact on biodiversity; integrating water, wastewater and solid waste management; introducing a series of certification programs for green and blue infrastructure; encouraging sustainable transportation (e.g., reducing use of private vehicles by ACG faculty and staff); implementing a Landscape Management Policy; adopting green procurement practices.

3.4 Governance

ACG strives to operate on the basis of governance best practices, is committed to healthy, enjoyable and sustainable ways of working, and cares for the wellbeing of all members of the ACG community. To these ends, ACG SS prioritizes the adoption of new or updated policies and protocols to ensure a just and inclusive organizational culture (e.g., gender, sexual orientation, disabilities, ethnicity, race, religion) and consistent, comprehensive and transparent internal reporting on sustainability performance as well as external benchmarking (e.g., AASHE- STARS and *Times Higher Education* Impact Ranking).

3.5 Community

ACG is committed to building good relations with the local community and society at large; to promoting sustainable development and practices in the local community and in Greece more broadly, and to creating bridges between education, research, and the local community by supporting faculty to work collaboratively with the local community for useful solutions to real problems towards a sustainable future. ACG SS calls for initiating new environmental awareness initiatives and campaigns in support of sustainable behaviors on campus and in the local community by connecting classes with community groups and by continuing close collaboration with local, regional and national authorities.

4 FINAL REMARKS

As described earlier, the ACG Strategic Plan has positioned sustainability as part of Goal Four. ACG SS, with its five pillars, makes it appropriate, if not essential, to position sustainability across all four goals in future versions of the ACG Strategic Plan. The Education and Research pillars in Goal One; Governance pillar in Goal Two; Community pillar in Goal Three, and Campuses – Facilities & Operations pillar in Goal Four. Thus, at this stage of the institutions and our global community's evolution sustainability emerges, including the existential concern of climate change, as a central and compelling issue requiring our most concerted and creative attention.

REFERENCES

E.U. (2023). *WMO*. Copernicus Climate Change Service European State of the Climate.

IPCC. (2023). Summary for Policymakers. In *Climate Change 2023: Synthesis Report. Contribution of Working Groups I, II and III to the Sixth Assessment Report of the Intergovernmental Panel on Climate Change* [Core Writing Team, H. Lee and J. Romero (eds.)]. (pp. 1–34). IPCC., DOI: 10.59327/IPCC/AR6-9789291691647

Richardson, K., Steffen, W., Lucht, W., Bendtsen, J., Cornell, S. E., Donges, J. F., Drüke, M., Fetzer, I., Bala, G., von Bloh, W., Feulner, G., Fiedler, S., Gerten, D., Gleeson, T., Hofmann, M., Huiskamp, W., Kummu, M., Mohan, C., Nogués-Bravo, D., & Rockström, J. (2023). Earth beyond six of nine planetary boundaries. *Science Advances*, *9*(37), eadh2458. https://aug.edu/news-&-events/news/the-alba-%E2%80%93-csr-hellas-esg-barometer. DOI: 10.1126/sciadv.adh2458 PMID: 37703365

The American College of Greece. (2024) website: https://www.acg.edu/

UN (2023) Global Sustainable Development Report 2023, Times of Crisis, Times of Change: Science for Accelerating Transformations to Sustainable Development

ENDNOTE

1 https://www.acg.edu/undergraduate/undergraduate-programs/school-of-liberal-arts-sciences/liberal-arts-sciences-majors/environmental-studies/

Chapter 11
Aurora University:
Campus Ecosystems in Action

Rebecca L. Sherrick
https://orcid.org/0009-0005-7031-371X
Aurora University, USA

ABSTRACT

In spring 2024, U.S. News & World Report published an article identifying 20 beautiful college and university campuses. Included were familiar institutional names — Duke, Harvard, Stanford. Also on the list was Aurora University, a private comprehensive college in the far western suburbs of Chicago, where this author served as president for over two decades. According to the magazine author, the beauty of a campus may sway college choice by making a positive first impression. (Claybourn, 2024) Indeed, this likely was true for Aurora. Prospective students and their families remarked frequently on the pleasing nature of the campus environment in post-visit evaluations. Particularly gratifying were the favorable comparisons made to the facilities of other institutions.

INTRODUCTION

With its lush landscaping, classic quadrangle, and architectural unity, the Aurora University campus did indeed make a memorable first impression. From its location on the far western edge of Chicagoland, the university competed with many excellent competitors for students. The natural beauty of the campus was a source of distinction, part of the "case" for enrollment. The story in U.S. News was not simply about first impressions, however. As one contributor to the magazine commented, "the campus acts as ground zero for social interactions and community engagement, whether it's students gathering for outdoor events, studying, socializing, or recreating, campus

DOI: 10.4018/979-8-3693-5454-4.ch011

green spaces bring students together and foster a sense of community." (Claybourn, 2024) This certainly was true at Aurora. The campus was a prime setting for many aspects of community life. From new student orientation to Homecoming weekends, activities fairs, graduation rehearsal picnics, and commencement itself, the campus was integral to the student experience.

It took more than two decades to create an environment worthy of such recognition. During these years, the institution also operated a second campus on the shores of Wisconsin's Geneva Lake. Both properties were in states of significant disrepair when the story recounted in this narrative began. A new generation of university leaders tackled problems in both settings fueled by strong beliefs in the importance of environmental sustainability. They were fueled by an understanding that providing an inspiring campus setting is akin to offering a sound general education curriculum, an attractive range of majors, and a talented faculty and staff. All were vital to fulfillment of the Aurora mission of transforming lives through education and to the vision of being worthy of emulation. Through the efforts presented below, university leaders impacted the welfare of current and future students in a meaningful way.

The narrative that follows explores the inter-connected strategies that the university relied upon to advance its distinctive campus redevelopment agenda. It begins with a brief examination of the ideas and influences that framed the thinking of leaders and turns next to an overview of the institutional landscape as the story began in 2000. The chapter then summarizes the processes and strategies used to introduce a new era of environmental responsibility at Aurora University. Several highly visible projects are presented next. An extended discussion of the changing treatment of environmental challenges in the university curriculum follows. The chapter turns next to a discussion of a large community-based collaborative project built upon a strong foundation of sustainable development thinking before concluding with final comments.

BACKGROUND

Influencers and Their Ideas

The thinking of university leaders was impacted by several different sources. The needs to satisfy external regulations and to acknowledge community sensibilities were always high priorities. Knowledge of best practices in other campus settings also contributed to development of goals and strategies. Finally, like the faculty members who fashioned Aurora's curriculum, senior administrators were influenced by the shifting views of prominent environmental champions over time, including Gifford Pinchot, John Muir, and Richard Louv.

As the Yale University-educated son of a prominent family, Pinchot enjoyed many early life advantages and advanced quickly to become the first chief of the U.S. Forest Service in 1905. A close ally of President Theodore Roosevelt, Pinchot advocated for the efficient professional management of natural resources through organizations such as the Bureau of Land Management and the Forest Service. He stressed the importance of striking a balance between human activity and nature and saw a role for government in maintaining this equilibrium. Public lands, Pinchot argued, belonged to all and were one of the greatest legacies of democracy.

If Gifford Pinchot sought to strike a balance through management, Scottish-born John Muir wanted to preserve and protect the environment from human activity. Pinchot saw the right kind of professional management as key to sustainable commercial use of natural resources. He was a conservationist and Muir a preservationist who found spiritual meaning and transcendence in nature and wilderness. Muir also influenced Teddy Roosevelt. The two camped together in the backcountry of the Sierras, an experience described by Roosevelt as transformative. As a founder of the Sierra Club and an advocate particularly for Yosemite, Muir is regarded as the father of America's national parks.

Iowa-born Aldo Leopold spent a childhood near the Mississippi River, preferring always to be outdoors rather than inside. He set his sights on earning a degree in forestry from Yale and subsequently served the U.S. Forest Service in Arizona and New Mexico. Early in his career, Leopold posted several important achievements, including preparation of the first comprehensive management plan for the Grand Canyon and compilation of an initial handbook on game and fish management. In time Leopold's career took him to Wisconsin and an appointment to the University of Wisconsin faculty. A milestone came when Leopold purchased 80 acres of sandy farmland in central Wisconsin that was largely barren due to logging, overgrazing, and recurrent fires. There he tested many of his own ideas and began the masterwork that would shape environmental thinking for years to come.

Pinchot and Muir offered two different views of the relationship between humanity and nature, Leopold advanced a third. His thinking was holistic, based in the understanding that humanity is part of the larger biotic community rather than set apart from it as conquerors or rulers. He argued that the natural world should be valued in its own right, not simply because of its benefits to humanity. In "A Sand County Almanac," published shortly after his sudden death from a heart attack, Leopold framed his thinking around what he called "the land ethic." He rejected any kind of utilitarian relationship between humanity and the wilderness or nature, arguing instead for a state of harmony.

As this chapter will demonstrate, Leopold's ideas played a prominent role in formation of the current Aurora University curriculum and in the author's own thinking. As the full magnitude of the global climate crisis became clear in the final decades of the 20th century, a second publication, "Our Common Future," had an equally significant impact. The book grew out of the work of the World Commission on Environment and Development, known often as the Brundtland Commission in tribute to its leader, a former prime minister of Norway with expertise in public health and science. The body was established in 1983 through an action by the United Nations secretary-general.

Its purpose was to unite nations in shared strategies designed to facilitate economic growth and environmental responsibility. Toward this end the commission advocated an intergenerational perspective. "Sustainable development," the Brundtland Commission asserted, "is development that meets the needs of the present without compromising the ability of future generations to meet their own needs." (Bruntland Commission,1987). Sustainable development, commission members contended, rests upon the three pillars of economic growth, environmental protection, and social equality. It was this holistic concept of sustainable development that compelled the decisions made by Aurora leaders.

Influence came from a third and final source, the work of environmental journalist Richard Louv, author of several books including "Last Child in the Woods," "The Nature Principle," "Vitamin N," and "Our Wild Calling." Through these publications, Louv examined the relationship between humanity and the natural world, based upon his understanding that less time is spent in nature now than in the past. This change, he argued, is responsible for both the behavioral and health problems evident among children and among some adults. Engagement with nature including animals, Louv suggested, will help humanity save itself and perhaps the planet too.

According to the journalist, the benefits of more time in nature may include improved well-being, increased creativity, improved social skills and learning capacity, sharpened mental acuity, a heightened ability to recognize patterns, and a sense of awe and wonder. Integral to his work is the juxtaposition between technology and nature and the argument that the two must be kept in balance. "The future will belong to the nature-smart," he asserted, "those individuals, families, businesses, and political leaders who develop a deeper understanding of the transformative power of the natural world and who balance the virtual with the real. The more high-tech we become, the more nature we need." (Louv, 2011)

Taken together, the environmental champions mentioned above shaped this author's thinking about campus redevelopment around several key principles. First, the university would pursue generational strategies designed to limit the need for repeated investment. Second, a high priority would be to create a learning environment rich in natural beauty. Finally, given Aurora's goals of strengthening educational quality and improving financial health, an aggressive growth plan would be the focus of the years ahead. As it implemented solutions guided by these principles, the university fulfilled its self-defined commitments to its mission, core values, and its larger vision of being worthy of emulation. Through this mindfulness, Aurora approached campus redevelopment with intentionality and a dedication to enriching student learning.

Two Campuses, One University

As a candidate for the presidency of Aurora, the author took an early-morning walk in spring of 2000 on the university's main campus. The goal was to find the music facility shown on the campus map. It turned out to be an unassuming brick building, constructed originally as an infirmary during an early-20th century influenza outbreak. The space was so limited that it was hard to imagine two grand pianos inside. Next door to the music building was a green metal structure that housed several departments and offices. Just across the drive was the maintenance building that delineated the southern boundary of the campus, while to the north was the newest building on the property, developed in 1987. The surrounding neighborhood consisted of small family homes.

A day later, the interview schedule moved 70 miles north for a visit to the university's Wisconsin location, George Williams College. The 150-acre property was adjacent to the University of Chicago's iconic Yerkes Observatory, scene of many of astronomy's most important breakthroughs during the 20th century. Transcending the boundary of the two was a small 18-hole golf course. The remainder of the property sloped dramatically toward the lake, dotted by small wood frame cottages and cabins grouped around a large, rambling dining hall, reminders that the facility long was home to summer camps and family and church conferences. Finally, along the lakefront were the historic buildings that once played important roles in the YMCA and YWCA movements.

The two campuses became part of the same university just weeks before the new president took office, the culmination of an affiliation and merger process begun almost eight years earlier. As the author would soon discover, both properties had deep roots in the 19th century. George Williams was founded first in 1886, when three YMCA leaders established the "Western Secretarial Institute," along the Geneva Lake shore. They were mindful of the association's rapid expansion westward and believed that a summer training center would help perpetuate cohesion among leaders and continuity within the growing organization.

The institute quickly became a popular destination for members of the growing number of collegiate Y associations as well as for faculty at YMCA colleges and universities. James Naismith, the inventor of basketball who served on the faculty of YMCA's Springfield College, frequently spent summers at the Geneva Lake facility.

As the Y movement spread across university and college campuses, the institute became even more important as a source of organizational unity and growth. It was during one of the collegiate gatherings, in fact, that a group of women in attendance founded an organization of their own, the Young Women's Christian Association.

The Western Secretarial Institute was one of many such properties in the area. In fact, within several decades of its establishment, more than 17 additional facilities would line the lakeshore. Why? In the aftermath of the Great Chicago Fire of 1871, wealthy families fled north to Geneva Lake in search of fresh air, clean water, and a beautiful natural environment. They brought with them a reform instinct rooted in the late-19th century belief that environment could shape — or even reshape — character. On their heels came celebrities, including Mary Todd Lincoln and several evangelists. Then tourists arrived. Quickly the area became known as a popular vacation and retreat destination.

A few of the new facilities were established to offer respite from the corruption and pollution of urban life. The Lake Geneva Fresh Air Association, founded by affluent Chicago women, was intended to benefit the children of their husbands' employees. The Eleanor Camp, meanwhile, was intended to provide single working women with a safe and suitable vacation destination of their own. There even was a camp inspired by a love of horses. The George Williams College campus was a product of this era. Over time, this distinctive heritage manifested itself in a deep loyalty to tradition and a passion for conserving and protecting the natural environment.

In time, the Geneva Lake training center affiliated with a Chicago-based institution of higher education with origins in the city's Central YMCA. Years later, on the anniversary of the birth of the YMCA's British founder, the organization was rechristened in his honor. George Williams' academic programs, designed to prepare graduates for careers in social services and youth development, were offered on a campus near the University of Chicago in Hyde Park. The property in Wisconsin, dubbed "College Camp" by this time, provided experiential learning, camp, and conference center programs.

In 1965, the college sought to escape neighborhood problems by relocating to the western suburbs and broadening its curriculum to compete with nearby liberal arts institutions, an effort that fell short. Unfortunately, the variable rate debt incurred to purchase and to develop the new campus became too heavy for the small college to shoulder. Trustees closed the Illinois campus in December of 1985 and sold the campus to satisfy obligations. Prominent academic programs moved quickly to other campuses so that students could complete their degrees from accredited institutions. Trustees then embarked upon the search to identify a new partner to operate the Wisconsin lakeside facility.

Aurora University claimed distinctive historical origins of its own. A full decade before George Williams founded the Young Men's Christian Association in London, evangelist William Miller carried his revival message up and down the East Coast of America. He preached from large muslin prophetic charts that portrayed Biblical events to buttress his idca that the Second Coming of Jesus Christ was imminent. Miller's calculations pointed to October 22, 1844, as the day. On the specified morning, his followers kept vigil on a day that would subsequently become known as the Great Disappointment.

Some Adventists recalculated. Others argued that a return was still imminent, a conviction that argued against planning too much for the future. Several factions emerged within the denomination with the Seventh Day Adventists becoming the most prominent. As other religious denominations pushed westward, many established seminaries and colleges. Adventists chose not to follow this practice. Some even contended that doing so would be heretical, a sign of disbelief. Not all agreed, particularly those who thought an educated clergy was important to the Adventist movement.

As the end of the 19th century neared, several small Bible colleges or seminaries opened. Among these was tiny Mendota Seminary. On a blustery January morning in 1893, three students waited at the front door of a building previously occupied by a Lutheran college just departed for Iowa. All were from the local community, as would continue to be the case for as long as the institution remained in Mendota. Even so, prospects were limited. In 1912, two decades after its founding, members of the college community boarded the train for a new home in nearby Aurora. Community leaders met them at the station and escorted them to a small property on the far western side of the community, where three partially constructed buildings waited on the new Aurora College campus.

For much of their history, neither college had a deep resource base. They prepared graduates for careers in modestly compensated disciplines. George Williams alumni excelled in non-profit leadership, while graduates of Aurora chose career fields such as nursing, teaching, and accounting. By the time the Downer's Grove campus of George Williams closed, the remaining facilities in Williams Bay and Aurora were quite different. They were separated by approximately 70 miles and existing in different states and locales. One was predominantly rural, occupying a prime location in a popular resort community, while the other sat in the middle of a modest residential neighborhood in the second largest city in Illinois.

It was not just the campuses that were different. In truth, there was limited mission congruence between the two, a reality that fueled skepticism among alumni and employees. Resources were strained for both organizations. Both were struggling financially. Aurora, meanwhile, had earned only seven years of accreditation from the Higher Learning Commission during its most recent comprehensive evaluation, considerably less than the standard ten years. Deferred maintenance was a major concern in both settings but was an especially serious issue in the highly competitive Illinois higher education marketplace.

These significant issues received minimal attention from trustees prior to the merger. Instead, as a new president prepared to take office, board members hurried to complete the process, though it would become evident later that some had significant reservations. In the short run, Aurora benefited from some immediate gains, but also inherited longer term issues. The Geneva Lake property itself was in considerable disrepair. It would prove easier to combine elements of the two academic programs than to blend the cultures of the two organizations.

Initial reservations on the part of Aurora faculty who favored liberal arts disciplines over professional preparation proved to be largely generational. Today, the social work program that relocated from Downer's Grove is among the largest and strongest offerings of the university. Moreover, the experiential emphasis long favored by George Williams now is evident throughout the university's curriculum. The campuses themselves remained fundamentally different with each having a culture shaped by history, tradition, and location. For all the problems, the opportunity to pursue a sustainability agenda was varied and rich. To this subject, the chapter now turns.

PHASE I: GETTING STARTED

Building Capacity

At the outset of a new administration, it was clear to the author that neither the Aurora nor the Lake Geneva property had the staffing, equipment, or expertise necessary for optimal operation. Apparent also was a resistance to new methodologies and practices and collaboration between the two campuses. Facilities were in states of significant disrepair. Finally, it was evident that minimal attention had been devoted to life safety and risk management. These issues became an immediate priority for the new administration. Following a fact-finding period, concerns were addressed by outsourcing all maintenance and custodial services, including grounds, to an experienced vendor.

The new model did not save money. As is common when resources are limited, both colleges had underfunded plant operations for years. Given the importance of upgrading the learning environment, difficult decisions were required to fund the shift. A recently leased facility in Chicago was shuttered and several Aurora campus functions that did not serve students directly were discontinued. The return on investment became quickly evident as the university experienced significant gains.

Outdated cleaning products and practices were replaced quickly with environmentally friendly alternatives. New product standards and operating policies were finalized, and a rigorous training and professional development program began.

With the new vendor relationship in place, the next crucial decision related to defining the division of labor between campus-based personnel and external professionals. The university took an innovative approach to this process. As small projects materialized, area architects, engineers, tradesmen, and construction professionals were invited to submit bids, which were reviewed carefully before interviews were conducted. The goal of this process was to communicate and align values around important considerations such as sustainability. Once work was completed, an outcomes review took place with the assistance of several experienced trustees. Through this intensive effort, Aurora selected the external providers who would join Aurora's multi-disciplinary team.

Sustainability Jump Starts

While much of the work just described took place behind the scenes, university leaders understood the importance of communicating shifting institutional values to members of the community and did so through several "jump starts." The first focused on recycling. Mismatched containers scattered about both settings were replaced by sturdy and attractive outdoor receptacles for trash and recycled materials. Offices and departments were encouraged to join in the effort as they received containers of their own. All were encouraged to prioritize digital communication over paper-intensive processes, and printer usage was limited through introduction of a new direct billing system. Student environmental clubs were established and championed these and other initiatives. Recycling was a good first project since virtually all understood its significance.

The second jump start targeted renovation of fast food venues on both campuses. Each was dingy and dark and evidenced hard wear. An early fundraising project targeted improvements to the two spaces. With dollars in hand, leaders made an unusual decision to schedule the work at times both campuses were occupied rather than over the summer months. As crews first gutted the spaces and then undertook the rebuilding process, regular updates were shared with the George Williams and Aurora communities on subjects ranging from improved exhaust and ventilation systems to the selection of materials. The goal was to educate members of the campus communities about the new emphasis on sustainability.

Clarifying Mission and Identity

The first phase of the university's emphasis on plant improvements and sustainability was almost complete. Only the work of clarifying mission and identity stood between the institution and development of new campus master plans. Here too, open communication and engagement were vital to shaping a shared understanding of educational purpose and agreement regarding the specific populations to be served in the years ahead. The first step in this effort was to craft a mission statement broad enough to encompass the programs and traditions of the two distinct colleges and their traditions. As faculty, staff, and trustees gathered to discuss this question, a consensus emerged quickly. Whether in Illinois or Wisconsin, participants agreed, the work of the university was inspired by a shared belief in the "transformative power of learning."

Opinions varied on questions of identity. Some in Aurora, for example, envisioned a "senior institution" that would educate upperclassmen, adults, and graduate students. These faculty were clustered heavily in professional programs where the preference was for recruiting transfer students who had completed general education requirements in other settings. Others, particularly those working in the liberal arts and sciences, favored a more traditional model built upon freshman recruitment. They believed that this was the best way to ensure that Aurora University graduates were truly prepared upon graduation.

The divide was significant and reflected to some extent the different natures of George Williams and Aurora. Initial conversations were divisive. In time, however, the two options merged around the argument that the best academic offerings for students were those where graduate and professional programs rested on strong undergraduate foundations. This was especially important, protagonists suggested, for an institution with a diverse student body and a large population of first-generation students. A strong general education experience would prepare them for the rigors of a professional program. Others noted that the university would require four years of tuition revenue to support the full costs associated with advanced offerings.

The question of identity was more complicated in Wisconsin, given the breadth of activities and programs offered by George Williams. At the time of the merger, Aurora used the campus as a base of operation for continuing education programming. For the most part, though, the work of the campus remained the same as it had been for years. Public school students came for outdoor education overnights, while others participated in adventure education experiences. When public schools were not in session, the emphasis turned to camps and conferences for groups of various sizes and composition, including children, youth and adults attending family reunions, large church gatherings, or specialty camps.

Summer months also saw the 18-hole golf course in operation as well active waterfront and mooring use. Most facilities were "rustic," to borrow from the facility's promotional literature with overnight accommodations varying in quality. Although one new lodge had been constructed following the 1992 affiliation between the two organizations, other residences were more basic and had badly outdated plumbing and poor ventilation. Partial winterization also was an issue. Several large dormitory-style buildings relied on plywood platforms as beds for youth and college-age residents. The need for modernization was readily apparent throughout the facility. University leaders opted to seek outside counsel as described next.

PHASE II: NEXT STEPS

Site Optimization – George Williams College

With both identity clarification and master planning in mind, the institution retained a respected national camp and conference center consulting firm from Arkansas to work closely with an internal task force to develop recommendations on both subjects. The intensive process unfolded over a year. The first suggestions focused on the optimal clientele to be served. Representatives of the firm noted that few adults would choose to stay at a facility that hosted large numbers of children. Next, they outlined the levels of investment required to make the life safety improvements necessary to accommodate minors in the lake environment. Finally, they argued that it was in the university's best interest to serve essentially the same audiences on its two settings.

For these reasons, the consultants suggested that George Williams College focus its programs on young adults and adults. Though not all on campus agreed with this analysis, the university accepted the recommendation and moved next to determining optimal capacity. This phase of the engagement focused on quality, marketability, and finances. Close attention was dedicated also to assessing practices in comparable settings. Here the consultants concluded the campus simply did not have sufficient resources to continue to host as many as 700 guests in appropriate accommodations. Instead, they advocated for reducing overnight capacity to a maximum of 200 occupants and for making the improvements necessary to increase the price point over time.

These first two recommendations established the context for entering a campus master planning process. This phase of the engagement began with an assessment of existing structures and identification of costs associated with bringing them to contemporary safety, comfort, and sustainability standards. Many of the small cottages and cabins used for guests were in rough shape. So too were the large dormitories constructed years before to house large numbers of summer employees. Throughout the campus, utility, electrical, and sanitary systems were badly out of date. And, as noted earlier, life safety was also a large issue.

This exercise was crucial to helping key individuals, including members of the Board of Trustees, understand the magnitude of facility shortcomings on the campus. For the most part, these leaders had been housed during visits in off-campus hotels or in the single new lodge. Many simply did not understand the poor condition of the facility. Expense numbers, accompanied by photos, diagrams, and text provided by experts familiar with standards in comparable settings, were sobering. This work marked an important turning point in the university's own considerations about the campus, resulting in the decision to demolish many of the worse structures.

Sustainable development principles impacted the consultants' remaining proposals as they advanced a concept for "zoning" campus redevelopment to optimize site potential. New residential and academic facilities should be developed, they argued, near the public street that ran along the crest of the hill that defined the lower campus. To maximize campus capacity while minimizing building footprints, they suggested that new buildings be at least two stories high with parking easily accessible from the adjacent street. The overall effect of this recommendation and the prior one relating to demolition of outdated properties was to open a wide swath of "park-like" green space through the center of the lower campus. Whereas before this area was densely occupied by small wooden buildings, upon implementation of these changes it would become dramatically different. The flow of water down the hill toward the lake would slow, allowing time for percolation and absorption, ideas designed to prevent lakefront erosion and to emphasize the importance of green space on the campus.

Recommendations relating to traffic and circulation on the campus also reflected sustainable development priorities. Several old Village of Williams Bay public streets transected the property. Many of these thoroughfares were in poor shape. Beneath them ran Village utilities that served the campus. Sink holes were a particular concern that grew worse over time, again due to rapid run-off and deteriorating infrastructure. To protect the integrity of the streets as long as possible and to avoid disruptions to campus operations, the Arkansas firm favored minimal traffic at the center of the property and proposed locating facilities serviced by large trucks, such as food service and maintenance, to the far boundaries of the campus.

Finally, members of the firm offered recommendations regarding the historic lakefront and the nearby buildings that were developed in the late-19th and early-20th century. This portion of the campus was truly distinctive, the consultants observed, due in part to the 26-mile public pathway that circles Geneva Lake. Local lore suggests that the path was first traversed by Native Americans and later by the workers who constructed the many mansions that sprang up after the Great Chicago

Fire. Here too the firm recommended eliminating as much impervious surface as possible to protect the shoreline. They also strongly discouraged development in this quasi-public area.

Upon approval by the Board of Trustees, the master planning recommendations took effect and guided future development. The plan was implemented in full with the last major project, the restoration of the lakefront itself, completed during the pandemic. The consultants' emphasis on sustainable development had a powerful impact on university leaders and those professionals who worked on the campus in the years after the plan was approved. For the most part, the university made positive changes by optimizing the use of existing structures, sometimes demolishing all but a few features and then rebuilding from the foundation upward. When work was required on a historic structure, preservation of original character and design were high priorities and minimal disruption of the natural environment was the goal.

Site Optimization – Aurora Campus

In Aurora, the consensus regarding the academic future of the institution and a strong enrollment effort propelled growth. The university met its short-term need for additional square footage by leasing nearby facilities, a surge space strategy that allowed decision-makers the time to evaluate the master plan inherited from the prior administration. The plan centered on construction of a multi-story academic building on the quadrangle, one of the few memorable, defining spaces on the small campus. The location was reflective of the tension among the university, its neighbors, and city leaders regarding campus growth.

This was not its only flaw. The plan failed to address the twin issues of stormwater management and parking that were growing concerns for university neighbors and city leaders. To put it simply, the university owned insufficient land to accommodate these inter-connected needs, a reality made more complicated by the fact that several different watersheds existed on the campus. Finally, the master plan did not take into consideration the condition and shortcomings of existing structures or the need to develop facilities common in other settings, such as a student union. Clearly a high priority of the new administration had to be persuading others that the university's long-term viability was contingent upon enrollment growth and campus expansion.

It took several years to complete this process, the first in a series of iterative master plans over the next 20 years. Three issues dominated negotiations each time Aurora attempted to expand its footprint. Neighborhood residents were vocal about the height and massing of buildings, apparently not worried that such limitations would force the university to seek additional land over time. Some even expressed concern about preserving the possibility of "seeing through" the university property from streets surrounding the campus. Neighbors and officials continuously pressured the administration to add more on-campus parking, while strenuously opposing the university proposal to solve this problem through construction of a parking garage.

As surface parking was added and impervious surface increased, storm water detention needs multiplied. Preventing flooding was a high priority for municipal engineers and city and county leaders. Memories of the 17 inches of rain that fell during a 24-hour period in 1996 and the resultant flooding on the west side of the city remained vivid in the minds of these officials. Most organizations responded to stormwater mandates in the Fox River Valley by including large above-ground detention areas in their development plans. Inevitably such facilities attracted large numbers of insects and geese, contributing to recurrent contamination of the local water supply.

University leaders evaluated the full range of strategies for complying with detention mandates, including sustainability and aesthetic implications. Given its limited land, financial constraints, and other ongoing concerns, university leaders hoped to find a solution that addressed multiple needs without creating new problems. Eventually they opted for the installation of massive detention vaults beneath parking lots and select new buildings. Although local ordinances provided for such a solution, municipal staff and officials were skeptical and acceptance came slowly. Eventually, however, the university gained approval for its plan and became the first organization in the county to develop below-grade detention.

Such systems came with hefty price tags. Still, they were preferable from the standpoint of optimizing site potential and the overall aesthetic of the campus environment. In future years, the university would consistently favor solutions that complied with external requirements and facilitated enhancement of the natural beauty of the property. These twin goals prompted development of other special features, including a dry creek that wound its way through one corner of the campus and a labyrinth fashioned partially from permeable pavers. Near another renovated building, the institution introduced a rain garden, otherwise known as a bioretention facility. The garden was developed in a depression designed to absorb and filter water. The new feature was planted with native shrubs and perennials that attracted birds, butterflies, and other insects. Its presence softened and brightened an area on the campus with dense development. Aurora took another step forward by adding a rooftop garden on the top of the STEM building that will be discussed later in the chapter.

The master planning processes in Wisconsin and Illinois were intensive efforts, which benefited from broad-based participation and required years to complete. Each new initiative triggered the need for research and examination of options. Neither facility had maps of essential utilities or fiber. It was not unusual to discover such during the construction process itself. Even the presence of three watersheds on the Aurora campus was not known in advance. For all their intensity, the two master planning efforts served the institution well and likely will continue to do so for several more years.

Communicating Values at the Front Door

Several years were required for the new administration to build capacity, to clarify institutional mission and identity, to develop master plans, to understand the extent of deferred maintenance needs, and to secure the necessary dollars. All told, over $300 million would be spent on plant improvements with support coming largely from grants, private gifts, and operating budget surpluses. Inevitably, projects included significant investments in sustainability. Often such positive efforts occurred behind the scenes in places not immediately visible to students, faculty, staff, and visitors. In select instances, however, Aurora chose to make visible statements to communicate its dedication to environmental responsibility and natural beauty.

The Aurora master planning project identified the need for a stronger entrance to the campus. Gone was the original horseshoe drive in front of "Old Main," leaving visitors uncertain about the location of a "front door." As university leaders corrected this shortcoming, they created one of the most memorable features of the Illinois campus. The project began with a review of examples at other institutions before turning to development of a distinctive design. In the place of a more traditional monument or gateway, the university created a dramatic sense of arrival by planting a bosque, an "entry forest" composed of 140 Dakota Pinnacle Asian Birch trees. Perpendicular to the bosque, Aurora developed an allee lined by 50 matched Greenspire Lindens leading to the newly constructed Welcome Center. The university continues to prize its status as a "tree campus." Even now, for every tree removed, two are planted.

The university's campus on Geneva Lake actually boasted two front doors, including the property's original 19th century entrance along the lakeshore. Still today, some in the local community prefer to arrive on campus by boat, tying up their classic wooden boats along the pier to attend concerts, lectures, or other events held on campus. Here too the institution chose to exemplify its twin commitments to sustainability and beauty. During a multi-phase restoration that unfolded during the pandemic, major changes were accomplished. During the first months of the effort, the goal was "addition by subtraction" through removal of the many impervious cement and wooden platforms and structures developed along the lakefront over a century to support waterfront activities.

Next deteriorating retaining walls and walkways were removed and the long pier itself was renovated. It too was an artifact of a much earlier time, one of three on the entire lake capable of accommodating one of the large late-19th and early-20th century steamers that transported visitors from a nearby train station to their destinations. During the next phase, the lakeshore itself was stabilized as boulders were lowered by cranes into place. Finally, the lawn that abutted the lakeshore was filled and cultivated. Native plantings were layered in place, all with a goal of holding the grade, absorbing water, and protecting the shoreline. A simple pergola marked the intersection between the pier and the stairway leading to the historic administration building.

Like the Aurora entry initiative, the lakefront restoration effort accomplished several purposes. First, it was informed by contemporary sustainability methodologies. Both projects directed attention toward slowing the flow of water to prevent erosion. As hard, impermeable surfaces were replaced by plantings, water was allowed to percolate naturally and to be absorbed by trees or plants. Second, the George Williams project celebrated natural beauty in a manner consistent with the ecological relationships celebrated by the lake community. To borrow from the thinking of conservationist Aldo Leopold, the new entry was a tribute to the whole biosphere, not simply to the humans who inhabited it. Even the local deer population was considered as plants were selected.

The two entryway projects were signature initiatives that drew the attention of the campus communities and the larger public. The university seized a long list of other visible opportunities to share its values. In campus dining, for instance, plastic containers were replaced by others made from recycled materials. Worn and mismatched outdoor furniture on both campuses was jettisoned. In its place, the university put seating arrangements of brightly colored polywood chairs and tables. During warmer months, when paired with vibrant umbrellas also crafted from recycled materials, they created perfect locations for conversations, while also minimizing annual maintenance.

SUSTAINABILITY IN THE CLASSROOM

University Curriculum

As the university's administration advanced its sustainability agenda, so too did the faculty. What began as a focus on environmental science became much broader over time. Initially, environmental science was housed exclusively in the Biology Department. On the Aurora campus, the curriculum had a heavy natural science and laboratory focus, while in Wisconsin the major was much more applied, a difference consistent with the identity and traditions of the two original colleges. The long-standing George Williams legacy of immersive learning and social entrepreneurship lent itself to development of the property's organic garden and hydroponics lab and to holding regular "lake studies" events for area school children.

In a sense, the two campuses acted from the same subject matter foundation before branching in different directions. Unfortunately, neither major attracted many students. Those who did focus in one of the sciences tended toward other concentrations within biology. This began to change when faculty from Biology, English, History, and Political Science/Public Policy fused the two variants to develop a popular new major in Environmental Studies and Sustainability that today enrolls a significant number of undergraduates. The program is purposely designed as a second major for those opting to marry interest in other disciplines with understanding and advocacy for the environment.

According to the senior history professor who directs the interdisciplinary program, the major finds its intellectual inspiration in the work of 20th century conservationist Aldo Leopold. Especially important is his landmark book, "A Sand County Almanac" that was published shortly after Leopold's death in 1948. At mid-century, only Rachel Carson's "Silent Spring" had a comparable impact on the environmental movement. Leopold argued on behalf of what he termed, "the land ethic." He rejected earlier ideas about the relationship between humanity and nature with their strong utilitarian or preservation priorities and chose to present the human species as part of the biotic community. Men and women existed within nature, according to Leopold, not alongside it.

The Aurora major draws heavily upon Leopold's thinking. In fact, a primary focus of the curriculum is to empower students to understand humanity as one of many species in the interdependent global web of life. Students are encouraged also to recognize that their decisions and actions have the potential to impact not only their own lives but those of subsequent generations. Through this lens, the program encourages students to problem solve as they chart responsible and sustainable paths into the future. Experiential engagement is a vital part of the program today as it was years ago on the Wisconsin campus. Learning comes alive through Citizen Science projects or through maintaining the university greenhouse, tending the roof garden, working in the hydroponics lab, or harvesting produce from the organic garden. Student organizations continue to play important roles in the educational process. Vegetables grown by some students in the campus garden are distributed by others who operate the food pantry.

STEM Partnership School

The university launched its boldest sustainability initiative midway through the author's term in office with the 2014 establishment of a STEM School on the Aurora campus. The effort, which drew on the expertise and resources of a long list of community organizations, had exactly the kind of comprehensive positive impact envisioned by the Brundtland Commission. It rested on the three pillars or principles of environmental, social, and economic sustainability. With its emphasis on the development and deployment of collaborative solutions to complex social problems, the effort demonstrated the significant leadership role a small, private university can play in leveraging the shared resources of a community to develop the kind of partnerships that accomplish multiple important purposes.

The project was the culmination of Aurora's long history of engagement with the West Aurora School District, efforts intended to enrich the learning experience of university and district students and to contribute in meaningful ways to the quality of life in the surrounding community. For many years, fourth graders from a neighborhood elementary school met in classrooms on campus. Select university faculty, meanwhile, served as "scholars in residence" in West Aurora facilities. And when the district developed a new facility, the university sometimes joined the effort by creating classrooms of its own within the building. The shared spaces which allowed teacher education students early opportunities for immersion in real public school settings.

In 2003, the institution received a large federal appropriation to expand this successful collaborative model. The grant helped fund construction of a major new academic building near the entry to the campus. The 45,000 square foot structure allocated space to multiple elementary classrooms and the College of Education. University classrooms and a large auditorium were also included along with the new high-tech Caterpillar Foundation Center for Teaching and Learning. The grant also allowed for establishment of a new university center for excellence, the Institute for Collaboration, which was the building's final occupant.

Many of Aurora's most important achievements in the years ahead would occur under the auspices of the institute, thanks especially to the leadership of the two highly accomplished retired school superintendents who served consecutively at its helm.

They were succeeded by a respected leader from the Illinois General Assembly. Together, the three furthered development of strong connections between the university and its neighbors throughout the Fox River Valley. Projects were varied. In response to elevated concern about juvenile crime rates, for example, institute leaders partnered with an area non-profit to develop a successful city-wide after school program, which was staffed by Aurora University students who gained valuable experience as well as compensation.

Equally important was the institute's leadership in research and grant writing. Even before the Institute was established, Aurora University consistently garnered external support through its partnerships with local school districts. Through the institute, grant writing efforts move to a new level. The university expanded upon its already impressive record by competing for a series of Illinois Mathematics and

Science Partnership (IMSP) grants. Aurora was among a handful of small, private institutions to secure such funding and benefited from its inclusion in the larger discussion surrounding excellence in mathematics and science instruction.

The state funding fueled expanded engagement between university professors in mathematics, natural science, and education and practicing area teachers. The overarching purpose of the dialog was to create new strategies to respond to recurring questions or problems. One such effort examined the sources of elementary and middle school student under-performance on standardized tests in mathematics. After observing in public school classrooms, Aurora faculty developed the hypothesis that student under-performance was more likely to take place in a class where the teacher him or herself was not adequately prepared in mathematics. As a result of a lack of mastery and confidence, they believed, teachers devoted insufficient time to the subject matter and did not engage student learners.

Faculty then worked with successful and effective teachers to develop a new master's degree curriculum in mathematics education for practicing teachers. Grant dollars supported costs associated with instruction and enrollment in the program as the effort moved into its next phase. Coursework emphasized content mastery and the development of new teaching methodologies. The results over time were impressive. As teachers grew stronger in mathematics and developed their own sense of ease, they devoted more time to the discipline in class and student attainment improved significantly. State of Illinois reviews scored the project at the highest level and praised its transformative impact.

The institute also championed a multi-partner response when a large local charitable foundation issued a "challenge for change," intended to spark collaboration throughout the community. Institute leaders convened colleagues from four area school districts and select non-profit organizations to identify a project. Eventually, the group settled on the shared goal of nurturing growth of a STEM culture throughout the city, beginning in the public schools. After winning first prize in the competition, the partners fashioned a specific vision that called for establishment of a grades 3-8 STEM School on the Aurora campus to serve students and teachers from the four participating public school districts.

The school's curriculum was highly innovative. Initial development took place during three graduate courses for practicing educators taught at Aurora University. Another team of professional educators then audited the curriculum to ensure alignment to Illinois Learning Standards. Corporate and non-profit partners next met with the university and public school educators for discussions held during the evening at the local Caterpillar plant. This part of the project was crucial, since one of the goals of the school was to encourage students to consider continued study and career choices in STEM disciplines. Toward this end, the multi-disciplinary participants created the "Caterpillar Framework," a model that integrated learning about industry processes, activities, and expectations into the curriculum.

Throughout the effort, partners came together to overcome challenges. One of the most significant early impediments was State of Illinois law that offered a narrow definition of a public school. This concept had been influenced heavily by powerful teachers' unions and those opposed to the charter school movement. Through the leadership of local legislators, the acumen of university lobbyists, and powerful public testimony, the partners convinced the General Assembly to pass the legislation necessary to move forward with establishment of the distinctive school despite the initial opposition of teachers' unions that eventually was overcome.

For the most part, the university president shouldered responsibility for securing the funding necessary to build and equip the new facility. The Governor's Office was helpful as was the local foundation that sponsored the original contest. Aurora University participated also alongside the City of Aurora. With dollars in hand, partners worked closely with a local architectural firm to scan for innovations in other public school districts and to develop a shared vision for a building that would serve multiple learning populations. The plan called for leveraging this initial fund raising with a second phase of the project focused on teacher professional development across all four districts and additional grant activity by the university.

When it opened in 2014, the STEM Partnership School was one of only three public school facilities in Illinois with a platinum LEED certification, a distinction that drew curious visitors each year that the facility was open. The building relied upon natural light to illuminate as much space as possible, thanks to the plentiful use of glass, light shelves, and adjustable fixtures. Visitors could see clearly into the elementary and middle school classrooms; just as younger students could easily catch glimpses of the college students moving through the corridors or across the quadrangle. Flooring was composed of recycled materials, creating soft and quiet passageways throughout the building. Rooftop solar panels and an adjacent wind turbine reduced energy consumption. Especially noteworthy were the teaching and learning resources literally built into the structure. Pipes were left exposed, and some walls had "portholes" that allowed students to see firsthand how the building worked. The school building itself functioned well for learners of all ages, thanks to the forethought involved in the selection of flexible furnishing and equipment that allowed a fourth graders to use a laboratory one hour and university students and their chemistry professor the next.

District and Institute leaders developed policies, procedures, and a budget, drawing upon best practices in other settings. A prominent school district attorney drafted procedures for governing the school. As the time approached for the school to open, conversations regarding logistics became more important. One of the local school districts, for instance, did not operate a transportation system, since their neighborhood-based model argued against buses. Unfortunately, given the distance from the east side of the city to the university campus on the west side, transportation assistance was necessary for students and families. After exploring a range of solutions, leaders from the district on the far east side of the city volunteered to arrange for their buses to pick up STEM School students.

Such solutions were common early in the project, a reflection of the emphasis on collaboration to solve problems. The vision came to life when the school opened in 2014, just as the final phase of technology innovations was set in place. The Partnership School did not serve gifted students. Instead, students were selected through district-based lotteries based upon their own interest and that of their families. As a result, students of varying abilities attended the STEM School, which was intended to serve as a microcosm of the whole. Teachers from participating districts also were selected based upon interest. University faculty and master teachers from the districts provided overall leadership.

STEM School students and teachers enjoyed an amazing array of classroom and "real world" activities. With the assistance of professionals from Scheck and Siress, they learned to develop prosthetic devices for amputees. A project supervised by experts from Waste Management focused on selecting a site for a new Kane County landfill. Students studied the supply chain under tutelage from Caterpillar staff, helped to research methodologies for retaining wall construction, communicated regularly with peers from other countries, implemented Citizen Science projects, and engaged the United Nations Sustainable Development Goals. And they consistently scored near the top of required state standardized tests. In fact, just as the pandemic began, a team of STEM Partnership School students took a top prize in the highly competitive Toshiba ExploraVision competition.

Sadly, the pressures of the pandemic forced the local school districts to withdraw from the partnership despite its noteworthy accomplishments. Anecdotal accounts suggest that STEM School students excelled the performance of peers in their home districts and across the state. Participating students and their families consistently were positive about the experience as were representatives of the various partner organizations. Like the teachers themselves, these volunteers frequently credited engagement with the school as an inspiring source of personal and professional renewal. They found meaning in helping to prepare a new generation to pursue similar careers. The same was true for university faculty members, who suggested that their own teaching was better for the experience.

From an overall perspective, the John C. Dunham STEM Partnership School was a resounding success, the product of an integrated sustainability strategy that offered social, environmental, and economic benefits. For instance, unlike many larger cities in Illinois, Aurora had multiple independent school districts that evidenced the socioeconomic character of the neighborhoods they served. The STEM School was the only public school in the community that served students from across the city. Its inclusive nature was distinctive. As students came together on the university campus from diverse homes, traditions, and neighborhoods, the possibility of developing a cohesive and holistic vision for the entire urban area came alive.

By virtually every measure and metric, the school was a success. Why, then, did school districts withdraw during the pandemic? From this author's perspective, several factors fueled this decision. First, the original intent of the model was to create a working laboratory to benefit all involved in the collaboration. Teachers would spend a year or two in the school to experiment with new ideas and test alternative methodologies before returning to their home districts to share what they had learned. The model hinged upon this concept of personnel rotation. The partnership school offered immersive professional development and curriculum innovation for teachers. In fact, this was the singular feature that attracted funders to the project, the belief that a working model could be an agent of significant school reform, curriculum and teaching innovation, and learner empowerment in the crucial STEM disciplines.

Unfortunately, the generation of school leaders who crafted and embraced this concept retired just as the facility opened. Despite the encouragement of university leaders, the planned personnel rotations did not happen. This, in turn, created a situation where the gap between the experience of those teaching in district settings and the STEM School grew ever wider. It is entirely likely that the same thing happened with student test scores. The districts' new leaders were hesitant and refused to provide internal comparative data. A model intended to be an innovative and affordable source of professional growth and renewal increasingly was perceived as disruptive.

The unwillingness of new district leaders to cooperate fully with the model developed by their predecessors by sharing information and complying with the governing model made it increasingly difficult to document positive impact with data. Qualitative information fell short of donor expectations concerned about return on investment. As a result, the foundation that itself launched the "Challenge for Change" competition withdrew its commitment to fund the second, even more expansive, phase of the project. Today the remarkable building itself is used by Aurora University natural science and health science educators with a continued emphasis on blending classroom and experiential learning.

Some of the school's teachers now hold important district leadership positions, and university faculty report that they continue to work closely with public school colleagues on smaller initiatives. One of the four districts involved in the partnership concept now has embarked upon development of its own mathematics and science academy. Most encouraging, however, is the change underway within the Chicago Public School District. Thanks to the continued leadership of the Institute for Collaboration's tireless and visionary executive director, a major new project is underway in the city's troubled North Lawndale neighborhood. Officials recently approved the funding necessary to create two Science, Technology, Engineering, Arts and Mathematics (STEAM) elementary schools and a STEAM high school. These new ventures are built upon the foundation poured in Aurora.

CONCLUSION

Climate change is an existential challenge. To paraphrase one observer, humanity is at once the source of the environmental crisis and its ultimate victim. This truth is especially clear to college-age Americans who consistently place environmental anxiety near the top of their lists of concern. (Palmer, 2024) For more than two decades between 2000 and 2023, Aurora University leaders took up the cause of sustainable development in behalf of current and future students and the communities surrounding its campuses. Their goal was to pursue several important purposes. First, they sought to improve the educational environment and the curriculum in hopes of inspiring students to create and pursues agendas of their own. Second, they evidenced their understanding of the inter-relationship between its campuses and surrounding communities by seeking mutually acceptable and beneficial solutions to challenges. Finally, thanks to these efforts, the institution established itself as a willing and intentional collaborative partner for all in the Fox River Valley of Illinois and Southeastern Wisconsin striving for the common good. The impact of this broad sustainable environmental development agenda will be measured over time as those who witnessed the transformation make decisions and take actions of their own.

REFERENCES

Louv, R. (2005). *Last Child in the Woods: Saving Our Children from Nature-Deficit Disorder*. Algonquin Books.

Louv, R. (2011). *The Nature Principle: Human Restoration and the End of Nature-Deficit Disorder*. Algonquin Books.

Louv, R. (2016). *Vitamin N: The Essential Guide to a Nature-Rich Life*. Algonquin Books.

Louv, R. (2019). *Our Wild Calling: How Connecting with Animals Can Transform Our Lives - and Save Theirs*. Workman.

Our Common Future. (1987). World Commission on Environment and Development.

Palmer, K. (2024, November 11). Recent Storms Stir Climate Anxiety Among College Students. *Inside Higher Ed.*http://www.insidehighered.com

KEY TERMS AND DEFINITIONS

Brundtland Commission (known also as World Commission on Environment and Development): Chaired by Norway's former prime minister Gro Harlem Brundtland. Organized by UN leader in 1983. Report published in 1987, known for its definition of sustainable development and identification of 3 pillars of sustainable development.

Conservation: View of early American environmentalists such as Gifford Pinchot who argued that nature and natural resources should be protected for the sake of sustainable human consumption.

Land Ethic: Concept advanced by environmentalist Aldo Leopold in *A Sand County Almanac*, published in 1949. Central argument is that humans are part of the biotic community and have an obligation to care for environment.

Preservation: Viewpoint argued by environmentalists such as John Muir. Natural areas and wilderness, he suggests, should be set apart and protected from humans. Logic behind National Park System.

Sustainable Development: Brundtland Commission definition argues for meeting needs of current generation without compromising capacity of future generations to meet their own needs.

Chapter 12
Sustainability at UC Merced

Daniel Okoli
https://orcid.org/0009-0003-1427-829X
University of California, Merced, USA

ABSTRACT

Sustainability has been at the heart of UC Merced's mission since its founding. From the outset, the campus was designed to embody a vision that balances environmental, social, and economic impacts. One of the most widely embraced definitions of sustainability comes from the Brundtland Commission's report (1987), Our Common Future, which describes it as: "Meeting the needs of the present without compromising the ability of future generations to meet their own needs." This philosophy aligns perfectly with UC Merced's approach—building a campus that not only serves today's community but also preserves resources and opportunities for those to come. UC Merced's commitment to sustainable development is evident across all aspects of campus life, making it a living example of how higher education can drive meaningful change for a more sustainable future.

INTRODUCTION

UC Merced began its journey in 2003 with the arrival of its founding faculty and graduate students. In 2005, the campus welcomed its first class of undergraduates and enrolled 875 students – graduate and undergraduate. As the newest addition to the University of California system and the nation's first 21st-century doctoral research university, UC Merced has experienced significant growth. By 2024, the campus community has expanded to more than 9,100 students, 445 faculty, and 1,625 staff members. The campus currently sits on over 1,000 acres of land of which

DOI: 10.4018/979-8-3693-5454-4.ch012

219 acres are developed and features 43 buildings covering more than 2.8 million gross square feet on its main campus and maintains presence at several off-campus locations. excluding off-campus facilities.

In 2002, UC Merced invited a panel from the Urban Land Institute (ULI) to review the campus development plans and recommend strategies to enhance sustainability for both the campus and the surrounding community. Concurrently, UC Merced commissioned UC Berkeley's College of Environmental Design to conduct an independent analysis of Phase One of the campus development, incorporating sustainability considerations. Completed in 2003, this study resulted in the 2004 release of "The Top Ten Action Items" to enhance sustainability at UC Merced.

This article will reference key recommendations from UC Berkeley's report, which advised that UC Merced should:

- Embrace a leadership role in sustainability,
- Set ambitious sustainability goals,
- Hire staff with expertise and passion for sustainability,
- Develop a sustainable land use and transportation plan,
- Pursue energy efficiency in campus buildings,
- Improve water quality,
- Create a solid waste management plan from the outset,
- Evaluate renewable energy options,
- Explore sustainable financing strategies,
- Integrate sustainability into the curriculum.

UC Merced has achieved these goals and more.

UC Merced was founded in California's Central Valley to expand educational opportunities in a region that has historically been underserved. With sustainability and accessibility at its core, the university is dedicated to addressing the area's low educational attainment levels and proudly serves a large population of first-generation college students. The campus is designated as a Hispanic-Serving Institution (HSI), Minority-Serving Institution (MSI), and Asian American and Native American Pacific Islander-Serving Institution (AANAPISI). Opportunities for students to join a higher education community centered on environmental responsibility were limited. That's why UC Merced's location was chosen with a focus on equity, recognizing that a sustainable future cannot be achieved without fairness.

In the 1980s, the San Joaquin Valley was the largest and most densely populated region in California without a UC campus. In response to growing enrollment and capacity limits at existing campuses, the UC Regents decided on May 19, 1988, to explore establishing a new campus in the area. UC Merced was founded in 2005 to expand educational opportunities in a region with limited educational resources, aiming to drive economic growth and advancement (Pelfrey, 2004)..

Merced was selected over other finalist locations in Madera and Fresno due to its central location between Fresno and Modesto, which improved statewide access to UC education. Additionally, the availability of land and water resources in Merced played a crucial role in the final decision.

Figure 1. Early campus development – 2004/2005

In a July 2001 letter, Chancellor Carol Tomlinson-Keasey (2001) outlined the goals for the new university, with sustainability as a central focus. She envisioned UC Merced as a university that would "set standards for the sustainable use of energy and other scarce resources and be a model of development in the San Joaquin Valley." Her leadership was described as "the key-enabling factor allowing the campus to embark on a path toward sustainability"(Brown et al., 2010, pp. 3–38). Since then, UC Merced has continuously set ambitious sustainability goals, achieving significant milestones as this article will demonstrate.

UC Merced's commitment to sustainability has earned it high recognition. In 2024, U.S. News & World Report ranked UC Merced among the top 30 public schools and top 60 national universities in the U.S., while the Wall Street Journal ranked it first in the nation for social mobility. Social mobility measures a university's effectiveness in serving disadvantaged students and supporting their success. Chancellor Juan Sánchez Muñoz stated, "Once again, UC Merced is being recognized as a leader in charting its own path—buoyed by a quest for knowledge and innovation that are hallmarks of the UC system, as well as its mission of facilitating access and opportunity for historically underserved populations" (Yniguez, 2024, p. 1). Equity, once again, remains central to UC Merced's mission.

As the first American research university of the 21st century, UC Merced aspires to be a model for the future of higher education. Sustainability initiatives touch all areas of the university, including curriculum, research, student engagement, operations, finance, and procurement. The following sections highlight some of UC Merced's ambitious initiatives and accomplishments.

LEADERSHIP IN SUSTAINABILITY

From the outset, UC Merced has been a leader in sustainable planning and environmental design. The campus has conserved over 26,000 acres of native vernal pool grasslands, which provide habitat for various species. The university demonstrates leadership in sustainability in three major ways:

1. **Education and Training:** UC Merced is committed to promoting social, economic, and ecological sustainability and to training global leaders in sustainability. This dedication is reflected across various fields of study and has gained global recognition.
2. **Public Service:** The university leverages its sustainability expertise to serve the local community and the broader state, using its location in the Central Valley—a critical area for sustainability efforts—to advance sustainable practices.
3. **Campus Operations:** UC Merced integrates sustainability into its planning, design, construction, and operations. The campus showcases cutting-edge sustainable practices, creating a living model of environmental stewardship.

UC MERCED'S SUSTAINABILITY POLICIES

UC Merced's vision of a sustainable campus has guided its development from the beginning. The university's approach is shaped by policies that reflect its commitment to the Triple Zero Initiative, along with sustainability principles in planning, design, construction, landscaping, and infrastructure.

The Triple Zero Initiative sets ambitious targets to achieve zero net energy, zero waste, and carbon neutrality. To meet the zero net energy goal, the campus aims to reduce energy consumption by 50%, exceed California's Title 24 energy efficiency standards by 30%, and optimize energy use according to LEED standards. The zero-waste target focuses on minimizing landfill waste through a robust system of waste reduction, recycling, and management of both construction and operational waste. The carbon neutrality objective aims to reduce emissions through various mitigation strategies, including planting trees and other carbon-offset measures.

UC Merced's sustainability policies extend into planning, design, and construction. The campus is committed to sustainable development by integrating environmental stewardship and striving for continuous improvement. All new buildings are designed to achieve at least LEED Gold certification, incorporating features like passive environmental systems, sustainable materials, and renewable energy to create a unique campus identity. Efforts to improve indoor environmental quality ensure maximum natural lighting and ventilation, while energy-saving technologies like thermal storage and solar power further reduce electricity demands. Energy monitoring systems are also installed in all new buildings to optimize operations, and water neutrality is pursued by assessing water usage and implementing reduction strategies.

Landscape and infrastructure planning play a vital role in UC Merced's sustainability strategy. The campus minimizes potable water use in landscaping by prioritizing recycled water for irrigation. Native plantings are favored to maintain compatibility with the local ecosystem, while heat island effects are reduced through extensive tree planting and shaded walkways. Additionally, roadways and parking lots are designed for effective stormwater management, further supporting UC Merced's environmental goals.

The 2009 Long Range Development Plan (LRDP) underscores UC Merced's commitment to sustainability in planning, design, construction, and operations. It aligns with the Triple Zero Initiative, striving for zero net energy, zero waste, and carbon neutrality. The plan seeks to achieve zero net energy through efficient building design and increased use of renewable energy, minimize waste to reach zero waste status, and balance emissions to attain net-zero carbon.

UC Merced's Zero Waste strategy emphasizes source reduction, recycling, and composting. This approach includes measures to minimize waste production from the start, along with well-labeled recycling stations and educational programs to promote proper recycling habits. The campus also prioritizes composting of organic waste, turning it into valuable resources and diverting it from landfills.

In a world increasingly focused on sustainability, UC Merced stands at the forefront of innovative environmental practices. The Triple Zero Initiative exemplifies the university's dedication to a sustainable future, using advanced technology, innovative strategies, and community involvement to reach its ambitious goals. This commitment sets a new standard for universities worldwide, demonstrating how cutting-edge methods and proactive engagement can lead to remarkable achievements in sustainability.

Figure 2. Zero waste-generation

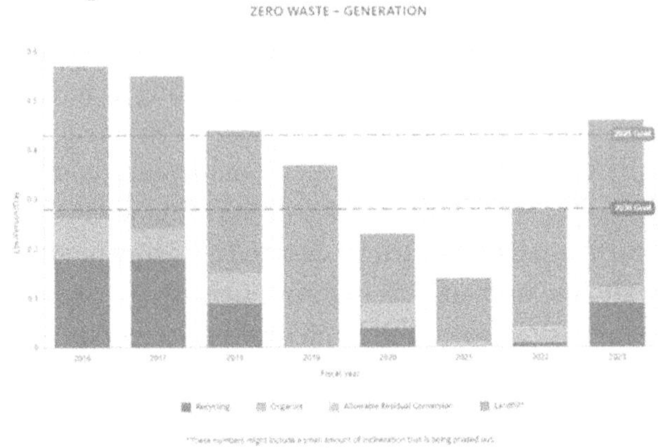

UC Merced observed an increase of 157 tons of waste sent to the landfill compared with fiscal year 2021–22. This was largely due to a return to normal business operations.

UC Merced diverted 33% of the waste generated on campus during the 2022–23 fiscal year (excluding construction and demolition waste), which represents a 2% increase in the waste diversion rate compared with the prior fiscal year. The total amount of waste diverted increased significantly, from 179 tons in 2021–22 to 269 tons in 2022–23.

Figure 3. Zero waste-diversion

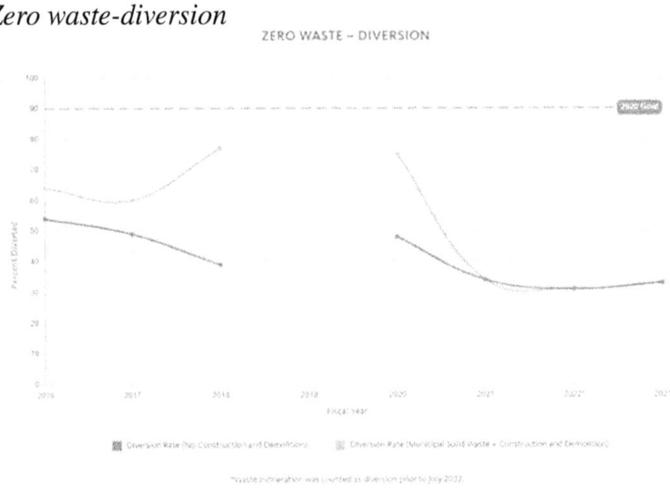

ZERO NET ENERGY CONSUMPTION

UC Merced works to achieve Zero Net Energy Consumption through a multi-faceted approach that combines energy-efficient building design, renewable energy integration, and advanced technology. The university focuses on reducing energy consumption by implementing sustainable building practices, such as energy-efficient insulation, lighting, and HVAC systems. Simultaneously, UC Merced invests in on-site renewable energy sources, like solar panels and wind turbines, to generate clean energy.

To further advance the Zero Net Energy concept, UC Merced continuously monitors and optimizes energy use through smart technologies and data analytics, ensuring efficient operation and minimizing waste. This comprehensive strategy aligns with the university's commitment to environmental responsibility, setting a sustainable example for the future.

Figure 4. Energy use intensity

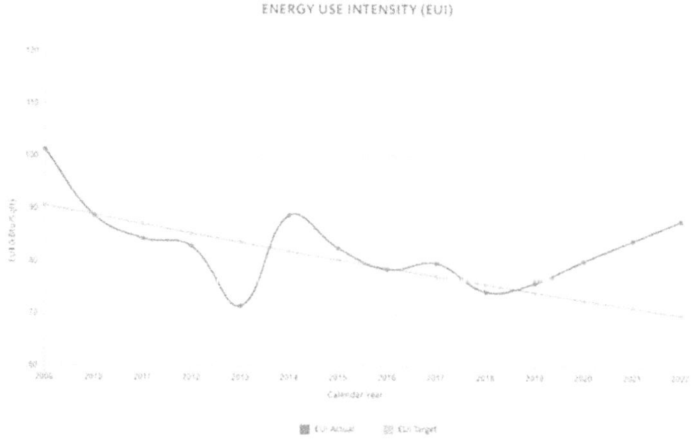

ENERGY USE INTENSITY (EUI) TRENDS AT UC MERCED

UC Merced experienced an increase in its Energy Use Intensity (EUI) in 2022. An analysis of current energy use data is underway, and one possible reason for the gap between the target and actual EUI is that the data has not yet been adjusted to reflect the energy-intensive buildings added during the 2020 Project.

In 2018, the University of California approved a policy requiring each campus to implement energy efficiency measures that reduce EUI—energy use per square foot—by an average of at least 2% annually. This goal aligns with the U.S. Department of Energy's Better Buildings Challenge.

Between 2018 and 2020, UC Merced's 2020 Project added approximately 1.2 million square feet of new buildings, including over 350,000 square feet of high-energy-use laboratory space. These new buildings have a combined policy-driven EUI target of 111 kBtu/sq. ft. The graph above displays the annual EUI targets and actual energy usage across all UC Merced buildings, including offsite and leased facilities. The divergence from the target EUI began in 2019, coinciding with the commissioning of these high-energy-use labs.

It is now clear that EUI targets need to be periodically adjusted to account for the energy demands of newly constructed buildings or changes in existing building use. Preliminary data suggests that the normalized EUI target for 2023 should be approximately 98 kBtu/sq. ft., making the actual EUI of 90 kBtu/sq. ft. compliant with UCOP Energy Policy.

ZERO NET GREEN HOUSE GAS EMISSIONS

UC Merced's Triple zero initiative sets a remarkable goal - achieving zero net greenhouse gas emissions. This means that the campus is committed to balancing the emissions it produces by reducing, offsetting, and adopting eco-friendly practices. Its aim is to minimize its carbon footprint and strive for a balanced state where we make as little impact as possible on the environment.

This commitment goes beyond words; it's a profound promise to combat climate change and lead in sustainability. UC Merced is pioneering innovative solutions, encouraging education, and fostering collaboration to redefine what it means to be truly green in the 21st century. Following UC Merced's achievement of carbon neutrality in 2018 as the first research university to do so, it has increasingly embraced its brand as a leader in sustainability in higher education.

Figure 5. Climate protection-emissions

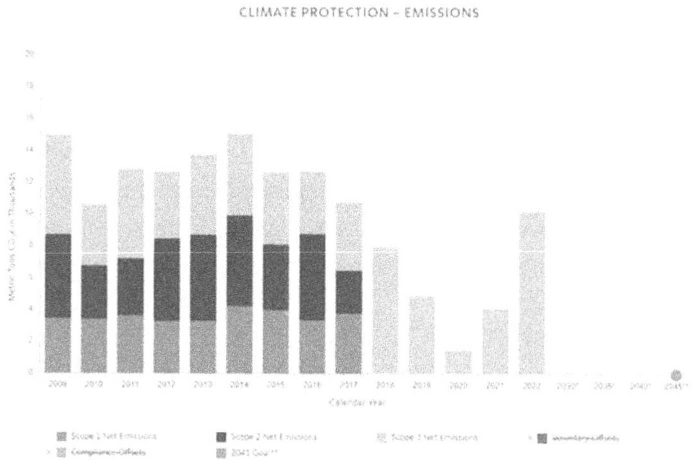

UC MERCED'S CARBON NEUTRALITY AND FUTURE SUSTAINABILITY EFFORTS

UC Merced remains a carbon-neutral campus for Scopes 1 and 2 emissions, with a renewed focus on investing in projects that support direct emission reductions in current fossil fuel systems. Planning is also underway to assess the electrification of future campus developments.

Natural Resource Attributes

The campus's ability to achieve its triple zero initiative goals relies on harnessing a variety of natural resources:

- **Solar:** Solar energy can be used for heating, lighting buildings, heating water, and generating electricity. In Merced County, average solar power potential ranges from 5.6 to 6.0 kW/m^2 per day, with the highest output occurring between March and October. (Source: Renewable Energy Atlas of the West, 2006)
- **Wind:** Wind turbines can capture intermittent wind energy, which can be relatively efficient. In Merced County, wind availability varies from 0 to 400 W/m^2. (Source: Renewable Energy Atlas of the West, 2006)

- **Biomass:** UC Merced is situated near some of the world's most fertile agricultural land. Currently, portions of campus property are used for livestock grazing to support organic milk production. Agricultural activities produce significant residues, which could be leveraged for energy production. Within a 30-minute radius of the campus, there is an untapped energy potential of 500,000 to 11,200,000 MMBtu, among the highest in the state.
- **Geothermal**: Geothermal energy, generated from heat stored beneath the Earth's surface, is limited in the San Joaquin Valley. However, geothermal heat pumps remain a viable resource throughout the region.

On-Site Solar Installations

UC Merced has installed on-site solar that provides approximately 25% of the campus's annual electrical energy needs and include:

- **1 MW PV Array – ground mounted system:** Installed in 2009
- **Additional 4.2 MW PV carport and 180 KW Systems**: Installed in 2018

The UC Clean Energy Program supplies the remaining 75% of on-campus electrical energy needs emission-free through the grid.

Figure 6. Energy-renewable energy use

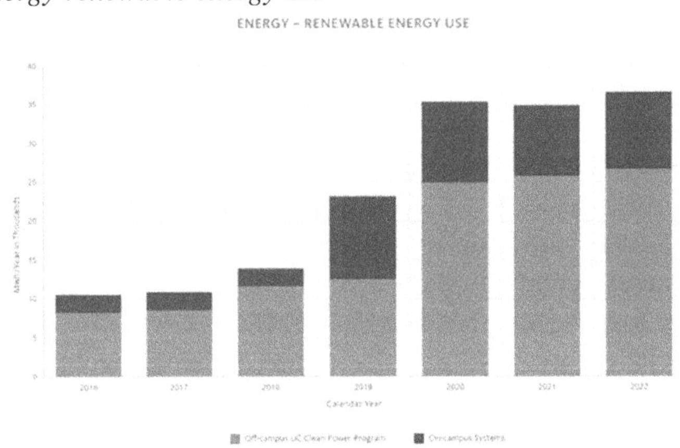

UC MERCED'S DECARBONIZATION INITIATIVE

Achieving the UC Carbon Neutrality Initiative and reaching net-zero greenhouse gas emissions without offsets will require accelerating the decarbonization of UC Merced's energy systems—an effort that is both technically challenging and capital-intensive. To address these challenges, UC President Michael Drake established a dedicated Task Force, and UC Merced initiated a decarbonization study as part of the Global Climate Leadership Council (GCLC) to guide emission reduction strategies.

The study's recommendations aim to phase out fossil fuel use on campus, strengthen climate action policies, and foster a climate-conscious culture. The initiative's deliverables include:

- A comprehensive plan to reduce greenhouse gas emissions by 90% or more by 2050, based on 2019 levels, with interim targets and decision support for decarbonization efforts.
- Cost estimates to inform funding requests, address climate justice issues, identify knowledge gaps for Net-Zero planning, and support climate and resiliency planning in academic contexts.
- A focus on climate justice, equity, and effective communication strategies as essential components of the decarbonization process.

Information on costs, savings, environmental justice considerations, uncertainties, and engagement strategies will be crucial for funding requests to transition the campus away from fossil fuels. Net-Zero planning must incorporate emission reduction goals, institutional constraints, risk management, and adaptability, with a focus on

UC Merced 2020 Project

The **2020 Project**, the largest social infrastructure project in the nation, expanded UC Merced's campus by 1.2 million gross square feet at a cost of $1.3 billion. This project provided essential research facilities, administrative buildings, student amenities, and residential halls to support UC Merced's growth.

The student population grew from 1,300 in 2007 to 6,300 in 2012, resulting in limited academic space, strained infrastructure, and a shortage of on-campus housing for students. The 2020 Project addressed these needs, enabling UC Merced to better accommodate its expanding community.

Figure 7. Growth over time at UC Merced

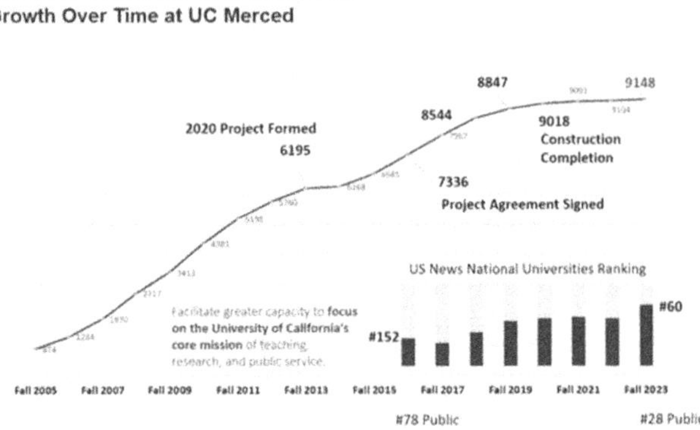

UC Merced hired the Urban Land Institute (ULI) to gather stakeholder input, assess financial readiness, and evaluate the campus's overall space and infrastructure capacity. ULI recommended adopting a strategy to design and construct sustainable facilities using optimal methods and construction types, in the shortest possible timeframe. This recommendation led to the **2020 Project**, a landmark initiative in higher education.

Development of the 2020 Project

Launched in 2012, the 2020 Project aimed to achieve four specific goals:

1. **Address Space Constraints:** The project sought to resolve space shortages across multiple facility types, including classrooms, residence halls, research spaces, athletic facilities, and student life areas. Traditional, incremental construction was insufficient to meet UC Merced's urgent space needs.
2. **Meet Urgent Timelines:** UC Merced needed to rapidly expand to meet established enrollment targets.
3. **Develop Comprehensive Infrastructure:** A holistic plan was needed to support housing, athletics, and academic areas.
4. **Avoid Deferred Maintenance:** By leveraging economies of scale, the project aimed to finance facilities based on life-cycle costs, thereby reducing deferred maintenance.

The 2020 Project utilized a DBFOM (Design, Build, Finance, Operate, and Maintain) strategy, partnering with a developer responsible for hiring architects, contractors, and operations/maintenance sub-consultants. This integrated delivery model ensured accountability for performance and shared risks related to finance, scheduling, and maintenance between the developer and the campus. The project spanned 136 acres of undeveloped land, doubling the campus's size by adding over 1.2 million gross square feet.

The project team, led by Plenary Partners Merced, included five prominent architecture firms, an infrastructure and sustainability consultant, a landscape architect, and a general contractor. The first phase was completed on schedule and within budget in 2018, with all 13 buildings achieving LEED certification.

The Public-Private Partnership (P3) approach enabled UC Merced to achieve necessary space expansion sustainably, with a financing model that accounts for life-cycle costs and reduces deferred maintenance for 30 years—an important factor in maintaining building efficiency as they age.

SUSTAINABILITY IN CURRICULUM

UC Merced integrates sustainability into its curriculum through courses, majors, and minors. In fact, a degree-granting environmental program is housed in each of the campus's schools: the Environmental System Science major in the School of Natural Sciences, the Environmental Engineering major in the School of Engineering, and the Environmental Humanities major and minor in the School of Social Science, Humanities and Arts. It is important to note that the Environmental Humanities program, which launched in Fall of 2024, is the first of its kind in the University of California system. This program aims to train the next generation of environmental communicators with career outcomes including science writers, environmental lawyers, governmental and non-governmental organization workers, and interpretive park rangers. In addition, there are more than 200 sustainability-related courses that students can take at UC Merced.

UC Merced operates as a "living laboratory", with all major buildings LEED-certified, showcasing green building practices. The LEED Laboratory course uses campus facilities as a teaching tool, where students conduct LEED certification audits of existing buildings. Students completing the course can take the LEED certification exams for professional accreditation. The Campus Community Garden also serves as a living lab, allowing students to explore sustainability through research and hands-on projects. UC Merced students have done class projects, carried out research or completed term papers on campus sustainability or how their unique lived experiences may shape their perspectives on sustainability as first generation or under-represented students (Wasson, 2022).

Food Initiatives

Food is a crucial aspect of the Central Valley's agricultural landscape, but food security remains a significant challenge for the UC Merced community. Food programs not only address food security and social justice concerns but also help reduce food waste.

The Global Food Initiative at the University of California aims to provide sustainable, nutritious food for a growing global population. It disseminates best practices, uses UC research to support communities, and influences food policy discussions at local, state, national, and international levels.

At UC Merced, programs like the Nutrition Education Strategy, Sustainable Food in Campus Dining, and Bobcat Pantry & CalFresh provide food assistance to students, staff, and faculty. Collaborating with the USDA and the Merced County Food Bank, the university offers a monthly food distribution program at no cost.

The Bobcat Eats Food Waste Awareness and Prevention Program has rescued over 2.3 million pounds of food from June 2018 to December 2021, providing nearly 2 million meals and reducing CO_2 emissions by over 2,400 metric tons. This effort engaged over 120 volunteers and student workers, with support from 60 food donors and 50 distribution locations.

The No Food Left Behind program uses a free texting service to connect the campus community to food resources. In 2022, 933 individuals registered, receiving an average of five weekly messages.

The Campus Community Garden, part of UC's Global Food Initiative, serves as a learning lab for sustainable food production and supplies fresh produce to those in need. Established in 2014, the garden addresses food security, health, and sustainability. Students and volunteers participate in weekly garden activities, contributing to the broader mission of food equity and sustainability.

Figure 8. Food

SUSTAINABLE FOOD PURCHASES

In the 2022-23 fiscal year, 24% of food and beverage purchases, totaling $1.7 million, met sustainability standards, and an equal percentage of purchases were plant-based. Although there was a slight decline compared to the previous year, UC Merced maintains a strong commitment to sustainable and plant-based food procurement. The decline follows the university's achievement of exceeding the systemwide goal of 25% in the 2021-22 fiscal year.

Student Involvement

Beyond the classroom, UC Merced students actively engage in campus sustainability efforts through various internships:

- **Green Office Interns**: These interns help certify offices for environmental performance, manage action plans, and enhance sustainability guides.
- **Garden Interns**: They oversee the management and maintenance of the campus garden, ensuring its smooth operation.
- **Green Lab Interns**: They verify labs' environmental performance, manage green lab action plans, and enhance sustainable lab guides.
- **Marketing Interns**: These interns manage social media, digital media, and the website to promote sustainability initiatives.
- **Special Project Interns**: These interns focus on sustainability research projects that align with their specific interests.

Research

Located at the intersection of global challenges related to water, energy, food, and the environment, UC Merced is dedicated to addressing these issues through faculty and student research. The campus institutes and research groups collaborate on complex topics such as energy efficiency, natural resource preservation, and environmental sustainability.

- **The Sierra Nevada Research Institute (SNRI)** studies natural resources, air, water, soil pollution, and climate change in the San Joaquin Valley and Sierra Nevada, aiming to provide solutions to environmental challenges.
    - **The Center for Climate Communication**, part of SNRI, analyzes and shares data to inform the public about climate and environmental issues. One project focuses on language use to better understand how

people discuss and perceive climate change, shaping effective policy communication.

    - **The Mountain Hydrology Research Group**, another SNRI initiative, conducts collaborative research in the Sierra Nevada and other western mountain regions. Their work analyzes water and biogeochemical cycles in forests, alpine areas, meadows, and streams.

- **UC Solar**, a multi-campus research institute based at UC Merced, focuses on developing advanced solar technologies, as well as studying the economic and policy implications of solar energy.

UC Merced's research community actively pursues projects aimed at advancing sustainability across various fields, from environmental sciences to renewable energy.

Four UC Merced researchers have received grants as part of a groundbreaking collaboration between the University of California and the State of California. These initiatives aim to:

- **Improve community readiness** for emergency evacuations using zero-emission vehicles.
- **Develop climate-resilient seed selection tools** to support reforestation efforts.
- **Create equitable water planning tools** for more sustainable water management.
- **Enhance support for maximizing the benefits** of California's methane reduction policy.

These projects were funded in response to a call for innovative solutions to address California's climate challenges.

Water Conservation

As the campus continues to expand, water usage has increased. To promote conservation, the university actively engages with students, raising awareness of methods they can use to reduce water consumption on campus.

Figure 10. Water

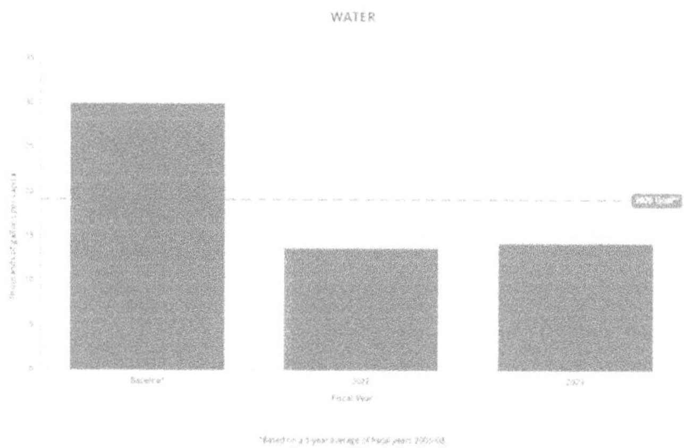

Transportation

Currently, 49% of UC Merced employees use alternative transportation methods for commuting. The campus has 34 electric vehicle charging stations, and 50% of all vehicles purchased in 2023 were either electric, plug-in hybrid, or eco-friendly.

Scope 3 emissions saw a significant increase, primarily due to improved data collection for more accurate calculations. Continued evaluation of current and past years' emissions will be crucial to ensure comprehensive accounting of all Scope 3 travel and commuting emissions.

Procurement

In the current reporting period, UC Merced made significant investments in sustainable products:

- **$324,000** spent on eco-friendly electronics (49% of total electronics spend)
- **$311,000** on environmentally friendly cleaning supplies (77% of total cleaning supplies spend)
- **$251,000** allocated for green indoor office furniture
- **$32,000** used for green office supplies (69% of total office supplies spend)

The University collected data on green spending in four product categories, following the Sustainable Procurement Guidelines, and reached out to suppliers for more detailed information. Green spending is defined as achieving basic or desired standards outlined in the UC Sustainable Procurement Guidelines. Suppliers provided data for electronics (5 suppliers), furniture (2), cleaning supplies (4), and office supplies (3). CalUSource offered category-specific data for UC Systemwide Spend Analytics.

Achievements in Sustainability

UC Merced continues to set ambitious sustainability goals, such as the Triple Zero Initiative, which aims for zero net energy use, zero waste, and zero net greenhouse gas emissions by 2020. While UC Merced became the first research university to achieve carbon neutrality in 2018, ahead of the 2020 target, efforts toward achieving zero waste and zero net energy use are ongoing.

The university has also embarked on a decarbonization strategy to create a fossil-free campus. Water conservation measures have been implemented in new construction, and plans are underway to enhance water management, including the use of greywater where appropriate.

The campus's location in Merced County, about two miles from the city of Merced, has implications for its carbon footprint. Unlike urban universities with closer access to amenities, UC Merced faces challenges in reducing vehicle trips. However, the recent annexation of the campus creates opportunities for development at its edge, enhancing student amenities and encouraging alternative transportation. Housing closer to campus will reduce the need for car trips and foster a more traditional campus environment.

UC Merced has made strides in addressing food security, integrating sustainability into the curriculum, and conducting extensive sustainability research—all aligned with its vision to be a model sustainable campus. The 2020 Project and other Public-Private Partnerships (P3), along with Power Purchase Agreements (PPA) for renewables, reflect the university's commitment to financial sustainability.

Recognitions and Awards

UC Merced's efforts have earned numerous awards, maintaining its status as a Platinum-rated institution under the Association for the Advancement of Sustainability in Higher Education (AASHE). In 2023, the campus received the Urban Land Institute's Americas Awards for Excellence in the Master Architect category for its significant infrastructure projects. The awards recognize excellence in areas such as design, planning, economic impact, management, community engagement, innovation, and sustainability. All major campus buildings have achieved LEED Platinum or LEED Gold except one.

REFERENCES

Brown, K., Daly, A., Engineering, T., Elliott, J., Higgins, C., & Granderson, J. (2010). *Hitting the Whole Target: Setting and Achieving Goals for Deep Efficiency Buildings*.

Brundtland, G. H. (1987). *Our common future world commission on environment and developement*. https://docenti.unimc.it/elena.borin/teaching/2019/21812/files/lessons-and-recommended-readings-1/Our%20Common%20future%201987.pdf

Pelfrey, P. A. (2004). *A brief history of the University of California*. Univ of California Press. https://books.google.com/books?hl=en&lr=&id=qbEwDwAAQBAJ&oi=fnd&pg=PT7&dq=a+brief+history+of+the+university+of+california&ots=lR0pGj7Avw&sig=ownjKS6d9rw75OfiARCcohJtJqM

Tomlinson-Keasey, C. & Merced. (2001). *Cover Letter to the Draft Long Range Development Plan: Public draft*. The University.

Wasson, A. (2022). UC Merced Through a Socio-Environmental Imagination Lens. *UC Merced Undergraduate Research Journal, 14*(1). https://escholarship.org/uc/item/0v48z732

Yniguez, S. (2024). *U.S. News Places UC Merced Among Top 30 Public Schools Following Top Social Mobility Ranking By Wall Street Journal | Newsroom*. https://news.ucmerced.edu/news/2024/us-news-places-uc-merced-among-top-30-public-schools-following-top-social-mobility-ranking

Chapter 13
Sustainability at Connecticut's Community Colleges:
Protecting the Environment Starts Locally

Antony L. Wormack
Naugatuck Valley Community College, USA

ABSTRACT

This chapter reviews recent environmental initiatives across Connecticut's 12 community college campuses and explores future opportunities as the newly formed Connecticut State Community College, established in July 2023 through the merger of the 12 campuses. It examines a range of sustainability efforts, historically rooted in environmental stewardship. These include academic programs, campus-wide green initiatives, hands-on learning, and community partnerships. Even prior to the merger, campuses collaborated through the Connecticut State Colleges and Universities System Office to implement statewide sustainability projects, achieving cost savings and operational efficiencies. More recently, the college has expanded its focus to include the United Nations' 17 Sustainable Development Goals, which encompass environmental, social, and economic dimensions of sustainability. In this context, the colleges have begun addressing issues like poverty, public health, employment, and education. As with all institutions, CTState faces several challenges that hinder progress including limited funding, institutional resistance, low student engagement, and stakeholder awareness. The chapter discusses each of these barriers and potential strategies to overcome them. As a new institution, Connecticut State Community College must adopt a systematic, statewide approach to sustainability leveraging important opportunities to scale energy efficiency initiatives, share ac-

DOI: 10.4018/979-8-3693-5454-4.ch013

ademic programs across campuses, and foster stronger ties with state universities.

INTRODUCTION

America's community colleges serve a critical role in higher education in our country. Their missions are comprehensive, serving the workforce development needs of local economies while offering the promise of gainful careers to local residents. Most community college students live and work in the communities served by their local college, and continue to make important contributions to their employers, families, and local communities after graduation. Community colleges also offer seamless transfer to public universities and specific articulation agreements with public and private institutions, allowing students to start their college education at their local community college, eventually graduating with a four-year degree from their school of choice. With non-credit and short-term training options, community colleges further extend their educational reach into their local community.

Sustainability, environmental conservation, and climate change are pressing global issues that require coordinated efforts across all sectors of society. Higher education institutions, including community colleges, play a critical role in addressing this challenge. Environmental stewardship and other issues of sustainability and community resilience occupy much of the consciousness of the higher education community and will continue to do so for decades to come. These institutions are increasingly integrating sustainability into their missions, curricula, operations, and community engagement strategies.

The history of Connecticut's 12 community colleges — as separate entities for decades and most recently as a single statewide community college — serves as a useful case study in how community colleges can address the global issue of sustainability by taking actions locally and in collaborations that can extend statewide. This chapter explores the various sustainability initiatives undertaken by Connecticut's community colleges, highlighting their impact, challenges, and future directions.

BACKGROUND

In 1965, the Connecticut General Assembly created the Connecticut Community College System, combining several existing campuses and future planned construction under a Board of Trustees separate from the four state universities and the University of Connecticut. Eventually the system included 12 campuses/separate colleges (Asnuntuck, Capital, Gateway, Housatonic, Manchester, Middlesex, Norwalk, Naugatuck Valley, Tunxis, Three Rivers, Quinebaug Valley and Northwestern). The system was to remain largely unchanged for 46 years.

In 2011-12, to support curricular alignment, promote student transfer and success, and achieve economies of scale, the 12 community colleges, the four state universities (Southern, Central, Western and Eastern) and Charter Oak College — the state's online college — were combined into the Connecticut State Colleges and Universities System (CSCU). With one Board of Regents, greater operational efficiency, closer campus collaborations, and increased financial support from the legislature were all envisioned.

In July 2023, the organizational structure of the state's higher education community was further refined, when the 12 community colleges became one community college with 12 campuses. Connecticut State Community College (CSCC) was born. (Connecticut State Colleges and Universities (CSCU), 2025)

As this chapter will discuss, in addition to significant sustainability efforts taking place at the 12 campuses, system-wide economies of scale, i.e. purchasing green electricity as a system, coupled with the potential of further statewide collaborations, give CSCC unprecedented opportunities to make an impact in the sustainability arena. The author hopes that this chapter can cast some light on the current achievements and future sustainability opportunities of the CSCC.

What follows begins with a review of recent past and present sustainability initiatives taking place on the 12 campuses of the new Connecticut State Community College. This discussion is centered on the historical definition of sustainability with its focus on environmental stewardship.

The chapter proceeds to a review of various statewide strategies that use economics of scale to achieve system level cost savings and energy efficiencies.

At this point the author will examine the opportunities presented by the United Nations' expanded definition of sustainability to include social justice and economic equity in addition to environmental considerations.

While Connecticut's community colleges and the combined single college — Connecticut State Community College — are making a significant contribution to the state's sustainability efforts, there are a number of challenges that must be addressed for CSCC to realize its vision to be fully responsive to the UN's Sustainable Development Goals. Some of those challenges are examined and possible solutions are advanced.

The final section of the chapter will look at the future, both in terms of sustainability planning currently taking place within the Connecticut State Community College and the larger Connecticut State Colleges and Universities System, as well as opportunities that exist for further development and innovation.

ENVIRONMENTAL STEWARDSHIP ON CONNECTICUT'S COMMUNITY COLLEGE CAMPUSES

In Connecticut, community colleges are doing a great deal to preserve the natural environment, reduce carbon emissions, and advance community resilience, social justice and economic equity. As the colleges have grown in scale over time, they have also prided themselves on serving students with quality educational programs and services.

In July of 2022, Wallet Hub published its findings, announcing that Connecticut had the best community college system in the country. With the merger of 12 community colleges into one Connecticut State Community College (CSCC), the resulting institution is the largest single community college in New England and the sixth largest in the nation. More than 8,000 degrees and other credentials are awarded annually to a student population of 64,000 students. (CSCU, 2025) This scale provides exciting opportunities for the college to collaborate across its 12 campuses and in partnership with the state universities and other statewide organizations to have even greater impact.

In this section of the chapter, the author will look at recent and current successful practices taking place on the 12 campuses of the Connecticut State Community College, breaking them down by Academic Initiatives, Campus Operations, and Community Partnerships and Outreach.

Academic Initiatives

Connecticut's community colleges have added sustainability to their curricula to prepare students for a world where climate considerations are paramount. This integration occurs in several ways: (1) dedicated programs (2) infusing sustainability into other academic disciplines, and (3) hands-on learning opportunities.

Dedicated Programs and Courses

A number of the community colleges offer specialized programs in environmental science, sustainable agriculture, and renewable energy. For instance, the Three Rivers campus offers a certificate in Environmental Engineering Technology: Environmental Health and Safety Management and an associate's degree program in Environmental Engineering Technology. Gateway offers a certificate and an associate's degree in Environmental Engineering Technology: Environmental Health and Toxicology. Naugatuck Valley offers a certificate and an associate's degree in Environmental Science: Environmental Biology. And Northwestern offers a certificate and an associate's degree in Environmental Science: Natural Resources. Manchester and Northwestern offer an associate's degree in Environmental Science: Sustainability.

Other programs that target environmental stewardship include Clean Water Management at Gateway, Water Management at Gateway, and Public Utility Management at Gateway. All 12 campuses offer a transfer associate's degree in Biochemistry.

Naugatuck Valley offers certificates and associate's degrees in horticulture and a certificate in landscape design. These programs are designed to prepare students for careers in sustainable agriculture, landscape design, and environmental stewardship, all while emphasizing hands-on learning through direct interaction with the college's diverse plant collections and greenhouses. These programs benefit from the campus having distinctive land holdings that provide students with significant hands-on experiences. Naugatuck's campus is renowned for its extensive and specialized horticultural gardens. The campus features a Sustainability Garden, a Rain Garden, and other themed gardens such as the Medicinal Garden and the Shakespeare Garden. These gardens serve as living laboratories where students and community members can engage with sustainable gardening practices and learn about various plant species and their ecological roles. The gardens are part of Naugatuck Valley's broader commitment to using sustainable designs and landscaping techniques (Naugatuck Valley Community College American Rose Trials for Sustainability).

Another notable feature on the Naugatuck Valley campus is the Tamarack Arboretum, which includes over 300 different species of trees and shrubs. This arboretum, along with the college's newly renovated hiking trails, provides a natural space for students and the public to explore and learn about native and non-native plant species, promoting biodiversity and conservation efforts in an urban setting (American Rose Trials for Sustainability).

Three Rivers offers a Nuclear Engineering Technician two-year associate's degree. In 2022, while still operating as Three Rivers Community College, the college's foundation and Millstone Nuclear Power Station announced the creation of a scholarship program to support the two-year associate degree. Sixteen full student scholarships are being funded for the program, which is the only one in New England and considered the best in the nation. In addition to scholarships, students in the program also participate in a three-month paid internship at Millstone and can transfer their associate degree credits to appropriate four-year bachelor's degrees. (CSCU, 2022)

Three Rivers' Energy Management and Environmental Engineering Technology degrees are two other programs that are designed to equip students with the knowledge and skills necessary to address the challenges of climate change and work in the growing green economy. By embedding sustainability into its curriculum, Three Rivers ensures that students are prepared to lead in fields related to environmental protection and energy efficiency.

Infusion into Existing Disciplines

Sustainability principles are being woven into a variety of CSCC academic disciplines. For instance, a CSCC program may include modules on environmental impact in business courses or discuss sustainable engineering practices in technical programs. This approach ensures that all students, regardless of their major, understand the importance of sustainability.

One example is the addition of electrical vehicle maintenance to Gateway's automotive technology program. In July 2024, it was announced that the National Science Foundation had awarded a three-year, $350,000 grant to the Connecticut State Community College-Gateway campus to support electrical vehicle technicians, with funds to support both students and local high school teachers. The grant will allow Gateway to develop and implement modernized hybrid and electric vehicle (EV) curriculum, including courses on vehicle safety, maintenance, diagnosis, and repair. These courses will lead to stackable credentials within existing programs, enhancing students' employability and meeting the National Institute for Automotive Service Excellence (ASE) electric vehicle technician/shop personnel electrical safety standards. (CSCU, 2024)

Hands-On Learning Opportunities

Experiential learning is emphasized through projects, labs, and fieldwork. The use of living labs — such as campus gardens or renewable energy installations — provides students with practical experience while advancing the institutions' sustainability goals.

At Three Rivers, the student Environmental Club maintains a pollinator garden as part of the U.S. pollinator movement. The campus also features three courtyards, where native varieties are planted to help counter the threat of invasive species. Support is also provided by students from the University of Connecticut's Master Gardener Program. In addition, the Environmental Club at Three Rivers conducts environmental projects such as a wildlife inventory to support the local community. At Naugatuck Valley, which offers certificate and degree programs in Horticulture, students also get the opportunity to work in the campus's greenhouses and gardens. Similarly, at Asnuntuck, the student Garden Club manages the campus's community garden.

Sustainable Campus Operations

Connecticut State Community College (CSCC) is adopting a number of sustainable practices in its campus operations to reduce the carbon footprint. Key areas of focus include:

Energy Efficiency and Renewable Energy

CSCC is investing in energy-efficient infrastructure and renewable energy sources on a number of fronts. All CSCC campuses have worked hard to upgrade and make their HVAC systems more energy efficient. In addition, all the campuses have either installed and/or are in the process of installing energy-efficient LED lighting throughout their buildings.

The Norwalk campus has made significant efforts to enhance energy efficiency. The campus has upgraded its facilities with energy-efficient lighting, heating, ventilation, and air conditioning (HVAC) systems. These upgrades are designed to reduce energy consumption and lower the college's overall carbon footprint. Additionally, Norwalk has implemented green building practices in the construction and renovation of its facilities, ensuring that new buildings meet or exceed standards for energy efficiency and environmental sustainability.

In the case of Three Rivers, the campus replaced an HVAC maintenance contractor with a full-time employee who can provide just-in-time service for the air conditioning, heating and ventilation systems on campus. Overall maintenance and utility bills have actually gone down as a result.

In summer 2022, Gateway Community College embarked on a solar energy project on the rooftop of its downtown New Haven parking garage. Seven hundred and thirty-two photovoltaic panels were installed to generate 400 kw, with a savings of $600,000 over 25 years. The project reflects Gateway's commitment to the environment, evidenced by the fact the campus is gold certified by the Leadership in Energy and Environmental Design (LEED) certification program of the U.S. Green Building Council. The certification reflects the campus's commitment to sustainability in building design and operation. The campus incorporates energy-efficient systems, sustainable materials, and design features that reduce energy consumption and minimize the carbon footprint of the facility. The building's features include energy-efficient lighting, advanced HVAC systems, and water conservation measures, all of which contribute to the college's overall sustainability goals. (CSCU, 2022)

Another campus committed to U.S. Green Building Council guidelines is Housatonic. For instance, the campus's Lafayette Hall expansion project included energy-efficient designs and materials aimed at reducing energy usage and environmental impact. In addition, an energy audit at Housatonic uncovered more than 50 strategies for reducing energy costs.

At Asnuntuck, motion sensors help manage electric lights and sensor faucets and toilets help to conserve water usage.

Connecticut State Community College has installed or has plans to install similar solar PV projects on the ground, on building roofs, and on parking garages across the state's 12 community college campuses. Some campuses already have the photovoltaic installations in place. Norwalk's solar panel installation contributes to students' education by providing students with hands-on learning opportunities with renewable energy technologies. Quinebaug Valley's solar presence is a visible demonstration of the campus's commitment to renewable energy, and the data on energy production is made available to the public, showcasing the tangible benefits of renewable energy. Asnuntuck has solar panels on two buildings with more installations planned.

The Northwestern campus conserves energy and reduces carbon emissions by making sensible use of its buildings. On Fridays, all buildings but one are closed, and it is used for the day's classes. Employees work remotely on Fridays, maintaining operations while lowering energy costs.

Waste Reduction and Recycling

Efforts to minimize waste and enhance recycling programs are common across the CSCC. Campuses have implemented a variety of programs to increase recycling and reduce waste. At Three Rivers, to reduce food waste, the campus cafeteria has gone tray-less and is investigating the replacement of plastic utensils with wooden utensils. Recycling programs are also in place on each campus to recycle plastics, metal, glass and paper recyclables.

The Norwalk campus is typical of what is taking place across the CSCC. The campus has implemented a comprehensive recycling and waste reduction program. It promotes recycling of paper, plastics, and other materials, with designated recycling bins located throughout the campus. In addition, Norwalk encourages the reduction of waste through various initiatives, such as promoting the use of reusable water bottles and reducing paper usage by encouraging digital communication.

The Capital campus in Hartford also has implemented a comprehensive recycling program to reduce waste on campus. Recycling bins for paper, plastics, and other materials are strategically placed throughout the campus to encourage proper disposal and recycling. In addition to recycling, Capital promotes waste reduction practices such as minimizing the use of single-use plastics and encouraging the use of reusable containers.

Water Conservation

In addition to academic programs such as Gateway's Clean Water Management Certificate and Water Management Certificate, campuses are also focusing on their own water usage. The Middlesex campus, for example, has implemented a water conservation program to use low-flow water fixtures across campus, while Asnuntuck uses sensor faucets to control water usage.

Sustainable Transportation

One major sustainable transportation project taking place within the Connecticut State Colleges and Universities System is the ongoing installation of 138 electric vehicle charging stations across system campuses; about half will be installed at the community colleges. The project's total cost is $1.3 million. Approximately $700,000 is funded from grant funds stemming from the federal emissions settlement with Volkswagen and the other portion is being funded from utility company rebates and CSCU project funds. This is one of many examples of how the CSCC can leverage its membership in the state college and university system to receive benefits that would not necessarily be available to individual, local campuses. (CSCU, 2017)

COMMUNITY ENGAGEMENT AND PARTNERSHIPS

Connecticut's community colleges often serve as hubs for local sustainability efforts, engaging with their communities to foster broader climate action.

Local Partnerships

Connecticut's community college campuses collaborate with local governments, businesses, and non-profits to promote sustainability. These partnerships can lead to joint initiatives such as community clean-up events, environmental education programs, and local sustainability fairs. One creative project at the Housatonic and Norwalk campuses involves hosting events for the annual Haitian Book and Art Festival. Sponsored by the Haitian American Professional Association of Connecticut, the festival is in its 25th year.

The Capital campus is actively involved in engaging with the Hartford community to promote sustainability. The college hosts events, workshops, and seminars that focus on various aspects of sustainability, including climate change, energy conservation, and sustainable living practices. These events are open to students, faculty, and the general public, providing opportunities for education and collaboration on environmental issues. Capital's partnerships with local organizations further enhance its ability to contribute to community-wide sustainability efforts.

On the Three Rivers campus, when nonprofits meet on campus, facility rental fees are often waived. In addition, the campus provides dedicated space to a nonprofit from the local community to offer English-as-a-Second-Language courses. This is in response to local conditions wherein 29 languages are spoken in the Norwich community. People learning English are then in a position to take on-the-job training to become gainfully employed as the campus furthers workforce development for the local community.

Student and Faculty Involvement

Students and faculty are actively involved in sustainability projects and advocacy across the Connecticut State Community College, ranging from Earth Hour to Climate Action Week. Campus organizations, such as environmental clubs or sustainability committees, often drive these efforts. For instance, Quinebaug Valley is involved in the annual Earth Hour event, where non-essential lights are turned off for one hour to raise awareness about climate change. This worldwide event connects Quinebaug students, faculty, and staff to people across the globe to combat climate change. Other campuses engage in beautification projects, food pantries, and other community activities. Three Rivers frequently convenes its Safety and Environmental Services Committee to provide planning and campus participation.

Norwalk students participate in projects that promote sustainability on campus, such as organizing recycling drives, advocating for energy conservation, and participating in environmental advocacy events. These activities provide students with practical experience in sustainability and empower them to make a positive impact on the environment.

Students at the Asnuntuck campus lead a number of sustainability initiatives through student clubs and other organizations. These student-led groups organize activities such as campus clean-up drives, awareness campaigns, and sustainability projects that contribute to the campus's overall environmental goals.

Public Outreach and Education

CSCC campuses host public lectures, workshops, and seminars on climate change and sustainability. By providing a platform for expert discussions and community dialogue, these institutions help to raise awareness and encourage local action. All campuses conduct outreach activities, including sustainability advocacy and promotion of related campus courses and events. Middlesex, for instance, hosts events, workshops, and seminars on topics related to climate change, energy efficiency, and sustainable living. These events are designed to educate both the campus community and the broader public about the importance of sustainability and the role that individuals and organizations can play in combating climate change.

SYSTEM-LEVEL PROJECTS YIELD LOWER ENERGY USAGE AND COSTS

While individual campuses within the CSCC must manage much of their budgets, the 12 community college campuses now benefit from being one single Connecticut State Community College. In turn, the college is part of the Connecticut State Colleges and Universities (CSCU), which includes the 12 community college campuses, Charter Oak College — the state's online institution — and the four state universities. For years, the CSCU has worked with public utility companies, other contractors, and the individual campuses to find opportunities for collective cost savings and reduced energy consumption.

Solar energy is a big initiative within the CSCC. Over recent years, 11 photovoltaic installations have been completed at seven campuses, with a total of 5,348 Kw of electricity generated. Over the 20-25 year period, the installations are estimated to save $9,416,173 in savings. An additional six installations are planned, generating 2,000-2,739 Kw for 10 campuses. (CSCU, 2017)

Fuel cells have also been installed with the support of the CSCU System Office. Locations include the Naugatuck Valley, Manchester, and Housatonic campuses, generating 1,060 Kw and resulting in a cost savings over 20-25 years of $14,631,280. (CSCU, 2017)

These bulk contracts and utility company rebates/incentives continue to be implemented when needed and where feasible. When the cost savings realized by the six additional solar installations are calculated, the long-term financial benefit of this system-wide approach to energy purchasing will be close to $30 million. (See Table 1 for details.)

Table 1. System-level electricity purchases and resultant savings

Done	Campus	Type	Location	Kw	$ Savings (20 years)
X	Manchester	PV	North	1,000	$2,054,512
X	Manchester	PV	South	1,000	$717, 322
X	Tunxis	PV	Roof	240	$1,108,405
X	Quinebaug Valley	PV	Site installation	730	$954,944
X	Asnuntuck	PV	Roof	640	$573,857

continued on following page

Table 1. Continued

Done	Campus	Type	Location	Kw	$ Savings (20 years)
X	Asnuntuck	PV	Parking lot canopy	240	$532,764
X	Housatonic	PV	Lafayette Hall	180	$1,869,511
X	Housatonic	PV	Parking lot canopy	720	$180,000
X	Middlesex	PV	Site installation	100	$310,048
X	Gateway	PV	Garage roof	249	$664,610
X	Houstonic	PV	Lafayette Hall roof	249	$450,000
	Subtotal			5,348	$9,416,173
TBD	Asnuntuck	PV	North parking lot Canopy	300-480	
TBD	Norwalk, West	PV	Parking lot canopy	300	
TBD	Norwalk, West	PV	Parking lot canopy	300	
TBD	Norwalk, East	PV	Parking lot canopy	300-460	
TBD	Naugatuck	PV	Parking lot canopy	200	
TBD	Naugatuck	PV	Site installation	600-999	
	Subtotal			2,000-2739	
X	Naugatuck	Fuel cell		350	$3,310,000
X	Manchester	Fuel Cell		250	$5,660,640
2025	Housatonic	Fuel Cell		460	$5,660,640
	Subtotal			1,060	$14,631,280
	Total Savings				$24,047,453

In addition, the CSCU System Office has led an effort to replace lighting at the CSCC campuses with LED fixtures. To date, LED lighting has been installed at seven campuses, with more installations planned. With $3,108,677 in utility company incentives, the LED installations will save $1,015,310 annually.

THE UNITED NATION'S SUSTAINABLE DEVELOPMENT GOALS: HAVING A BROADER IMPACT

Up to this point, the many examples of sustainability initiatives taking place on Connecticut State Community College campuses referenced in this chapter have dealt with a historical definition of sustainability focused on environmental stewardship. In 2015, as part of its 2030 Agenda for Sustainable Development, the United Nations published an expanded set of goals aimed at ensuring a sustainable future for the Earth and its inhabitants. The 17 goals include environmental objectives, e.g. Clean Water, Clean Energy, Climate Action, Land Use, but have also added social justice and economic equity to the UN's definition of sustainability. Ending poverty and hunger, improving people's health, expanding educational opportunities, and reducing inequality were identified as social justice/equity goals, while economic growth, job opportunities, sustainable cities, and industry infrastructure were stated economic goals.

This expansion of sustainability beyond the natural environment offers the Connecticut State Community College's 12 campuses increased opportunities to advance sustainability on their campuses and in their local communities. While the global community shares in the challenges of hunger, poverty, healthcare, and employability, it is the actions of local communities that collectively serve as mankind's response to these challenges. The following is a review of some current examples of social and economic equity initiatives in the CSCC that could be scaled up across the state and beyond.

Hunger

Food insecurity is a constant issue on college campuses and certainly exists in the communities where CSCC campuses are located. At the Three Rivers campus, to address food insecurity an active food pantry gets heavy usage five days a week. Donations have come from United Way of Southeastern Connecticut and Stop-n-Shop grocery stores. To support healthy food choices, an industrial freezer and refrigerator was purchased to offer fresh and frozen foods. In addition, one day a month, a mobile food pantry visits courtesy of the United Way of Southeastern Connecticut. At Asnuntuck, the campus food pantry also benefits from a partnership with Stop-n-Shop. While these efforts are focused on campus food insecurity, most larger communities in Connecticut have food pantries and soup kitchens. Community college students could easily organize themselves to assist local efforts to fight hunger, benefiting from donations from the grocery community and staffing pantries with student volunteers.

Expanded Access to Educational Opportunity.

In July of 2024, the CT State-Asnuntuck campus received a $75,000 grant from the Metallica Scholars Initiative. Included in the funding are 30 scholarships for $1,000 each to support students in robotics, welding and advanced manufacturing programs. Equipment, career readiness activities, advanced manufacturing training and OSHA certification are also part of the program. Since 2019, the legendary rock band has donated $10.5 million in 25 fields of study. Through Metallica's "All Within My Hands" foundation, and in collaboration with the American Association of Community Colleges, students in more than 60 community colleges in all 50 states are benefitting from the program. (CSCU, 2024)

A major initiative of Connecticut State Community College is to make a college education affordable to students from underrepresented populations. Providing free tuition as space is available is the keystone of the initiative, with $45 million serving 22,000 students since 2020. In FY25, the Connecticut General Assembly has earmarked $28.5 million for the program.

In fall 2023, the Louis Stokes Alliance for Minority Participation (LSAMP) awarded a $1 million, three-year grant to support STEM students and programs at Connecticut State Community College's Gateway, Housatonic and Norwalk campuses. The grant will support students who complete associate degrees and continue on to four-year programs in science, technology, engineering or mathematics (STEM). Underrepresented students majoring in computer science, engineering, physics, biology, mathematics and chemistry will benefit from new STEM centers at Gateway and Norwalk, which will replicate the STEM center at Housatonic. Over the three-year period, 135 students will benefit from the grant. (CSCU, 2023)

More than 1,800 prison inmates from nine Connecticut correctional institutions are attending courses at Asnuntuck, Housatonic, Middlesex, Quinebaug and Three Rivers. Program areas include business, manufacturing and human services. National data suggests that postsecondary education contributes to 48% less recidivism among inmates, and each dollar spent on education for inmates saves $4-5 in prison costs.

In fall 2024, the National Science Foundation awarded Connecticut State Community College-Naugatuck Valley campus a $2 million grant over six years to support underrepresented populations in STEM careers. The grant will provide $5,000-$15,000 a year for 60 students for career advising, tutoring, field trips, and educational workshops. The grant is a follow up to a five-year, $650,000 NSF grant to support 82 STEM students. The grant's goal is to support student articulation to four-year degrees in a variety of STEM fields. Overall, Connecticut State Community College educates 5,600 STEM students annually through the Connecticut College of Technology, established in 1995 between the 12 community colleges and eight public and private colleges and universities. Goals include workforce development through the seamless transfer and articulation of credits and course experiences across high schools, community colleges and four-year institutions. (CSCU, 2024)

Work and Economic Growth

In November of 2024, a partnership with Pursuit Aerospace has resulted in a 3,000-square-foot manufacturing center at the Connecticut State Community College-Capital campus. It will provide students pursuing careers in computer numeric control manufacturing with one central location as they articulate from technical high school to community college to four-year engineering degrees. Connecticut State Community College is training 2,400 students a year at 11 centers in robotics, mechatronics and precision manufacturing. These training programs support industry infrastructure as well as promoting economic growth and providing students with gainful employment opportunities.

Another workforce sector — this one with labor shortages — is nursing and allied health. Connecticut State Community College educates 3,800 nursing and allied health students annually at six campuses. In addition to entering the workforce upon completion of their licensing exam, graduates of the nursing programs can transfer their associate degree to four-year nursing programs at a number of colleges and universities. Over the past five years, the college has awarded nearly 3,000 degrees in nursing at its Capital, Gateway, Naugatuck Valley, Northwestern, Norwalk, and Three Rivers campuses. Approximately 90 percent of CT State nursing graduates find employment in Connecticut within two years of graduating, with an average starting salary of $73,352. They benefit from clinical placements at local health care facilities throughout the state, including Hartford HealthCare.

In early 2023, Quinebaug Valley and Capital Community Colleges received $120,000 from the Connecticut Health and Educational Facilities Authority (CHEFA) to expand their allied health (certified nursing assistant, medical assistant) and nursing training programs.

Small businesses are also the beneficiaries of the community colleges of Connecticut. In spring and summer 2023, a series of workshops and seminars ranging from human resources to fund-raising, e-commerce, cybersecurity, finance and social media were offered at various locations across the state in support of local small businesses. Northwestern Community College and Housatonic Community College were the facilitators of the program.

Improved Public Health

Many academic programs in the CSCC produce trained professionals who directly impact public health in communities across Connecticut. Students in the Dental Hygiene program at CT State-Tunxis campus are conducting dental clinics at health centers, clinics, nursing facilities and also on a field study to the Bahamas. In the past year, 22 students have served more than 1,700 patients, including children in Hartford Public Schools and the University of Connecticut Health Dental Clinic. The only public program in Connecticut, Tunxis' Dental Hygiene program historically has 100% job placement.

CSCC also addresses health issues on its own campuses. A recent Connecticut survey indicated that 67% of community college students self-identify with mental health issues, mirroring national data. In response, Connecticut State Community

College has added several supports to provide improved and expanded mental health services. These include hiring seven additional mental health counselors and a contracted partner to provide telehealth 24/7. With the help of more than $600,000 in grants awarded in 2021 from the Governor's Emergency Education Relief (GEER) Fund, the Manchester and Tunxis community college campuses have created wellness centers. Other campuses have been able to add part-time temporary positions to support students. In addition, all 12 CSCC campuses have been able to purchase mental health screening and referral platforms. Each campus now has onsite mental health professionals to assist students with academic, personal, career, social injustice, and COVID-related issues. Peer education programs, employee training and connectedness programming are also part of the initiative.

When the more than 300 academic degrees and certificates offered by the CSCC are considered against the United Nation's expanded definition of sustainability, it is clear that Connecticut's community college system is having significant local impact across the state in advancing environmental sustainability, social justice, and economic development and equity. (See Table 2 for a select list of sustainability-related academic programs) Whether human services, business development, healthcare, social work, advanced manufacturing or other career pathways, CSCC is making a difference in the communities it serves.

Table 2. Sustainability-related academic programs – CT State Community College

Program Title	Campus(es)	Type
Arborist Training	Capital	Other
Architectural Design Technology	Capital, Norwalk, Three Rivers	Degree
Automotive Technology Management	Gateway, Naugatuck Valley	Degree
Aviation Maintenance Technology	Asnuntuck, Gateway	Degree
Biochemistry Degree	All campuses	Transfer
Biotechnology Certificate	Middlesex, Naugatuck Valley	Certificate
Biotechnology Degree	Capital, Middlesex	Degree
Basic Life Support	All campuses but Manchester and Housatonic	Other
Biomedical Engineering Technology	Gateway	Degree
Business Studies Degree	All campuses	Transfer
Certified Nurse's Aide	All campuses but Asnuntuck	Certificate
Civic Engagement	Tunxis	Certificate
Clean Water Management	Gateway	Certificate
Community Health Worker	Capital, Housatonic, Gateway, Three Rivers	Other
Community Interpreter	Middlesex	Certificate

continued on following page

Table 2. Continued

Program Title	Campus(es)	Type
Computer Science: Mobile Programming	Capital, Norwalk, Gateway	Degree
Construction Management	Capital, Norwalk	Degree
CPR	All campuses but Housatonic	Other
Criminal Justice	All campuses but Three Rivers and Quinebaug Valley	Degree
Cyber/Homeland Security Certificate	Capital, Naugatuck Valley, Tunxis	Certificate
Cyber/Homeland Security Degree	Naugatuck Valley, Tunxis	Transfer
Dental Assistant Certificate	Asnuntuck, Manchester, Gateway, Tunxis, Norwalk, Three Rivers	Certificate
Dental Hygiene Degree	Tunxis	Degree
Diagnostic Medical Sonography	Gateway	Degree
Disability Specialist	Manchester	Certificate
Disability Specialist	Manchester	Degree
Drug and Alcohol Recovery Counselor	Gateway, Manchester, Naugutuck Valley	Certificate
Drug and Alcohol Recovery Counselor	Gateway, Manchester, Naugutuck Valley	Degree
Early Childhood Education Degree	All campuses	Degree
Early Childhood Education Certificate	Asnuntuck, Capital, Gateway, Housatonic, Middlesex, Norwalk, Tunxis, Quinebaug Valley	Certificate
EKG Technician	Asnuntuck, Capital, Gateway, Middlesex, Norwalk, Tunxis, Naugatuck Valley, Three Rivers	Other
EMT	Asnuntuck, Manchester, Middlesex, Northwestern, Norwalk, Tunxis, Quinebaug Valley	Other
Environmental Engineering Tech: Env. Health and Safety Management	Three Rivers	Certificate
Environmental Engineering Tech	Three Rivers	Degree
Environmental Science: Environmental Health and Toxicology	Gateway	Certificate
Environmental Science: Natural Resources	Northwestern	Certificate
Environmental Science: Env. Biology	Naugatuck Valley	Degree
Environmental Science: Environmental Health and Toxicology	Gateway	Degree
Environmental Science: Natural Resources	Northwestern	Degree
Environmental Science: Sustainability	Manchester, Northwestern	Degree
Fire Technology/Administration	Gateway, Naugatuck Valley	Degree

continued on following page

Table 2. Continued

Program Title	Campus(es)	Type
Firefighter 1 and 2	Gateway	Certificate
Fundamentals of Marine Service	Middlesex	Other
Health Information Technology	Middlesex	Degree
Health Science	Capital	Degree
Health Care Administration	Middlesex	Degree
Horticulture	Naugutuck Valley	Certificate
Horticulture	Naugutuck Valley	Degree
Human Services: Gerontology	Asnuntuck, Capital, Gateway, Middlesex, Naugatuck Valley	Certificate
Human Services: Mental Health	Asnuntuck, Capital, Housatonic, Middlesex, Naugatuck Valley	Certificate
Human Services Degree	All campuses	Degree
Human Services: Mental Health	Asnuntuck, Capital, Housatonic, Middlesex	Degree
Human Services Management	Asnuntuck, Capital, Housatonic, Middlesex	Degree
Human Services: Social Work	All campuses	Transfer
Introduction to Medical Cannabis	Three Rivers	Other
Medical Assistant Certificate	Asnuntuck, Capital, Housatonic, Northwestern, Norwalk, Three Rivers, Quinebaug Valley	Certificate
Medical Assisting	Asnuntuck, Capital, Housatonic, Northwestern, Norwalk, Three Rivers, Quinebaug Valley	Degree
Medical Interpreter	Asnuntuck, Gateway	Certificate
Medical Lab Technician	Quinebaug Valley	Degree
Natural Science/Math	Gateway, Tunxis	Degree
New Media Production: Corporate	Middlesex	Degree
Nuclear Engineering Technician	Three Rivers	Degree
Nursing	Capital, Gateway, Naugatuck Valley, Northwestern, Norwalk, Three Rivers	Degree
Occupational Therapy Assistant	Manchester	Degree
Ophthalmic Assistant	Asnuntuck, Housatonic	Other
Paramedic Certificate	Capital	Certificate
Paramedic Studies	Capital	Degree
Paramedic Studies: EMT instructor	Capital	Degree
Patient Care Technician	Asnuntuck, Capital, Gateway, Housatonic, Middlesex, Tunxis	Other
Pharmacy Technician	All campuses but Norwalk	Other
Phlebotomy Technician	All campuses but Capital and Housatonic	Other

continued on following page

Table 2. Continued

Program Title	Campus(es)	Type
Certified Phlebotomy Technicians	Asnuntuck, Quinebaug Valley	Certificate
Physical Therapist Assistant	Naugatuck Valley, Norwalk	Degree
Physical Therapy Aide	Norwalk	Certificate
Pre-Nutrition Transfer Degree	Gateway, Housatonic	Transfer
Public Utility Management	Gateway	Certificate
Public Utility Management	Gateway	Degree
Radiation Therapy Degree	Gateway, Manchester	Degree
Radiography: Computed Tomography	Middlesex	Certificate
Radiography: MRI	Middlesex	Certificate
Radiography; Mammography	Middlesex	Certificate
Radiography	Capital, Gateway, Manchester, Middlesex, Naugatuck Valley	Degree
Registered Medical Assistant	Tunxis	Other
Respiratory Care Degree	Manchester, Naugatuck Valley, Norwalk	Degree
Sociology Studies Transfer Degree	All campuses	Transfer
Surgical Technology Degree	Gateway, Manchester	Degree
Technology Studies Degree	All campuses except Capital	Degree
Therapeutic Recreation	Manchester, Middlesex	Certificate
Therapeutic Recreation	Manchester	Degree
Water Management	Gateway	Certificate

IMPACTING SUSTAINABILITY LOCALLY—IT COMES NATURALLY

The global community shares a planet, an atmosphere, water and air systems. Even so, ecosystems are more regional — the Amazon rain forest, the Mississippi River and its environs, the Sahara desert. The three areas of sustainability described in this chapter—environmental, social, and economic—are all impacted more directly at the local level, but ordinary people doing the right thing. As individuals, we can best impact sustainability locally. Collectively we can work best as communities to impact climate change, economic inequality and educational access.

Environmental Equity

Recycling, reducing food waste, and purchasing green electricity starts with local initiatives. In Connecticut, community colleges educate local students who can witness the fruits of their environmental stewardship in their local communities, the same communities served by the 12 individual college campuses.

Economic Equity

Economic growth, educational access, and industry partnerships start at the local level, no matter how large in scale they become. Community colleges are well positioned to address these issues through their academic programs, commitment to access, and partnerships with local businesses, government, and the nonprofit sector.

Social Equity

Poverty. Hunger. Homelessness. Health care. These are global issues, no doubt. But while we send checks to charities to feed children in sub-Saharan Africa, we need to look around us and see to it that the children in our own communities are feed, clothed, housed. Again, these huge issues should be addressed locally; in the case Connecticut, a relatively small state geographically, we can measure the progress being made throughout the state's 169 townships.

With 12 campuses serving those 169 townships with their villages, towns and cities, there is a plethora of opportunities for Connecticut's community colleges to impact sustainability in their own backyards. Collectively they can impact environmental, social and economic equity goals throughout Connecticut. And such efforts, built upon each other across our nation can become a strong response to the UN's 17 Sustainable Development Goals.

CHALLENGES AND OPPORTUNITIES

Community colleges play a crucial role in higher education, yet they face unique challenges in developing and sustaining environmental conservation and sustainability initiatives compared to four-year universities. The following is a short list of challenges facing community colleges like the CSCC, with comments in each instance of how CSCC could address the challenges.

Funding and Resources

Community colleges typically operate with tighter budgets than four-year institutions. They often rely on state funding or a local property tax base, and like other tax-supported institutions, must be conscious of anti-taxation sentiments. In addition, securing grants and external funding is crucial but often competitive and challenging, and campuses may not have the resources to apply for federal funding and competitive grants.

Response

The Connecticut State Community College's new scale — largest in New England, sixth largest in the nation — and its partnership with other institutions in the Connecticut State Colleges and Universities System presents many opportunities for additional funding from state, federal, and private sources. In addition, grant writing expertise is available that would otherwise not be accessible. As well, CSCU System level projects such as the LED lighting project, EV charging station project, and bulk purchasing of green electricity save money that can be reallocated for other sustainability projects. In the end, if sustainability is a priority in Connecticut, investing in sustainable practices in the community colleges makes sense and must continue to be pursued by state leaders.

Institutional Resistance and Lack of Support

Some institutions may face resistance to change from stakeholders who prioritize other aspects of college operations. Overcoming this resistance requires strong leadership and a clear demonstration of the benefits of sustainability. Many four-year colleges have dedicated offices for sustainability, which are responsible for implementing and managing green initiatives. This level of staffing and support does not exist on local campuses of the Connecticut State Community College; support instead is provided as time is available in facilities departments and other offices. This makes it difficult to develop new initiatives, especially given some of the other challenges being outlined here.

Response

The National Wildlife Federation (NWF) recommends state-level sustainability staffing to provide the leadership needed. "Appoint a high-level champion within each college (and at the system level) to facilitate sustainability planning, assessment, and reporting; coordinate partnerships with employers; identify job demand and skill needs; and articulate project- and work-based learning to defined sustainability competencies." (NWF, 2015)

Lack of Student Engagement

While many students are passionate about sustainability, engaging all students in these efforts can be challenging. Without a residential community, and with a commuter population that turns over every two years or less, it is harder to create a campus culture around sustainability on a community college campus compared to a four-year, residential campus. In addition, students have full-time jobs off-campus and don't typically develop a long-term relationship with the campus community.

Response

Short-term, yet high-impact initiatives, e.g. composting, community gardens, have the most chance of success on a local community college campus. As well, a statewide campaign involving all 17 institutions in the Connecticut State Colleges and Universities, with local campus goals and partnerships with local municipalities and nonprofits, could serve as motivation for statewide participation in broader projects that are manageable and measurable. An example could be a statewide recycling and/or waste management competition among the 12 community college campuses. It should also be noted that any progress is positive, and that individual campus initiatives can be celebrated locally if not replicated statewide. At Naugatuck Valley, Electrical Engineering students are conducting an "Unplug It" campaign to turn office refrigerators off during winter and spring breaks. They are also monitoring electricity usage to gauge success. Small in scope, this project shows how engaging students in active sustainability programs can inspire the rest of the campus and pave the way for replication on other campuses.

Limited Engagement and Awareness Among Stakeholders

Community colleges serve a diverse population, including non-traditional students. Given the short-term (two years or less) commitment of community college students, the fact that all community college in Connecticut commute to campus, and the likelihood that students are working full time off campus makes it difficult for students to commit to issues unrelated to their focus on work. Certainly, a lack of student engagement on campus can diminish the effectiveness and sustainability of environmental programs.

Response

For years, sustainability has been couched as a worldwide issue ("global warming"). In actuality, sustainability is ultimately local. Energy consumption takes place locally, food waste is generated locally. Recycling takes place locally. That means that any efforts on a CSCC campus to reduce energy consumption, promote recycling, use alternative energy, and practice conservation of natural resources directly impacts the local community where students live. Given the UN's expanded definition of sustainability to include economic issues (job development, training, sustainable cities), the Connecticut State Community College's focus on serving the state by staying connected to local communities is also on point in helping create a sustainable future.

Insufficient Research and Data on Best Practices

Community colleges have a mission grounded in the present, meeting current workforce needs of local communities. By intention, community colleges do not have a strong research function. Innovative sustainability strategies depend on new knowledge and information.

Response

One of the value-added benefits of the Connecticut State Community College's membership in the Connecticut State Colleges and Universities System is the potential collaboration with state university researchers — faculty and students committed to finding new green, clean solutions to environmental challenges. These types of collaborations do not occur by chance; they are intentional and should be a major planning goal of sustainability leaders on the state's two- and four-year campuses.

Lack of Public Awareness of Current Sustainability Efforts

With adequate staffing necessary to monitor, measure, and record sustainability successes on CSCC campuses — perhaps located at the CSCC state office — there are several recognition programs that the CSCC could join. In doing so, the resulting awards and public awareness could be translated into additional funding and other public support. Each year, The Princeton Review releases its Guide to Green Colleges, a document based on school data and student surveys. In the 2024-25 guide, there were no community colleges among the nation's Top 50 schools, a list that included private liberal arts colleges (Williams), national universities (Princeton), public land grant institutions (University of Vermont, Arizona State), and state universities (Colorado State-Fort Collins). Out of the total set of 511 "green colleges" only nine were community colleges. In 2024, 1,582 of the 4,414 institutions of higher education were community colleges—36%. ("research.com", 2024) At that rate of representation, there should have been several community colleges on the Top 50 list, and 184 among the 511 green colleges in the United States. Instead, there were only nine — community colleges were underrepresented 20-fold.

Another organization that recognizes sustainability efforts among colleges and universities in the United States, Canada and beyond is the Association for the Advancement of Sustainability in Higher Education (AASHE), which ranks institutional progress through its Sustainability Tracking, Assessment and Rating System (STARS). Of the 805 members of AASHE, only 74 are community colleges. At 9%, that is better representation than the Princeton Review's Green Colleges program but is still far less than it should be. Establishing a monitoring program at CSCC would be a pioneering act, but could help inform best practices while increasing public awareness and approval of CSCC sustainability efforts. It might even encourage other community colleges to join the Princeton Review an AASHE recognition programs.

FUTURE DIRECTIONS AND OPPORTUNITIES

Looking ahead, Connecticut's community college campuses can continue to advance their climate sustainability efforts by focusing on several key areas:

Planned Progress

The Connecticut State Community College is still in its early stages as a single entity, having been founded in July 2023. Since then, the college has been busy reorganizing itself, greeting new leadership, dealing with the never-ending task of managing challenging budgets, and working through the New England Association of Colleges and Schools (NEASC) accreditation process as an institution undergoing a substantive change. In 2025, NEASC will conduct a focused site visit, and the college will begin its first Strategic Planning cycle.

While challenging, these times should be seen as a tremendous opportunity for the CSCC to plan for its sustainability future. The CSCC is clearly committed to sustainability in its broadest definition (environmental, social, and economic). Among its guiding principles, "The college will prioritize resources toward diverse student needs, equity, inclusivity, and the advancement of environmental, social, and economic sustainability." (Maduko, J., 2023, p. 13)

Based on this stated commitment to sustainability, we can anticipate that serious attention will be given to including feasible, sensible sustainability initiatives as part of the emerging Strategic Plan. Innovation, collaboration, and accountability should be hallmarks of such planning. While some projects will cost money, others simply involve staff time, campuswide commitments, and innovative thought. And the CSCU System has already demonstrated that careful statewide planning can result in significant cost savings. Among options for investment is system-level staffing to help facilitate campus-level initiatives, manage statewide projects such as LED installations, advocate for resources and funding, and create tools to monitor, measure, and report sustainability successes. Finding opportunities for collaboration between Connecticut State Community College and the four Connecticut State Universities is another important planning strategy.

Scaling Successful Models

Expanding successful sustainability programs and practices to other institutions can amplify their impact. With 12 campuses and sister schools in the form of the four Connecticut state universities, CSCC has considerable opportunity to replicate successful projects across the state's higher education community. In addition to localized programs and activities, the Connecticut State Colleges and Universities have the ability to act in unison through economies of scale, purchasing alternative energy in bulk; installing LED lights, solar panels and EV charging stations; sharing curriculum tied to sustainability; and sharing best practices, thereby impacting the statewide community in the process.

It is important to make the point that success does not always mean that every CSCC campus must be involved in a given solution. The campuses range widely in size, serve different populations (Gateway, Capital and Housatonic serve three large cities while Northwestern serves rural Litchfield County), and feature different academic programs. Any initiative that works — whether for 1, 2, or 12 campuses — is worthy of consideration. Naugatuck Valley and Manchester have been collaborating with Cascade Energy on a Strategic Energy Management project, conducting training and energy audits at their respective campuses. The pilot may be extended to other campuses, but regardless, is a step forward and one that will yield important lessons for the future. In addition, the state community college and state universities central offices can serve as facilitators for sharing resources as newer energy technologies are discovered.

Enhanced Research and Innovation

Investing in research and innovation can help community colleges develop new solutions for sustainability challenges. Encouraging student and faculty research in this area can lead to practical advancements and increased institutional visibility. Given the applied learning inherent on the community college campuses, it may be that CSCC must look to its partners in the four Connecticut State Universities to conduct research and share findings across the Connecticut State Colleges and Universities system. This would seem to be a natural partnership and one surely to have the support of state leaders.

Strengthening Policy Advocacy

Community colleges can play a role in advocating for policies that support sustainability at local, state, and national levels. By engaging in policy discussions and lobbying efforts, these institutions can help shape a more supportive environment for climate action. In Connecticut, sustainability already is a state priority. The Connecticut Climate Action Plan, led by the Governor's Council on Climate Change, seeks to reduce greenhouse gas emissions, improve water and waste management, and promote increased sustainability across all state facilities and institutions. The Connecticut State Colleges and Universities' Climate Initiatives are intended to complement and support the state's plans. Using Quinebaug Valley's Climate Action Plan as a model, future CSCC planning could include Climate Action Plans for all 12 CSCC campuses, as well as one for the system. In this environment, given the broader definition of sustainability found in the UN's Sustainable Development Goals, and with the potential local impact of the 12 community college campuses, Connecticut is poised to move forward with a coordinated approach to sustainability.

Academic Alignment

Beginning with environment-focused curriculum, there are many exciting opportunities for Connecticut's educational institutions to align courses and programs. Starting at the high school level with career pathways and dual credit programs, a focused effort could be made to link sustainable education all the way through the community colleges to four-year degree programs. More than simply transfer of credits, such alignment could create skills thresholds that would allow students to base successive learning on existing skill ladders. Sharing innovative curriculum across the CSCC and the CSCU can only help students on individual campuses to prepare for green economy careers and lives as environmental stewards.

Expanded Local Partnerships

With stronger relationships among the 12 CSCC campuses and an articulated, planned approach to local implementation of broader global sustainability initiatives, each CSCC campus is in a position to work with local municipalities and nonprofits to move ahead on a shared vision. Food security, housing, workforce development, educational access, economic development and more can be addressed within a broad definition of sustainability, motivating students, bridging organizational boundaries, and focusing on a community's leading social issues.

One statewide coalition of local partners that could be approached is Sustainable Connecticut, a nonprofit group that certifies municipalities and other organizations against a rubric of best practices. Founded in 2017 at Eastern Connecticut State University, the organization achieved 501(c)3 status in 2022. The group offers best practices, technical assistance and funding support to help towns and cities mitigate climate change, reduce greenhouse gas emissions, support affordable housing, promote public health and more. Since founding, Sustainable Connecticut has funded 354 projects ($2.3 million in crowdfunding and $2.1 million in matching funds) ranging from social justice workshops, to gardening projects, public art, food insecurity, environmental education, clean rivers and more. Of 134 communities registered with Sustainable Connecticut, 63 are certified, including 37 at the Bronze level, 24 at the silver level, and two at the Gold level. Linking the Connecticut State Community College's 12 campuses with active Sustainable Connecticut projects would connect students with sustainable projects in their home communities, benefiting local residents and communities. (SustainableCT, 2025).

CONCLUSION

Connecticut's community colleges are making significant strides in promoting climate sustainability through curriculum integration, sustainable operations, community engagement, and overcoming challenges. Their efforts not only contribute to addressing climate change but also prepare students to be future leaders in sustainability. As these institutions continue to evolve and adapt, they will play an increasingly crucial role in fostering a more sustainable future. By leveraging their unique strengths, the scale of the CSCC, and the partnerships that exist and that can be developed within the Connecticut State Colleges and Universities, Connecticut's community colleges can lead by example and inspire broader societal change in the fight against climate change, while promoting social justice and economic equity.

As the Connecticut State Community College embarks on its first Strategic Plan as a single college, several fundamental assumptions should be included in the planning process.

(1) Local stakeholders — both on- and off-campus — must be given ample opportunity to provide input. Students, faculty, staff, as well as community leaders and groups must share their needs, their experiences, their resources and commitments.
(2) With economies of scale, statewide workforce needs, and other statewide stakeholders must also be given opportunities to craft a Connecticut sustainability plan.
(3) Thought must be given to developing a sustainability governance structure that balances local needs, state priorities, and the potential for sharing resources.

REFERENCES

Maduko, J. (2023). *CT State Transitional Strategic Plan,* May 1, 2023, p. 13; "www .ctstate.edu"

National Wildlife Federation. (2015). *Greenprint: A Plan to Prepare Community College Students for Careers in the Clean Economy,* December, 2015; Cross Mwabe, G., Lamback, S., and Wright, M. editors.

Research.com. (2024). *U.S. College Statistics: 2024.*

Connecticut State College and Universities. (2017). *Energy Master Plan*, August 2017. (Also see 2021-24 Eversource and UI Project, Connecticut State Colleges and Universities)

State, C. Colleges and Universities. (2022). *Board Agenda,* June 23, 2022, pp. 264-267.

State, C. Colleges and Universities. (2022). *"GCC Garage Rooftop Solar Panels Provide Renewable Energy, Cost Savings,* November 28, 2022, "www.ctstate.edu/newsroom"

State, C. Colleges and Universities. (2023). *CT State's NSF Grant to Help Increase Underrepresented Students in STEM Fields, October 11, 2023,* "www.ctstate.edu/newsroom"

State, C. Colleges and Universities. (2024). *CT State Gateway Awarded $350,000 NSF Grant to Train Technicians for Evolving Needs of Automotive Industry*, July 18, 2024, "www.ctstate.edu/newsroom"

State, C. Colleges and Universities. (2024). *CT State Asnuntuck First in Connecticut to be Selected for Metallica Scholars Initiative's Workforce Education Program," July 22, 2024,* "www.ctstate.edu/newsroom"

State, C. Colleges and Universities. (2024). *CT State Naugatuck Valley Awarded $1.99 million NSF grant to Support STEM Students,* September 11, 2024, "www .ctstate.edu/newsroom"

State, C. Colleges and Universities. (2025). *Get to Know Us*, www.cscu.edu

Sustainable CT. (2025). "sustainablect.org"

Chapter 14
Geographic Advantage, Visionary Philanthropy, and Strategic Growth: Virginia Wesleyan University's Rise as a Sustainability Leader in Higher Education

Scott D. Miller
Virginia Wesleyan University, USA

Kelly Cordova
Virginia Wesleyan University, USA

ABSTRACT

As a result of strategic leadership from all levels of campus, the university's geographic advantage, and visionary philanthropy, Virginia Wesleyan University has grown into a higher education leader in environmental sustainability. The university fosters community engagement and inspires responsible environmental action at every level of the institution through targeted fundraising from individuals and grant-making organizations. Its commitment to environmental education and stewardship extends beyond the classroom by integrating green practices into academic programs and campus initiatives from Virginia Beach, Virginia, to Tokyo, Japan.

DOI: 10.4018/979-8-3693-5454-4.ch014

BACKGROUND

Virginia Wesleyan University is located in Virginia Beach, a beach resort town in southeastern Virginia with a population of approximately 450,000, making it the largest city in Virginia. Virginia Beach is a unique coastal city where the natural beauty of the Atlantic Ocean and Chesapeake Bay meets the strength of maritime industry. Located on the edge of rich ecosystems teeming with diverse marine life, the area is home to thriving ecological treasures like Back Bay National Wildlife Refuge and First Landing State Park. Yet, Virginia Beach is also a hub of economic activity, driven by its close proximity to the world's largest naval base and the Port of Virginia, one of the busiest and most significant ports on the East Coast. This balance between preserving its vibrant environment and supporting key industries makes Virginia Beach a dynamic and vital city, and a wonderful home for a liberal arts university like Virginia Wesleyan.

Over the past 60 years, the university has continually evolved to serve the ever-changing needs of the region. Chartered in 1961 and first opening its doors at its current location in 1966, the university was originally founded by the United Methodist Church as Virginia Wesleyan College in Norfolk, Virginia ("Our History," n.d.). Accreditation by the Southern Association of Colleges and Schools Commission on Colleges (SACSCOC) was obtained in its first year of eligibility (1970) and the college graduated its first class of 41 pioneering students that same year ("Our History," n.d.)

As regional and global concerns about climate change became ever more prevalent during the 21st Century, Virginia Wesleyan introduced The President's Environmental Challenge Grant program in 2007 as part of a larger campus sustainability initiative designed to heighten awareness of environmental issues among faculty, staff and students. These efforts included the creation of the Environmental Sustainability Council, Marlins Go Green (the university's mascot is a marlin), and the awarding of President's Environmental Challenge Grants.

For example, one such grant was awarded in 2011 to Professor Phil Rock to create a worm farm on campus for vermicomposting, which is faster than traditional composting (Baker, 2011). The resulting soil has many advantages including increased healthy microbial activity, improved water holding capability and decreased susceptibility to pests and disease (Baker, 2011). The university used the resulting soil in the flower beds and gardens on the campus as an environmentally friendly and less costly alternative to traditional fertilizers while combating the loss of topsoil, a primary environmental concern (Baker, 2011).

Other efforts at the time included a Master Landscape Plan, which was designed with the concept of sustainability through the use of native plantings on campus (Shearin, 2010).

In 2007, campus Dining Services and Facilities Management teamed up to create compost tea, through retaining and composting appropriate food waste. The subsequent "tea," a rich, liquid fertilizer full of nutrients, was used across campus on flower and shrub beds (Shearin, 2010).

As these examples demonstrate, there was definite interest and support on campus for sustainability initiatives, but these initial efforts were rather small-scale and occurred largely on the fringes of campus.

Meanwhile, before I began my presidency at Virginia Wesleyan, I served as president at Bethany College in West Virginia (2007-15), Wesley College in Delaware (1997-2007), and Lincoln Memorial University in Tennessee (1991-97). When I served as the president of LMU, I was just 31 years old at the time of my appointment, making me the youngest college president in the nation.

During my presidency at Wesley College, I first became involved in what was then the American College and University Presidents' Climate Commitment (ACUPCC) in 2006. The original group of 12 college presidents formulated the commitment plan, and 50 additional presidents, including me, joined them in Washington, D.C., for the first meeting. Thanks to the ACUPCC, I was able to help advance a proactive environmental agenda at Wesley College, which was starting to gain momentum nationwide. By 2008, over 400 colleges and universities across the United States had signed on to the Commitment.

The ACUPCC rebranded in 2015 to become the Presidents' Climate Leadership Commitments. The program is centered on three commitments: "a Carbon Commitment (focused on reducing greenhouse gas emissions), a Resilience Commitment (focused on climate adaptation and building community capacity), and a Climate Commitment that integrates both" ("About Us," 2024).

With extensive experience in shaping climate initiatives at prior institutions, I joined Virginia Wesleyan then-College in 2015, driven by a deep passion for harnessing higher education's potential to combat the climate crisis. Having served as president at three other institutions, I felt uniquely prepared to lead Virginia Wesleyan in addressing one of the most pressing challenges of our time: revolutionizing how universities confront environmental change. Positioned at the intersection of academia and a coastal environment, Virginia Wesleyan offers a unique vantage point to pioneer solutions in the fight against climate change, as we stand on the edge of an ecosystem on the verge of irreparable change.

In one of my first meetings with faculty at Virginia Wesleyan, Dr. Elizabeth Malcolm, Processor of Ocean and Atmospheric Sciences and Director of Sustainability, shared her environmental aspirations for Virginia Wesleyan. Already a member of the faculty for 10 years, Dr. Malcolm had a rather aggressive plan to make Virginia Wesleyan a leader in environmental education and sustainability. Her plan included a state-of-the-art educational facility, a rigorous program that matched the tenets of a liberal arts education, a new mission statement, an environmentally and socially responsible investment policy, and protection of the institution's 300 acre park-like campus. Together, she and I acted on the agenda - she with academics and me with donors, board members, and other stakeholders.

Eventually, as part of the overall agenda to enhance the reputation of the institution, Virginia Wesleyan College became Virginia Wesleyan University, effective July 1, 2017. This change moved Virginia Wesleyan to Level III status with SACSCOC which allowed for the addition of graduate programs and an online degree program.

This change in name also included a change in mailing address. The university's campus, which has straddled the Norfolk/Virginia Beach city lines since its development, officially moved to a Virginia Beach mailing address on July 1, 2017. This is significant as it allowed the university to further our environmental agenda by capitalizing on our location and the intrinsically powerful aesthetic and environmental connotations of "the beach."

Symbolically, "the beach" conjures breathtaking images of marine life and sandy coastlines, while emotionally, "the beach" inspires thoughts of relaxing or having fun in idyllic weather while enjoying the scenic views at one of nature's finest playgrounds. But in reality, the beach is also a place where anyone can witness the direct effects of climate change. Due to rising sea levels, every present-day beach is also an ecological battleground, in which a fragile ecosystem feels as if it could all be lost to the ocean at any time.

This sobering feeling is based in fact: a first-of-its-kind satellite survey has determined that "24 percent of Earth's sandy beaches are eroding," representing a combined coastline distance of nearly 50,000 miles (Merzdorf, 2018). By 2050, the "sea level along contiguous U.S. coastlines could rise by as much as 12 inches," further eroding and disrupting not just beaches but entire metropolitan areas (Younger, 2023). Additionally, the health of beach regions directly affects the health of the people who live in and visit them: in 2022, "over 350 beach waters in the U.S. were found to be unsafe for swimmers" on at least one quarter of the days they were tested (America, 2023). These data points underscore the crucial role that beaches play in both ecological and societal health.

It is in this setting that a new environmental vision for Virginia Wesleyan was born — one developed from the unique collaboration of the university's philanthropic spirit, its particular geographic advantage, and its innovative leadership at many levels of campus.

VISIONARY PHILANTHROPY

The rise of Virginia Wesleyan as a leader in environmental stewardship in higher education is thanks in large part to visionary philanthropists who care deeply about the university, its students, and the greater community. At a time when only "2% of global philanthropy is focused on climate action" (Schueman, 2023), Virginia Wesleyan has raised over $150 million dollars in the last ten years for environmental education and green initiatives on campus and beyond.

The start of this incredible fundraising largely began with the construction of the Greer Environmental Sciences Center, named in honor of former Virginia Wesleyan president William T. Greer, Jr., who started the institution on a path to sustainability. The announcement served as a declaration to society and the larger academic community that Virginia Wesleyan is resolved to be a premier liberal arts institution in the field of environmental science.

Although originally announced as an anonymous gift in 2015, the Greer Environmental Sciences Center was funded by Jane P. Batten, a name that is synonymous with growth and prosperity at Virginia Wesleyan. Every aspect of the campus—academic, student life, athletics, endowment, and facilities—has been touched by the visionary leadership and philanthropic support of Jane and Frank Batten.

Frank Batten, Sr., was an internationally known businessman and co-founder of The Weather Channel, the country's first nationwide 24-hour cable weather channel. His media company, Landmark Media Enterprises, once owned nine daily newspapers, more than 50 weekly newspapers, television stations in Las Vegas and Nashville, and a national chain of classified advertising publications ("The Batten Legacy," n.d.). Until 2008, Landmark Communications (now Landmark Media Enterprises) was one of the country's largest privately held media companies. Eventually, Mr. Batten sold multi-system cable TV company "TeleCable in 1995 to TCI for $1 billion and The Weather Channel in 2008 to NBC Universal and two private equity firms for nearly $3.5 billion" ("Newspaper Exec," 2009).

Mr. Batten declined President Clarke's offer to join Virginia Wesleyan's Board of Trustees. But when President Clarke invited Mr. Batten's wife, Jane, to join the Board, it changed the trajectory of Virginia Wesleyan forever.

A native of Norfolk and alumna of Hollins University, Jane Parke Batten's interest in Virginia Wesleyan was kindled when her daughter, Betsy, enrolled as a freshman in 1978 ('The Batten Legacy," n.d.). Mrs. Batten joined the Board of Trustees in 1981 and served as Board Chair from 1995-1998 ("The Batten Legacy," n.d.). At the conclusion of her term as Chair, Mr. Batten made the lead contribution for the construction of the Jane P. Batten Student Center, a state-of-the-art $22 million multi-purpose facility which opened in 2002 ("The Batten Legacy," n.d.).

In addition to the student center, the Battens have funded a number of endowed faculty positions, student scholarships, and an outstanding faculty award. Frank Batten was recognized with a Doctor of Humane Letters, honoris causa, in 2002 ("The Batten Legacy," n.d.). Jane Batten was honored with the Doctor of Laws, honoris causa, in 2006 and became a Trustee Emerita in 2015 ('The Batten Legacy," n.d.). Mr. Batten passed away on September 10, 2009 ("Newspaper Exec," 2009).

With her incredible history of philanthropy and service to Virginia Wesleyan, which spanned the tenure of President Greer (1992-2015) ("Presidents of Wesleyan," 2015), Mrs. Batten then chose to fund the Greer Environmental Sciences Center and surrounding gardens upon his retirement.

The announcement of the Greer Environmental Sciences Center in 2015 was the first large-scale commitment by the university to its green initiatives. Completed in 2017, the 44,000 square-foot state-of-the-art facility offers wonderful opportunities for students while also encouraging regional collaboration with like-minded organizations including the Chesapeake Bay Foundation, the Elizabeth River Project, Lynnhaven River NOW, the Norfolk Botanical Garden, and many more ("Greer Environmental Sciences Center," n.d.).

The Center is designed around the Earth's four spheres—the atmosphere, hydrosphere, lithosphere, and biosphere—and features a multitude of indoor and outdoor learning spaces that emphasize "science on display." This wide variety of relevant situational learning opportunities promotes hands-on experiences, interactive learning, and interdisciplinary research.

The Center also includes four large teaching laboratories dedicated to specific aspects of earth science, such as ecology and marine biology, earth science, atmospheric science, oceanography, and analytical chemistry, which is equipped with advanced instrumentation. Dedicated research labs further enhance students' learning experiences, including the Chesapeake Bay fish lab, geology and hydrology lab, nutrient analysis lab, a trace metal analysis clean room, and an environmental imaging suite. Additionally, there is a geographic information system (GIS) classroom.

Sustainability is woven into the facility's design with strategically placed features easily accessible to students. These include a geothermal system for heating, cooling, and domestic hot water, an array of photovoltaic panels, an extensive green roof, and a greenhouse with research gardens. The surrounding landscape features native plantings, wildlife habitats, vegetated bioretention basins, and constructed wetlands for managing stormwater runoff.

The Greer Environmental Sciences Center achieved Leadership in Energy and Environmental Design (LEED) Gold Certification for compliance within a rigorous rating system created by the U.S. Green Building Council by proving measurable strategies and solutions aimed at achieving high performance in energy efficiency, material choice, water efficiency and savings, and site enhancement (Zork, 2018). LEED certification is a globally recognized symbol of sustainability achievement.

Fittingly, the Greer Environmental Sciences Center dedication ceremony was held on Thursday, September 7, 2017 ("Greer Environmental Sciences Center Dedicated," 2017). In Virginia Wesleyan tradition, the first Thursday of September is recognized as Founders Day, as that was the week the institution first held classes in 1966. So while Virginia Wesleyan was celebrating its Founders Day by dedicating its new facility, in so many ways September 7, 2017, also became a new founding date for the institution. With the opening of the Center, Virginia Wesleyan marked the beginning of a new chapter in its history under new leadership, a new name, a new address, and a vision for the future that capitalized on its brand new facility located in a pivotal coastal community.

The construction of the Greer Environmental Sciences Center was the biggest step to date in Virginia Wesleyan's commitment to studying, understanding, and protecting the environment. Since its opening, the university's commitment to the environment has only continued to grow.

Creation of the Batten Honors College

Halfway through the two-year process of constructing the Center, I entered discussions with Jane Batten about an academic program that might complement this soon-to-be-completed state-of-the-art facility. Mrs. Batten suggested that an honors college with a focus on the environment in a location like VWU's would develop into a national niche program for the university. After months of developing the model, she agreed to fund the program with annual contributions for the first four years.

The Batten Honors College was announced in 2016 and recruited its first class to begin in Fall 2017, the same time the Greer Environmental Sciences Center opened. The opening of the Batten Honors College was greeted with tremendous success–30,000 inquiries per year, incredibly high selectivity, and positive reviews from students, parents, and the media.

The vision of the Batten Honors College is to recruit high-achieving students from around the world and equip them to be problem solvers who are environmentally-conscious, strategic-thinking global citizens, ethically-bound to bettering their community and the world. Of those accepted into the program, 80 total scholars are offered full tuition and the other 80 total scholars are offered three-fourths tuition scholarships.

The mission of the Batten Honors College is "to prepare academically advanced and socially engaged students to lead society's professions and institutions with innovative approaches and ethical values that steward an environment threatened by global degradation and climate change" ("Mission and Vision," n.d.)

Mrs. Batten knew that bright, young, and motivated individuals would be the ones to solve the environmental issues of today and the future, and she trusted the Virginia Wesleyan liberal arts education to equip these students with the analytical, synthetic, and resolution-oriented skills they need to address the root causes of environmental degradation and develop solutions to a plethora of other challenges facing the world.

To accomplish this, the Batten Honors College explores diverse disciplines from the humanities, social sciences, mathematics, and natural sciences through an interdisciplinary and thematic curriculum that educates and graduates lifelong learners who are interested in shaping the future and taking action to improve the world. Its curriculum encourages independent thinking and self-discovery while challenging students to view the world from multiple perspectives and to integrate theory and practice. Each student has a fully-funded experience in the three priority areas of the program: the environment, global engagement, and leadership. Their robust learning experiences include study-area opportunities, special lectures and conferences (both on and off-campus), and general education courses in environmental sciences. And no matter their majors, all VWU students benefit from access to the national award-winning Greer Environmental Sciences Center.

By 2021, the impact of the Batten Honors College program had grown significantly, and the results were convincing. The inaugural cohort graduated with the following quality indicators: 3.7 GPA, 92% graduation rate, and 100% placement. Based on these numbers, Mrs. Batten committed to endow the program.

Additionally, the university established early assurance agreements that guarantee graduate admission with the Nicholas School of the Environment at Duke University, the Frank Batten School of Leadership and Public Policy at the University of Virginia, and the Virginia Institute of Marine Sciences of the College of William & Mary, which provides Batten Honors College students with built-in pathways to continue their education to create a better, greener world.

Inspiring New Programs in the Batten Honors College

After Mrs. Batten's commitment to endow the Batten Honors College in 2021, the university announced the Louis and Prudence Ryan Environmental Research Symposium and Ryan Prize for Excellence in Environmental Research for scholars in the Batten Honors College. The symposium asks scholars to conduct independent research on one of the United Nations' seventeen Sustainable Development Goals ("Batten Honors College Symposium," 2021). Under the mentorship of VWU faculty and community leaders, student scholarship featured in the symposium explores complex environmental challenges in a global context.

The symposium is named with gratitude to Louis and Prudence Ryan. We sought out the Ryans for funding such a program for several reasons: (1) Their main areas of philanthropic focus are education and the environment (Bulletin, 2024); (2) Louis Ryan has been a member of the Virginia Wesleyan Board of Trustees since 2007; and (3) Louis Ryan worked as General Counsel for Landmark Communications under Frank Batten; thus, it is fitting that this symposium is forever tied to the Batten name. The symposium was created by an endowment to support the research and fund its prizes.

The first Ryan Environmental Research Symposium was held in April 2023. During the symposium, students present their research as a senior capstone project. A committee of Batten Professors and student representatives select the three best projects to be presented live during the symposium, but all projects are displayed during the poster session ("Batten Honors College Symposium," 2021). The Ryan Symposium is a premiere public event that brings together scholars, mentors, and community leaders for the awarding of the Ryan Prize for Excellence in Environmental Research.

The projects are judged on four criteria: (1) potential impact of the project, (2) addressing environmental complexities, (3) strength of evidence, and (4) communication of the value of the project ("Batten Honors College Symposium," 2021).

In a speech at the first Ryan Environmental Research Symposium, Mr. Ryan stated, "What a good education does is teach you how to think. And what I mean is thinking critically, creatively, and holistically. Doing that is really necessary to guarantee the survival and thriving of humanity" ("Earth Angels," 2023, p. 14). He felt that environmental research was the perfect place to both learn those skills and practice them.

Batten Honors College scholars were up to the challenge, showing interest in a variety of topics and offering real-world solutions to help solve complex environmental problems. In the symposium's first year, the top three research projects were "Reducing Campus Food Waste"; "Combating Fraudulent Pharmaceuticals in Kenya"; and "Urban Tree Planting. Outlining How to Diminish Urban Stormwater Runoff Utilizing Community Partnerships."

Abby Peterson's "Urban Tree Planting" project was awarded the first-ever Ryan Prize in Environmental Research ("Abby Peterson Wins," 2023). A native of Wisconsin, Peterson joined the Batten Honors College at Virginia Wesleyan in 2019 ("Abby Peterson Wins," 2023). Initially a self-proclaimed climate change skeptic, her perspective evolved significantly throughout her time in the Batten Honors College. By her senior year, Peterson was not only deeply committed to environmental issues but also determined to make a tangible impact. Her project was meticulously researched, and she took it a step further by implementing her plan in the Chimney Hill community of Virginia Beach ("Abby Peterson Wins," 2023). Peterson secured grants to fund the tree-planting project, turning her academic work into real-world environmental action ("Abby Peterson Wins," 2023).

STRATEGIC GROWTH

The environmental risks facing the region surrounding Virginia Wesleyan underscore an urgent, critical need to prepare the next generation of environmental leaders and stewards. With the visible impacts of climate change in Virginia Beach, including rising sea levels and coastal erosion, VWU is positioned at the forefront of addressing these challenges. In fact, a recent study from Penn State University states that "sea levels will rise an additional 1 to 6 feet in the next 50 years (by 2070), submerging property and permanently reshaping Virginia Beach's coastline" ("Community Climate Outlook," 2022).

As the Coastal Virginia area becomes a focal point for climate resilience efforts, the university has the opportunity to play a pivotal role in educating students to take immediate, impactful action.

With this in mind, Virginia Wesleyan's commitment to the environment had to be bigger than one building or one program. The university needed to dedicate itself to its climate commitments on a campus-wide scale, involving more students, faculty, and staff in this urgent problem facing the region and the world. With the guidance from colleagues at Allegheny College, Virginia Wesleyan established its version of the Environmental Guiding Principles in 2019.

The idea behind the Environmental Guiding Principles is to embed these values in all policy and operational activities that will withstand the test of time. Presidents change. Deans change. Boards change. The purpose of the Environmental Guiding Principles are to reflect a larger commitment to ethical conduct and social responsibility through environmental stewardship and education.

Virginia Wesleyan's Environmental Guiding Principles state:

We seek to be a campus community:

- *In which the University is an environmental leader in all aspects of institutional functions, including planning, operations, purchasing, and maintenance;*
- *In which environmental efforts encourage interdisciplinary and collaborative teaching and research;*
- *In which diversity, inclusion, and equity are valued and recognized for their role in supporting and creating a vibrant, sustainable, healthy, and resilient community;*

- *In which we acknowledge our obligation to live responsibly as part of a forested ecosystem that mantles our campus and promotes a rich learning environment;*
- *In which we build a campus community that promotes health and wellness efforts in harmony with the natural world; and*
- *In which University leadership and expertise contribute to Coastal Virginia's local and regional environmental efforts.*

To advance our goals, we will:

- *Incorporate environmental and social concerns as a priority in University decision making, planning, and investing;*
- *Consider social, environmental, and economic impacts of Virginia Wesleyan's policies, practices, and growth;*
- *Train the next generation of environmental stewards and professionals of local and regional distinction;*
- *Foster the University's national-niche programs in environmental sciences and the Batten Honors College by encouraging investments from internal and external resources;*
- *Seek practices and procedures that protect, enhance, and regenerate the environment;*
- *Promote the University as a living laboratory for sustainability, contributing to the education, outreach, and research mission of the University;*
- *Use participatory processes in developing Virginia Wesleyan's policies and practices; and*
- *Fulfill our responsibility as good neighbors by working with the community to foster social equity, environmental justice, and healthy natural systems.*
- *Our decisions and actions will be guided by the Virginia Wesleyan University mission statement, reflective of the University's values and resources. As a learning institution, we recognize that achieving environmental sustainability will be an evolving practice.*

("Environmental Guiding Principles," 2020)

The VWU Board of Trustees adopted the Environmental Guiding Principles during their Winter Board Meeting in February 2020. Since then, the Environmental Guiding Principles have helped to shape the decisions of the university in multi-faceted ways.

One example would be that the Environmental Guiding Principles inspired the university to update its mission statement to include its commitment to the environment. Virginia Wesleyan University's mission statement is "an inclusive community grounded in the liberal arts and sciences and dedicated to scholarship, service, lifelong learning, and environmental stewardship, Virginia Wesleyan University inspires students to build meaningful lives through engagement in Coastal Virginia's dynamic metropolitan region, the nation, and the world" (Our mission, n.d.).

This mission statement, along with the Environmental Guiding Principles, has inspired support across the university's constituency groups - students, faculty, staff, and donors - and has resulted in multiple accolades for the university.

For example, the president's residence, DeFord Manor, was named a River Star Home by the Elizabeth River Project ("DeFord Manor Recognized," 2022). The Elizabeth River Project is an independent non-profit that works with partner organizations to restore the Elizabeth River, a tributary of the Chesapeake Bay (Bulletin, 2024).

DeFord Manor was opened in 2022 and named with gratitude to Bob and Nancy DeFord, another example of philanthropy inspiring philanthropy ("Dedication of DeFord Manor," 2022).

Thanks to the DeFords' support, the home was constructed with the university's Environmental Guiding Principles in mind. The landscape includes about 460 trees, shrubs, and plants, as well as nearly a mile of underground stormwater chambers and a bio pond ("DeFord Manor," n.d.). The driveway and circle are constructed with permeable materials that absorb water, helping to prevent runoff, flooding, and groundwater contamination ("DeFord Manor," n.d.). This system filters surface water, preventing harmful chemicals from vehicles from polluting storm drains or groundwater, helping the home to earn this distinction ("DeFord Manor," n.d.).

A Mission-Driven Endowment

But the most important mission-driven, campus-wide change inspired by the Environmental Guiding Principles came in 2021. The success of the Batten Honors College was well-documented by that point. Inspired by this success and the many positive changes on campus, many of which were green initiatives, the original donor committed an additional $60 million to the Batten Honors College. Of the total given to the program since 2017, $80.3 million was for the endowment ("State of the University," 2021). Of the 147 honors colleges in the country, the Batten Honors College of Virginia Wesleyan University is believed to have one of the largest endowments among the private college programs.

There were three conditions of the endowment gift: (1) Gift anonymity, (2) $80.3 million for the endowment of the honors college with a guarantee that the mission of the honors college remain focused on the environment and (3) that all endowment funds be invested in an environmentally friendly way. Shortly after investing the initial $80.3 million, with the Environmental Guiding Principles in mind, the

Virginia Wesleyan Board of Trustees voted unanimously to move the entirety of the university's endowment to environmentally-friendly investing ("Values-Aligned Investments," n.d.).

To do this, the university turned to its outsourced Chief Investment Officer, Brown Advisory of Baltimore, Maryland, and a member of the Intentional Endowments Network. The Intentional Endowments Network is a peer-learning network that encourages and supports higher education institutions to adopt investment strategies that create an equitable, low-carbon, and regenerative economy ("About the Intentional Endowment Network," n.d.). Sixty higher educational institutions nationwide partner with over 200 financial institutions committed to environmentally friendly investing "(About the Intentional Endowment Network," n.d.). VWU was an early member of the Intentional Endowments Network, joining in 2018. The network helps institutions like VWU and Brown Advisory in creating roadways for sustainable investing (and divesting).

The higher education sector, "educating 20 million students per year in the US" ("About the Intentional Endowment Network," n.d.), is a formidable force, and Intentional Endowments Network recognizes its potency. Spearheaded by influential figures in higher education, there's a tangible drive to leverage the immense assets of educational institutions towards creating a sustainable economic landscape, and Virginia Wesleyan is proud to be a part of it.

The decision to move the university's entire endowment to environmentally-friendly investments was monumental. Inspired by philanthropy and aptly led by our Board of Trustees, the decision shows the university's deep and lasting commitment to its mission, its Environmental Guiding Principles, and the fragile coastal environment it calls home.

Student Support

One of the most rewarding outcomes of the university's mission statement and Environmental Guiding Principles is the students' deep engagement with and impassioned promotion of environmental stewardship and sustainability initiatives in both their academic programs and extracurricular activities. Through these efforts, students have authentically connected with the university's beautiful and diverse location and created initiatives that reflect their affinity for this region.

Among my favorite examples are VWU students engaging in the Chesapeake Bay Foundation's Oyster Gardening Program, where they care for oysters to help restore the Bay's ecosystem (Mayfield, 2019). In April 2024, the "Environmental Champions" panel brought together young VWU students and alumni dedicated to tackling environmental challenges both during their studies and in their careers ("Environmental Champions," 2024). In Fall 2023, the junior-level Management,

Business, and Economics students organized the bi-annual Marlin Business Conference with the theme, "A Greener Tomorrow," highlighting environmentally-friendly career opportunities and sustainable business practices ('Spring 2023," 2023).

Most notably, the university's membership in the Intentional Endowments Network allowed scholars from the Batten Honors College to participate in the 2022 IEN Student Corporate Engagement Competition, which is a first-of-its-kind student investment competition. In the competition, students pitch an investment in a publicly-traded company and include in that recommendation a shareholder engagement strategy focused on addressing inequality and the climate crisis, as they are systemic risks that threaten a healthy society, and in turn, healthy portfolio returns (Frank, 2022).

Building on what they had learned about the environment through the Batten Honors College, VWU's team chose to study Ford Motor Company for the competition (Frank, 2022). The team noted that Ford's corporate commitment to Electric Vehicle production, changes to U.S.-based manufacturing, and commitment to reduce waste and emissions made it a worthy consideration for investment ("Batten Honors College Scholars," 2022). The team also proposed a shareholder strategy that included the development of EV-Lithium-Ion battery recycling program through Ford's factories and dealerships ("Batten Honors College Scholars," 2022).

VWU's team competed against nine other student teams from colleges and universities across the U.S. and Canada, including Arizona State University, Central Washington University, New York University Stern School of Business, Presidio Graduate School, Ryerson University, University of California Berkeley Haas School of Business, and University of Minnesota (Frank, 2022).

During the finals, the top two teams in both the graduate and undergraduate categories livestreamed their presentations as they presented their final deliverables and discussed their strategies to improve company business performance and impact on society (Frank, 2022).

Thanks to their Virginia Wesleyan liberal arts education rooted in environmental stewardship and climate action, the Batten Honors College team, composed of five sophomores and juniors, took first place in this national competition.

As the culmination of engaging in campus environmental initiatives and practices, VWU established the Sustainability Leaders program to honor "green" bachelor degrees at Virginia Wesleyan. The program recognizes students who have completed volunteer service and coursework focused on real-world issues of sustainability and the environment. Those deemed a Sustainability Leader receive a special cord (with colors dyed from recycled food waste scraps, of course) to wear during Commencement when they graduate.

Academic Programs Showcasing Commitment

Building on the success of student-led environmental programs on campus and strong leadership in environmental education from Dr. Malcolm and others, the impact of these programs has expanded beyond university grounds. VWU students now participate in courses and research abroad, empowering them to apply their sustainability efforts and knowledge to address broader community and global challenges while preparing them to thrive in environmental science careers.

In one of the first Batten Honors College study-away courses, students explored pressing environmental challenges in Alaska. Through collaboration with local experts and native Tlingit elders, students delved into topics such as climate change, glaciers and glaciation, geology, ecology, and sustainability, with a strong emphasis on understanding the complexity of these challenges and the interdisciplinary approach required to find sustainable solutions ("Exploring Sustainability," 2017).

As opportunities expanded to all Virginia Wesleyan students, one student group explored various habitats and geological features of Costa Rica, such as a biological station, a volcano, and wildlife sanctuaries to study biodiversity and sustainable agricultural practices (January Term, 2024). Another student participated in a National Science Foundation Research for Undergraduates program in Belize, studying how frequent flooding affected infrastructure and impacted a community ("Bridging Academic," 2023). One student surveyed freshwater ecosystems with the Department of Environmental Protection in Maryland (Wampler, 2023, p. 6). The list goes on and on.

As one of its niche programs, Virginia Wesleyan has developed a collaborative campus with Lakeland University of Wisconsin that is located in neither Virginia nor Wisconsin, but rather in the heart of Tokyo, Japan. This campus operates as "Lakeland University Japan and Virginia Wesleyan University Global." The unique and exclusive agreement between the two institutions was finalized in the spring of 2020, thanks to a donor who was inspired by the environmental and academic progress of Virginia Wesleyan.

In June 2024, VWU offered a course called "Urban Ecology in Japan" to give students a hands-on exploration of urban environments and their human impact. The three-week program, which took place at the Tokyo campus, examined urban ecosystems and focused on issues like water, air, energy, and public health ("Exploring Urban Ecology," 2024). Field visits included the Tokyo Port Wild Bird Park, which showcased green spaces in urban settings, and the Shinagawa waste incineration plant, which demonstrated Tokyo's advanced waste management ("Exploring Urban Ecology," 2024). Students also studied the urban heat island effect through thermal imaging ("Exploring Urban Ecology," 2024). A cross-country trip via bullet train allowed comparisons between urban and rural ecosystems in cities like Kyoto, Osaka, and Hiroshima ("Exploring Urban Ecology," 2024). The course emphasized global education, sustainability, and cross-cultural understanding, deepening students' appreciation of the interconnectedness of ecological and human systems.

Japan is a nation of over 6,800 islands, so climate change and rising seas levels are at the forefront of their policies and action plans. By operating a campus in Japan, Virginia Wesleyan continues to grow into a higher education leader in environmental sustainability thanks to its strategic leadership, visionary philanthropy, and geographic advantage beyond Coastal Virginia.

Perhaps one of the longest-running environmental research and climate action projects of Virginia Wesleyan students and faculty is a heat-mapping study in Virginia, led by Dr. Malcolm. This program directly connects the university's mission to bettering the environment in its Coastal Virginia home. In the summer of 2021, faculty and students from member schools of the Virginia Foundation for Independent Colleges, a consortium of 17 private colleges and universities in the Commonwealth, organized citizen science research to measure the urban heat islands across Virginia Beach. They engaged over 200 volunteers including VWU students, staff, alumni, and community members to participate in the project ("Measuring Urban Heat," 2022).

Faculty and students then analyzed the collected data using a geographic information system. This analysis created a map of heat risk index, which combines the heat data with demographic data such as asthma and kidney disease, that can be used to identify neighborhoods facing the highest health risk from extreme heat ("Measuring Urban Heat," 2022). This risk map, along with maps of tree cover and impervious surfaces, are now used to identify neighborhoods to target for tree planting initiatives, which provide shade and carbon removal benefits ("Measuring Urban Heat," 2022). These maps were shared with the City of Virginia Beach and Lynnhaven River Now, a local environmental group, for use in future projects ("Measuring Urban Heat," 2022).

Dr. Malcolm presented results from the project along with Dr. Karin Warren (Randolph College) and Dr. Michael Wolyniak (Hampden-Sydney College) at the 2022 Higher Education Climate Leadership Summit in a session titled "Virginia Heat Watch: Heat Mapping to Assess Vulnerability and Address Inequity in Virginia Communities" ("Measuring Urban Heat," 2022). Shortly after the project's conclusion, Dr. Malcolm applied for a grant from the Virginia Department of Forestry to continue its work in urban heat mapping. In March 2024, Virginia Wesleyan was awarded a $222,000 grant to combat urban heat stress in Portsmouth, Virginia, a close neighbor of Virginia Beach and a key area of disparity discovered during the initial heat mapping program (Hafner, 2024). The two-year project, led by VWU faculty, focuses on reducing rising temperatures in disadvantaged areas with low tree coverage and impervious surfaces, emphasizing VWU leadership in this area.

Furthering Campus Engagement

Back on campus, Virginia Wesleyan has continually demonstrated its commitment to sustainability, innovation, and community engagement through a series of partnerships and initiatives that integrate environmental stewardship into the heart of the institution. These collaborations not only enhance VWU's role as a leadership in sustainability in higher education but also expand its impact on all students.

VWU served as a host institution for the 2021 Climate Action Pursuit, a leadership summit focused on climate priorities, led by Second Nature and the Intentional Endowments Network. VWU participated alongside other institutions, addressing key themes like reducing greenhouse gas emissions, aligning endowment investments with sustainability, and advancing diversity, equity, and inclusion. The event, which included faculty, staff, and students, emphasized collaboration to achieve environmental and social justice goals throughout the year.

Expanding its connection with the Intentional Endowments Network, VWU hosted the organization's Fall Forum in October 2023. The Forum brings in guests and participates from all over to build consensus in investment policies for values-aligned investment, promoting racial equity in portfolios, and upholding the sanctity of investment autonomy ("Virginia Wesleyan University Hosts," 2023). VWU was featured in a panel which included VWU alumni and Trustee Gary Bonnewell, a personal wealth advisor; Rowan Stuart, a Batten Honors College scholar and member of the award-winning Brown Advisory Student Investment Fund team; and me. The group shared their insights and experiences in the development of VWU's endowment incorporating Environmental, Social, and Governance (ESG) criteria into its portfolios ("Virginia Wesleyan University Hosts," 2023). As a student, Mr. Stuart spoke to what this initiative means to students and young people and how this momentum is carried forward by the student body in extracurricular activities.

In other initiatives and partnerships, Virginia Wesleyan welcomed Dr. Alexander Siegmund, a German sustainability scholar, to campus in August 2023. This visit not only created a platform for cross-cultural exchange but also provided students with exposure to international perspectives on sustainability practices, enriching the campus community with global insights and encouraging a diverse approach to ecological challenges ('VWU Hosts," 2023).

Further emphasizing its commitment to environmental sustainability, VWU has expanded its partnership with the Port of Virginia, an essential hub for the state's economy. This partnership aims to incorporate sustainability into logistics and operations, aligning with global trends that promote environmentally responsible business practices ("VWU Develops Custom," 2022). Through collaboration, VWU and the Port of Virginia engage in educational opportunities and workforce development initiatives that prepare students for careers in the maritime industry while integrating principles of environmental sustainability. This partnership is a powerful example of how higher education and industry can work together to create a sustainable future while driving economic growth.

Virginia Wesleyan's Environmental Institute: Summer Scholars is a week-long residential environmental science program designed for top-performing, rising ninth-grade girls. Participants take part in authentic environmental science, technology, engineering, and math (E-STEM) experiences alongside environmental science professionals in academia, industry, business, government, and non-profit organizations in order to better understand the vast number of opportunities for women in E-STEM fields to make a positive difference in the world ("Summer Scholars," 2018). The program received the 2020 Inspiring Programs in STEM Award from INSIGHT Into Diversity magazine, the largest and oldest diversity and inclusion publication in higher education ("VWU Receives," 2020).

All of these programs show the commitment of VWU to offer environmental programs that engage the campus community at every level.

Accolades Acknowledging Efforts

The VWU community, including faculty, staff, and students, is dedicated to creating a greener campus and contributing to a healthier planet. Visitors to the campus immediately notice the emphasis on environmental stewardship, reflecting its broader commitment to ethical conduct, social responsibility, and environmental stewardship.

Virginia Wesleyan has earned recognition locally, regionally, and nationally for its sustainability practices. In 2019, Virginia Wesleyan was recognized as National Conservationist of the Year by the Chesapeake Bay Foundation (Mayfield, 2019). The university was the 2023 inductee to the Elizabeth River Project and Inside Business's River Star Hall of Fame ('VWU Recognized," 2022). Virginia Wesleyan has also been consistently named by *The Princeton Review* as one of the most environmentally responsible colleges in the U.S. and Canada in its *Guide to Green Colleges* ("Green Colleges," 2024).

Virginia Wesleyan University Dining Services is one of a handful of Virginia colleges and universities to receive the green restaurant designation. To qualify, restaurants (or dining halls) must minimize the use of disposable food service products, recycle grease, recycle and reduce waste, use water efficiently and conserve energy.

Thanks to a grant and inspiration from the Greer Environmental Sciences Center, the university installed a solar SmartFlower, a cutting-edge photovoltaic system that tracks the sun, right outside of the Greer Center. The dual-axis system allows the modular fan to follow the sun across the sky throughout the day, always maintaining the optimal angle and generating 40 percent more energy than a traditional fixed solar panel (Yoakum, 2018). Installed in August 2018, Virginia Wesleyan was the second institution of higher education in the nation to install this revolutionary solar system (Yoakum, 2018), which powers Honors Village for students in the Batten Honors College and serves as a hands-on learning tool.

Virginia Wesleyan was honored for its commitment to effective urban forest management as a 2022 and 2023 Tree Campus, a recognition by the Arbor Day Foundation. VWU achieved the distinction by meeting Tree Campus Higher Education's five standards, including maintaining a tree advisory committee, a campus tree-care plan, dedicated annual expenditures for its campus tree program, an Arbor Day observance, and student service-learning project ("Tree Campus," n.d.). Currently, there are just 411 campuses across the United States with this recognition ("Tree Campus," n.d.).

In keeping with its urban forest management, in June 2024, the university proactively addressed the threat posed by the invasive emerald ash borer, which has devastated ash trees across North America ("No Need to Fear," 2024). While the beetle has not yet been detected in the Coastal Virginia area, VWU partnered with the Virginia Department of Forestry to protect its large stands of ash trees. The university was selected as a 2024 Biocontrol Release Site, receiving shipments of a wasp species that targets the beetle ("No Need to Fear," 2024). With faculty and students actively involved in this conservation effort, the initiative enhances both ecological preservation and hands-on education in environmental science ("No Need to Fear," 2024).

Because of all these initiatives and more, Virginia Wesleyan completed the Advancement for Sustainability in Higher Education (AASHE) Sustainability Tracking, Assessment & Rating System (STARS) reporting process. STARS is a "transparent, self-reporting framework for colleges and universities to measure their sustainability performance," that is "intended to engage and recognize the full spectrum of higher education institutions, from community colleges to research universities" ("The Sustainability Tracking," n.d.). VWU earned a STARS Silver rating in recognition of reported accomplishments in campus sustainability in its first submission and review ("Virginia Wesleyan University STARS Report," 2023). Like the mission statement and the Environmental Guiding Principles, the rating helps the university frame its next steps in creating a greener, more sustainable campus.

Opportunities for the Future

Looking ahead, Virginia Wesleyan University is well-positioned for significant growth in its sustainability journey. As the campus evolves, the university is prioritizing eco-friendly upgrades to reduce the carbon footprint of its aging facilities, such as solar and green roofs and energy-efficient HVAC systems. Currently, 40 percent of the university's energy comes from sustainable sources, but further long-term investment in more advanced renewable energy systems, such as wind or geothermal energy, could enhance the university's energy independence. The inclusion of EV charging stations at the new Virginia Museum of Contemporary Art (slated to open on campus in 2026) also signals a broader commitment to supporting clean transportation options while putting the opportunity for change directly in users' hands, which helps raise their awareness of how their individual actions affect our shared environment. In time, this EV charging station initiative may be expanded across campus to further encourage electric vehicle adoption among VWU faculty, students, and visitors.

Another exciting growth area lies in the university's academic programs. Through the statewide Batten Environmental Education Initiative, the university is exploring ways to offer free environmental science and sustainability courses to all students, regardless of their major, reflecting a progressive and inclusive vision for society-wide ecological education.

To further this initiative, VWU will continue to partner with local businesses and environmental organizations to create internship and job placement pipelines, ensuring that students have practical experiences aligned with their coursework. In a region teeming with relevant eco-focused career opportunities, from off-shore wind projects to nearby solar farms, VWU is poised to support the growing green job market in its region and beyond.

Looking outward, VWU alumni are already driving positive environmental change through professional, community, and individual efforts. The university has identified that it can further strengthen the ecological influence of its alumni network by creating mentorship programs that connect current students with graduates working in environmental fields. Highlighting success stories—such as VWU alumni who are addressing sea-level rise on Tangier Island or working with the Elizabeth River Project— will inspire a new generation of environmental advocates while also showing that their academic education can lead to the opportunity to create practical, tangible, and measurable environmental improvements.

Collaboration is also key to VWU's proactive influence within the region. By expanding institutional partnerships with regional environmental organizations to offer joint research opportunities, workshops, and public outreach programs, VWU will continue to embrace its role as a sustainability leader in the community. Revitalizing campus and community events like "Generations Uniting to Address Climate Change," held in both 2019 and 2023, brought together a panel of experts from science, policy, and communication backgrounds who discussed the climate crisis's evolution, impacts, and solutions for an audience of students and community members ("Celebrating Earth Week," 2023). The event was offered for free and open to anyone looking to be inspired by change makers from across the generations.

Events like these ensure that the university is not only seen as a leader in environmental education and stewardship, but also that the conversations and programs it initiates are helping to actively inspire broader change.

CONCLUSION

As Virginia Wesleyan University continues to champion environmental stewardship at both the institutional and personal level, it has the opportunity to lead by example during higher education's ongoing fight to achieve greater nationwide sustainability—especially in a region as ecologically significant as Virginia Beach. Nestled between the Chesapeake Bay and the Atlantic Ocean, VWU is ideally positioned to address the severe environmental challenges that are unique to coastal communities. In a city renowned for its natural beauty, wetlands, and diverse ecosystems, the university's location underscores its responsibility to chart an environmentally conscious path that other institutions and individuals can follow, not just for the region but for the world.

Virginia Beach, with its vulnerability to rising sea levels and climate change, highlights the need for institutions like VWU to integrate sustainability into every facet of campus operations and curriculum. By doing so, VWU educates students about the real-world impact of environmental degradation while providing them the tools they need to develop innovative, practical, replicable, and scalable solutions. The university's coastal setting also serves as a living laboratory where students can engage in hands-on learning experiences, conducting research that directly benefits the local environment and strengthens global ecological resilience. VWU's collaborative campus in Tokyo, Japan, further reinforces its dedication to empower students to think globally while acting locally when addressing climate issues.

The philanthropy of alumni and other benefactors who are personally invested in combating the dangers of climate change has played a pivotal role in advancing VWU's mission. Their gifts have allowed the university to expand its sustainability initiatives and offer programs that grow its prestige and influence, making VWU an even more magnetic destination for students who are inspired to change — and protect — the world. This generous donor support has enabled investments in environmental research and academic programs, sustainable campus practices, state-of-the-art facilities, and study-away opportunities. These efforts not only benefit Virginia Wesleyan students, but also help to contribute to the broader global conversation on sustainability, which is a discussion that individuals in all industries and walks of life should be having far more often.

By fostering a collaborative culture of environmental responsibility, Virginia Wesleyan is ensuring that its students graduate with a lifelong commitment to sustainability. This commitment, deeply rooted in both the university's mission and its location, and fully supported by university leadership at all levels as well as outstanding philanthropists, prepares VWU students to lead and innovate in ways that will have a lasting positive impact on the environment. As a coastal institution, VWU's choices today resonate locally and globally, establishing the university as a leader in environmental efforts in modern higher education and influencing how future generations will adapt to the challenges of environmental stewardship while developing solutions to protect and sustain the planet.

REFERENCES

Abby Peterson wins Ryan Environmental Prize for Excellence in Research. (2023, April 27). Virginia Wesleyan University. https://www.vwu.edu/about/news-and-events/features/story/winners-name-wins-ryan-environmental-prize-for-excellence-in-research

About the Intentional Endowment Network. (n.d.) Intentional Endowment Network. https://www.intentionalendowments.org/about_ien

About us. (2024, August 14). Second Nature. https://secondnature.org/about-us/

America, E. (2023, July 6). *Beach pollution puts swimmers at risk nationwide – report.* YubaNet. https://yubanet.com/usa/beach-pollution-puts-swimmers-at-risk-nationwide-report/

Baker, L. (2011, March 11). *A way with worms.* Virginia Wesleyan University. https://www.vwu.edu/about/sustainability/a-way-with-worms.php

Batten Honors College scholars take first place in IEN national competition. (2022, May 2). Virginia Wesleyan University. https://www.vwu.edu/about/news-and-events/features/story/batten-honors-college-scholars-take-first-place-in-ien-national-competition

Batten Honors College symposium and prize to encourage environmental exploration. (2021, December 22). Virginia Wesleyan University. https://www.vwu.edu/about/news-and-events/features/story/new-program-announced-for-the-batten-honors-collegenbsp

Bridging academic and real-world challenges. (2023, November 28). Virginia Wesleyan University. https://www.vwu.edu/about/news-and-events/features/story/bridging-academia-and-real-world-challenges

Bulletin, B. (2024, June 4). *Resilience lab, designed to float when it floods, opens on the Elizabeth River.* Chesapeake Bay Magazine. https://www.chesapeakebaymagazine.com/resilience-lab-designed-to-float-when-it-floods-opens-on-the-elizabeth-river/

Celebrating Earth week: April 17-21. (2023, March 24). Virginia Wesleyan University. https://www.vwu.edu/about/news-and-events/features/story/celebrating-earth-week-april-17-ndash-21

Community climate outlook VIRGINIA BEACH VA. (2022, August 16). Pennsylvania State University. https://www.marisa.psu.edu/misc/outlooks/2022-01-11/VA/Virginia_Beach_VA.pdf

Dedication of DeFord Manor. (2022, May 12). Virginia Wesleyan University. https://www.vwu.edu/about/news-and-events/features/story/dedication-of-deford-manor

DeFord Manor. (n.d.) Virginia Wesleyan University. https://www.vwu.edu/about/president-and-leadership/deford-manor.php

DeFord Manor recognized as 'River Star Home'. (2022, July 25). Virginia Wesleyan University. https://www.vwu.edu/about/news-and-events/features/story/deford-manor-recognized-as-river-star-home

Earth Angels. (2023). *Virginia Wesleyan University Magazine, Fall 2023,* 14–15.

Environmental champions encourage others to take action. (2024, April 22). Virginia Wesleyan University. https://www.vwu.edu/about/news-and-events/features/story/environmental-champions

Environmental Guiding Principles. (2020). Virginia Wesleyan University. https://www.vwu.edu/about/sustainability/environmental-guiding-principles.php

Exploring sustainability in Alaska. (2017, June 14). Virginia Wesleyan University. https://www.vwu.edu/about/news-and-events/features/story/exploring-sustainability-in-alaska

Exploring urban ecology in Japan. (2024, September 10). Virginia Wesleyan University. https://www.vwu.edu/about/news-and-events/features/story/exploring-urban-ecology-in-japan

Frank, A. (2022, May 3). *Bard & Virginia Wesleyan win top honors in national collegiate investment competition.* IEN. https://www.intentionalendowments.org/bard_and_virginia_wesleyan_win_top_honors_in_national_collegiate_investment_competition

Green colleges. (2024). The Princeton Review. https://www.princetonreview.com/college-rankings?rankings=green-colleges

Greer Environmental Sciences Center dedicated on Founders Day. (2017, September 7). Virginia Wesleyan University. https://www.vwu.edu/about/news-and-events/features/story/greer-environmental-sciences-center-dedicated-on-founders-day

Greer Environmental Sciences Center. (n.d.). Virginia Wesleyan University. https://www.vwu.edu/academics/school-of-mathematics-and-science/greer-environmental-sciences-center.php

Hafner, K. (2024, July 16). *How hot does it get in Portsmouth? New mapping project aims to find out.* WHRO. https://www.whro.org/environment/2024-07-16/how-hot-does-it-get-in-portsmouth-new-mapping-project-aims-to-find-out

January term 2024: Students dive into hands-on learning, intensive field experiences. (2024, January 18). Virginia Wesleyan University. https://www.vwu.edu/about/news-and-events/features/story/january-term-2024-students-dive-into-hands-on-learning-intensive-field-experiences

Mayfield, D. (2019, August 9). *Virginia Wesleyan wins top conservation award from Chesapeake Bay Foundation.* The Virginian-Pilot. https://www.pilotonline.com/2017/12/06/virginia-wesleyan-wins-top-conservation-award-from-chesapeake-bay-foundation-2/

Measuring urban heat across Virginia. (2022, March 29). Virginia Wesleyan University. https://www.vwu.edu/about/news-and-events/features/story/measuring-urban-heat-across-virginia

Merzdorf, J. (2018, July 18). *Study of the world's beaches shows threats to protected areas-NASA Science.* NASA Science. https://science.nasa.gov/science-research/earth-science/study-of-the-worlds-beaches-shows-threats-to-protected-areas/

Mission and vision. (n.d.) Virginia Wesleyan University. https://www.vwu.edu/academics/batten-honors-college/mission-vision.php

Newspaper exec who founded Weather Channel dies. (2009, October 30). 13NewsNow. https://www.13newsnow.com/article/news/local/newspaper-exec-who-founded-weather-channel-dies/291-375839746

No need to fear these wasps, unless you're an emerald ash borer. (2024, August 1). Virginia Wesleyan University. https://www.vwu.edu/about/news-and-events/features/story/no-need-to-fear-these-wasps-unless-yoursquore-an-emerald-ash-borer

Our history. (n.d.). Virginia Wesleyan University. https://www.vwu.edu/about/our-history.php

Our mission. (n.d.). Virginia Wesleyan University. https://www.vwu.edu/about/our-mission.php

Presidents of Wesleyan. (2015). Virginia Wesleyan University. https://www.vwu.edu/about/president-and-leadership/presidents-of-wesleyan.php

Schueman, L. J. (2023, December 13). *How philanthropy can support climate justice.* One Earth. https://www.oneearth.org/how-philanthropy-can-support-climate-justice/

Shearin, M. Z. (2010, June 24). *Tulip poplars populate campus.* Virginia Wesleyan University. https://www.vwu.edu/about/sustainability/tulip-poplars-populate-campus.php

Shearin, M. Z. (2010, September 22). *Being seen being green.* Virginia Wesleyan University. https://www.vwu.edu/about/sustainability/being-seen-being-green2.php

Spring 2023 Marlin Business Conference: A Greener Tomorrow. (2023, April 13). Virginia Wesleyan University. https://www.vwu.edu/about/news-and-events/features/story/spring-2023-marlin-business-conference-a-greener-tomorrow

State of the university: Record fundraising and endowment growth. (2021, August 20). Virginia Wesleyan University. https://www.vwu.edu/about/news-and-events/features/story/state-of-the-university-record-fundraising-and-endowment-growth

Summer scholars. (2018, August 7). Virginia Wesleyan University. https://www.vwu.edu/about/news-and-events/features/story/summer-scholars

The Batten legacy: Frank and Jane Batten. (n.d.). Virginia Wesleyan University. https://www.vwu.edu/academics/batten-honors-college/the-battens.php

The sustainability tracking, assessment, and rating system. (n.d.). STARS: A program of AASHE. https://stars.aashe.org/

Tree campus higher education. (n.d.). Arbor Day Foundation. https://www.arborday.org/programs/tree-campus-higher-education/

Values-aligned investments. (n.d.). Virginia Wesleyan University. https://www.vwu.edu/about/sustainability/values-aligned-investments.php#:~:text=Virginia%20Wesleyan%20University%20invests%20in,%2Dcarbon%2C%20and%20regenerative%20economy

Virginia Wesleyan University hosts IEN Fall Forum: a platform for sustainable investments and societal impact. (2023, October 23). Virginia Wesleyan University. https://www.vwu.edu/about/news-and-events/features/story/virginia-wesleyan-university-hosts-ien-fall-forum-a-platform-for-sustainable-investments-and-societal-impact

Virginia Wesleyan University STARS Report. (2023, January 19). https://reports.aashe.org/institutions/virginia-wesleyan-college-va/report/2023-01-19/

VWU develops custom sustainability course with Port of Virginia. (2022, October 5). Virginia Wesleyan University. https://www.vwu.edu/about/news-and-events/features/story/vwu-expands-partnership-with-port-of-virginia

VWU hosts German sustainability scholar. (2023, August 1). Virginia Wesleyan University. https://www.vwu.edu/about/news-and-events/features/story/vwu-hosts-german-sustainability-scholar

VWU receives 2020 Inspiring Programs in STEM Award. (2020, August 17). Virginia Wesleyan University. https://www.vwu.edu/about/news-and-events/features/story/vwu-receives-2020-inspiring-programs-in-stem-award

VWU recognized by Elizabeth River Project. (2022, December 6). Virginia Wesleyan University. https://www.vwu.edu/about/news-and-events/features/story/vwu-recognized-by-elizabeth-river-project

Wampler, J. (2023, November 17). Surveying freshwater ecosystems. *Excelsis. (Fall 2023),* 6.

Yoakum, R. (2018, September 4). *Solar SmartFlower making power for Virginia Wesleyan University.* Williamsburg Yorktown Daily. https://wydaily.com/latest/local/2018/09/04/solar-smartflower-making-power-for-virginia-wesleyan-university/

Younger, S. (2023, August 9). *NASA study: Rising sea level could exceed estimates for U.S. coasts.* NASA Jet Propulsion Laboratory. https://www.jpl.nasa.gov/news/nasa-study-rising-sea-level-could-exceed-estimates-for-us-coasts/

Zork, C. (2018, February 13). *Three LEED projects contributing to Virginia's ranking on the Top 10 States.* U.S. Green Building Council. https://www.usgbc.org/articles/three-leed-projects-contributing-virginia-s-ranking-top-10-states

Compilation of References

8th Fire Solar. (n.d.). Why solar thermal. Retrieved December 2, 2024, from https://www.8thfiresolar.org/why-8th-fire

Abby Peterson wins Ryan Environmental Prize for Excellence in Research. (2023, April 27). Virginia Wesleyan University. https://www.vwu.edu/about/news-and-events/features/story/winners-name-wins-ryan-environmental-prize-for-excellence-in-research

About the Intentional Endowment Network. (n.d.) Intentional Endowment Network. https://www.intentionalendowments.org/about_ien

About us. (2024, August 14). Second Nature. https://secondnature.org/about-us/

Abreu, M., & Grinevich, V. (2024). The entrepreneurial university: Strategies, processes, and competing goals. *The Journal of Technology Transfer*, *49*(6), 1991–2034. DOI: 10.1007/s10961-024-10085-7

Adamo, N., Al-Ansari, N., & Sissakian, V. (2021). Review of climate change impacts on human environment: Past, present and future projections. *Engineering*, *13*(11), 605–630. DOI: 10.4236/eng.2021.1311044

Akins, E., Giddens, E., Glassmeyer, D., Gruss, A., Kalamas Hedden, M., Slinger-Friedman, V., & Weand, M. (2019). Sustainability education and organizational change: A critical case study of barriers and change drivers at a higher education institution. *Sustainability*, *11*(2), 501. DOI: 10.3390/su11020501

Altbach, P. G. (2016). *Global Perspectives on Higher Education*. JHU Press. DOI: 10.56021/9781421419268

America, E. (2023, July 6). *Beach pollution puts swimmers at risk nationwide – report.* YubaNet. https://yubanet.com/usa/beach-pollution-puts-swimmers-at-risk-nationwide-report/

American Association of Colleges & Universities. (2024). Trending topic: What is liberal education. Retrieved December 2, 2024, from https://www.aacu.org/trending-topics/what-is-liberal-education

Aramburuzabala, P., & Cerrillo, R. (2023). Service-learning as an approach to educating for sustainable development. *Sustainability (Basel)*, *15*(14), 11231. DOI: 10.3390/su151411231

Arbo, P., & Benneworth, P. (2007). Understanding the Regional Contribution of Higher Education Institutions: A Literature Review. OECD Education Working Papers, No. 9. OECD Publishing, Paris.

Armstrong, M., & Stewart-Gambino, H. W. (2009). Building curricular diversity through a "social movement": How faculty networks support institutional change. *Journal of Curriculum Theorizing*, *25*(1). Advance online publication. DOI: 10.63997/jct.v31i1.512

Aronoff, K. (2024, September 30). Climate change does not care about your borders. *The New Republic.* https://newrepublic.com/article/186526/hurricane-helene-asheville-climate-havens

Arrhenius S. (1896). On the influence of carbonic acid in the air upon the temperature of the ground. Philosophical Magazine and Journal of Science Series 5: 237-276.

Arthur, M. M. L. (2016). *Student activism and curricular change in higher education*. Routledge. DOI: 10.4324/9781315611129

Association for the Advancement of Sustainability in Higher Education (AASHE). (2024). Sustainability Tracking, Assessment & Rating System (STARS). Accessed July 20, 2024. https://reports.aashe.org/institutions/participants-and-reports/

Association for the Advancement of Sustainability in Higher Education. (2020). Sustainable Campus Index. Retrieved from https://www.aashe.org/resources/sustainable-campus-index/

Astor, M. (2022, July 1). As federal climate-fighting tools are taken away, cities and states step up. *The New York Times*. https://www.nytimes.com/2022/07/01/climate/climate-policies-cities-states-local.html

Baker, L. (2011, March 11). *A way with worms.* Virginia Wesleyan University. https://www.vwu.edu/about/sustainability/a-way-with-worms.php

Baker, J., & Spracklen, D. (2019). Climate benefits of intact Amazon forests and the biophysical consequences of disturbance. *Frontiers in Forests and Global Change*, *2*(47), 47. DOI: 10.3389/ffgc.2019.00047

Ball, P. (2004). *The Elements: A Very Short Introduction. Very Short Introductions.* OUP Oxford. DOI: 10.1093/actrade/9780192840998.001.0001

Barber, C. (2024). America has a $250 billion problem: Microplastics have invaded our bloodstreams and may increase the risk of heart attack and stroke. *Fortune.* May5th, 2024. https://fortune.com/well/article/microplastics-health-effects/

Barlett, P. F., & Chase, G. W. (2012). Curricular innovation for sustainability: The Piedmont/Ponderosa model of faculty development. *Liberal Education, 98*(4), 14–21.

Barlett, P. F., & Chase, G. W. (Eds.). (2013). *Sustainability in higher education: Stories and strategies for transformation.* MIT Press. DOI: 10.7551/mitpress/9418.001.0001

Barlett, P. F., & Rappaport, A. (2009). Long-term impacts of faculty development programs: The experience of Teli and Piedmont. *College Teaching, 57*(2), 73–82. DOI: 10.3200/CTCH.57.2.73-82

Batten Honors College scholars take first place in IEN national competition. (2022, May 2). Virginia Wesleyan University. https://www.vwu.edu/about/news-and-events/features/story/batten-honors-college-scholars-take-first-place-in-ien-national-competition

Batten Honors College symposium and prize to encourage environmental exploration. (2021, December 22). Virginia Wesleyan University. https://www.vwu.edu/about/news-and-events/features/story/new-program-announced-for-the-batten-honors-collegenbsp

Beach, A. L., Sorcinelli, M. D., Austin, A. E., & Rivard, J. K. (2016). *Faculty development in the age of evidence: Current practices, future imperatives.* Routledge.

Benbow, R. J., Lee, C., & Hora, M. T. (2021). Exploring college faculty development in 21st-century skill instruction: An analysis of teaching-focused personal networks. *Journal of Further and Higher Education, 45*(6), 818–835. DOI: 10.1080/0309877X.2020.1826032

Bittner, G. D., Denison, M. S., Yang, C. Z., Stoner, M. A., & He, G. (2014, December 4). Chemicals having estrogenic activity can be released from some bisphenol A-free, hard and clear, thermoplastic resins. *Environmental Health : A Global Access Science Source, 13*(1), 103. DOI: 10.1186/1476-069X-13-103 PMID: 25477047

Bongaerts, J. C. (2022). The Humboldtian Model of Higher Education and its Significance for the European University on Responsible Consumption and Production. *BHM. Berg- und Hüttenmännische Monatshefte, 167*(10), 500–507. DOI: 10.1007/s00501-022-01280-w

Boyle, K. (2024). The importance of sustainable agriculture: The future of farming. *Lead The Change.* https://leadthechange.bard.edu/blog/the-importance-of-sustainable - agriculture-the-future-of-farming

Brand, D. (2000). *At the full and change of the moon: A novel.* Grove Press.

Brenan, M. (2024, July 15). U.S. confidence in institutions mostly flat, but police up. *Gallup News.* https://news.gallup.com/poll/647303/confidence-institutions -mostly-flat-police.aspx

Brennan, J. (2008). Higher Education and Social Change. *Higher Education Quarterly, 62*(2), 384–386.

Bridging academic and real-world challenges. (2023, November 28). Virginia Wesleyan University. https://www.vwu.edu/about/news-and-events/features/story/bridging-academia-and-real-world-challenges

Brown, K., Daly, A., Engineering, T., Elliott, J., Higgins, C., & Granderson, J. (2010). *Hitting the Whole Target: Setting and Achieving Goals for Deep Efficiency Buildings.*

Brundiers, K., Wiek, A., & Redman, C. L. (2010). Real-world learning opportunities in sustainability: From classroom into the real world. *International Journal of Sustainability in Higher Education, 11*(4), 308–324. DOI: 10.1108/14676371011077540

Brundtland, G. H. (1987). *Our common future world commission on environment and developement.* https://docenti.unimc.it/elena.borin/teaching/2019/21812/files/lessons-and-recommended-readings-1/Our%20Common%20future%201987.pdf

Bullard, R. D., Mohai, P., Saha, R., & Wright, B. (2007). *Toxic Wastes and Race at Twenty: 1987–2007. A Report Prepared for the United Church of Christ Justice and Witness Ministries.* United Church of Christ.

Bulletin, B. (2024, June 4). *Resilience lab, designed to float when it floods, opens on the Elizabeth River.* Chesapeake Bay Magazine. https://www.chesapeakebaymagazine .com/resilience-lab-designed-to-float-when-it-floods-opens-on-the-elizabeth-river/

Burrage, A., Carter, T., & Goldstein, A. (2022, March). *The future of academic research on climate solutions* (p. 3). Second Nature. https://secondnature.org

Bush, V. (1945). *Science, the endless frontier* (Reprinted 1960 by the National Science Foundation). https://archive.org/details/scienceendlessfr00unit/page/n3/mode/2up

Cabrera, A. (2024). A Message from President Cabrera. Georgia Institute of Technology, https://strategicplan.gatech.edu/about/cabrera-message

Cairo, A. (2011). Student engagement in campus sustainability. *Facilities Manager*, March/April. https://files.eric.ed.gov/fulltext/EJ920384.pdf

Cameron, L. A. (n.d.). Rural electrification administration in Minnesota. MNopedia, Minnesota Historical Society. Retrieved December 2, 2024, from https://www.mnopedia.org/thing/rural-electrification-administration-minnesota

Campbell, J. (1995). *Understanding John Dewey: Nature and cooperative intelligence*. Open Court.

Campbell, J. L. (2005). Where do we stand? Common mechanisms in organizations and social movements research. In *Social Movements and Organization Theory* (pp. 41–68). Cambridge University Press., DOI: 10.1017/CBO9780511791000.004

Carson, R. (1962). *Silent spring*. Houghton Mifflin.

Celebrating Earth week: April 17-21. (2023, March 24). Virginia Wesleyan University. https://www.vwu.edu/about/news-and-events/features/story/celebrating-earth-week-april-17-ndash-21

Cerenvak, S. (2023). *Black gathering*. Duke University Press.

Chism, N. V. N., Holley, M., & Harris, C. J. (2012). 9: Researching the impact of educational development: Basis for informed practice. *To Improve the Academy*, *31*(1), 129–145. DOI: 10.1002/j.2334-4822.2012.tb00678.x

Cho, R. (2020). Recycling in the U.S. is broken: How do we fix it? *State of the Planet*.https://news.climate.columbia.edu/2020/03/13/fix-recycling-america/

Clayton, P., & Schwartz, W. A. (2019). *What is ecological civilization: Crisis, hope, and the future of the planet*. Process Century Press.

Climate Smart Minnesota. (n.d.). About. Retrieved December 2, 2024, from https://www.climatesmart-mn.org/about

Cohen, S. (2014). Understanding how Americans view the environment. *Huffpost*, May 24. https://www.huffpost.com/entry/understanding-how-america_b_5020669

Cohen, A., & Kisker, C. (2010). *The shaping of American higher education: emergence and growth of the contemporary system* (2nd ed.). Jossey-Bass.

Collazo Expósito, L. M., & Granados Sánchez, J. (2020). Implementation of SDGs in university teaching: A course for professional development of teachers in education for sustainability for a transformative action. *Sustainability (Basel)*, *12*(19), 8267. DOI: 10.3390/su12198267

Community climate outlook VIRGINIA BEACH VA. (2022, August 16). Pennsylvania State University. https://www.marisa.psu.edu/misc/outlooks/2022-01-11/VA/Virginia_Beach_VA.pdf

Compagnucci, L., & Spigarelli, F. (2020). The Third Mission of the university: A systematic literature review on potentials and constraints. *Technological Forecasting and Social Change*, *161*, 120284. DOI: 10.1016/j.techfore.2020.120284

Condé, M. (1989). *Crossing the mangrove*. Penguin Random House.

Condon, S. (2015, November 10). "GOP debate fact check: Was Marco Rubio right about welders vs. philosophers?" CBS News. https://www.cbsnews.com/news/republican-debate-fact-check-was-marco-rubio-right-about-welders-versus-philosophers/

Connecticut State College and Universities. (2017). *Energy Master Plan*, August 2017. (Also see 2021-24 Eversource and UI Project, Connecticut State Colleges and Universities)

Coops, N. C., Marcus, J., Construt, I., Frank, E., Kellett, R., Mazzi, E., Munro, A., Nesbit, S., Riseman, A., Robinson, J., Schultz, A., & Sipos, Y. (2015). How an entry-level, interdisciplinary sustainability course revealed the benefits and challenges of a university-wide initiative for sustainability education. *International Journal of Sustainability in Higher Education*, *16*(5), 729–747. DOI: 10.1108/IJSHE-04-2014-0059

Cortese, A. (2010). Higher education's purpose: A healthy, just, and sustainable society. *Inspired Economics*, 1/28/2010.

Cox, M. D. (2004). Introduction to faculty learning communities. *New Directions for Teaching and Learning*, *2004*(97), 5–23. DOI: 10.1002/tl.129

Cox, M. D. (2013). The impact of communities of practice in support of early-career academics. *The International Journal for Academic Development*, *18*(1), 18–30. DOI: 10.1080/1360144X.2011.599600

Cross, I. D., & Congreve, A. (2021). Teaching (super) wicked problems: Authentic learning about climate change. *Journal of Geography in Higher Education*, *45*(4), 491–516. DOI: 10.1080/03098265.2020.1849066

Crownhart, C. (2024, March 13). Methane leaks in the US are worse than we thought. *MIT Technology Review*.https://www.technologyreview.com/2024/03/13/1089725/methane-leaks-oil-gas/

Dabbagh, N. (2019). Effects of PBL on critical thinking skills. The Wiley Handbook of Problem-Based Learning, 135-156. DOI: 10.1002/9781119173243.ch6

Darby, A., & Willingham, L. (2022). Faculty motivation in service-learning based on expectancy x value theory. *Journal of Experiential Education*, *45*(3), 337–353. DOI: 10.1177/10538259211058292

Davidson, C. N. (2017). *The New Education: How to Revolutionize the University to Prepare Students for a World in Flux*. Basic Books.

Dedication of DeFord Manor. (2022, May 12). Virginia Wesleyan University. https://www.vwu.edu/about/news-and-events/features/story/dedication-of-deford-manor

DeFord Manor recognized as 'River Star Home'. (2022, July 25). Virginia Wesleyan University. https://www.vwu.edu/about/news-and-events/features/story/deford-manor-recognized-as-river-star-home

DeFord Manor. (n.d.) Virginia Wesleyan University. https://www.vwu.edu/about/president-and-leadership/deford-manor.php

Deren, M. (1983). *Divine horsemen*. McPherson.

Design Curial. (2012, February 20). University of Minnesota Morris' Welcome Centre wins LEED Gold. Progressive Media International Limited.

Desmangles, L. (1992). *Faces of the gods: Vodou and Roman Catholicism in Haiti*. University of North Carolina Press.

Dewey, J. (1916). *Democracy and education: An introduction to the philosophy of education*. Macmillan Publishing.

Dewey, J., Pautz, M. C., & Diede, M. K. (2024). How do we Address Faculty Burnout? Start by Exploring Faculty Motivation. *Innovative Higher Education*, *49*(3), 521–539. DOI: 10.1007/s10755-023-09685-2

Diani, M., & McAdam, D. (Eds.). (2003). *Social Movements and Networks: Relational Approaches to Collective Action*. Oxford University Press. DOI: 10.1093/0199251789.001.0001

Dmochowski, J. E., Garofalo, D., Fisher, S., Greene, A., & Gambogi, D. (2016). Integrating sustainability across the university curriculum. *International Journal of Sustainability in Higher Education*, *17*(5), 652–670. DOI: 10.1108/IJSHE-10-2014-0154

Downs, R. (2010, January 28). Morris biomass plant sees setback. *The Minnesota Daily*. https://mndaily.com/226520/uncategorized/morris-biomass-plant-sees-setback/

Drury, S., Briscoe-Smith, A., Longo, N., Napoli, L., Winslow, R., Farley, M., & Weaver, L. (2024). *Better Discourse: A Guide for Bridging Divides in Challenging Times*. Campus Compact.

E.U. (2023). *WMO*. Copernicus Climate Change Service European State of the Climate.

Earth Angels. (2023). *Virginia Wesleyan University Magazine, Fall 2023,* 14–15.

Educational Partnerships for Innovation in Communities Network. (n.d.). *EPIC-N*.https://www.epicn.org/

EIA. (2023). *Carbon Dioxide Emissions Coefficients*. U.S. Energy Information Administration. Washington: DC. https://www.eia.gov/environment/emissions/co2_vol_mass.php

Ekins, E. (2017). *The state of free speech in America: Attitudes about free speech, campus speech, religious liberty, and tolerance of political expression*. Cato Institute, 10/31/2017

Emanuel, K. (1998). *What we know about climate change*. MIT Press.

Engineers, K. F. I. (n.d.). University of Minnesota Morris Welcome Center. Retrieved December 2, 2024, from https://www.kfi-eng.com/project/welcome-center-u-of-m-morris/

Environment America. (2019, April 4). New rankings compare colleges and universities on progress toward clean energy. https://environmentamerica.org/media-center/new-rankings-compare-colleges-and-universities-on-progress-toward-clean-energy/

Environmental champions encourage others to take action. (2024, April 22). Virginia Wesleyan University. https://www.vwu.edu/about/news-and-events/features/story/environmental-champions

Environmental Guiding Principles. (2020). Virginia Wesleyan University. https://www.vwu.edu/about/sustainability/environmental-guiding-principles.php

EPA. (2023). *Quantifying Methane Emissions in Landfilled Food Waste*, Environmental Protection Agency. Washington: DC. October 2023, https://www.epa.gov/land-research/quantifying-methane-emissions-landfilled-food-waste

EPA. (2023). *Recycling Basics and Benefits*. Environmental Protection Agency. Washington: DC. https://www.epa.gov/recycle/recycling-basics-and-benefits

EPA. (2024). *Electric Power Sector Basics*. Environmental Protection Agency. Washington: DC. https://www.epa.gov/power-sector/electric-power-sector-basics

Etzkowitz, H. (1983). Entrepreneurial scientists and entrepreneurial universities in American academic science. *Minerva, 21*(2-3), 198–233. DOI: 10.1007/BF01097964 PMID: 11611566

Etzkowitz, H., & Leydesdorff, L. (2000). The dynamics of innovation: From National Systems and "Mode 2" to a Triple Helix of university–industry–government relations. *Research Policy, 29*(2), 109–123. DOI: 10.1016/S0048-7333(99)00055-4

Exploring sustainability in Alaska. (2017, June 14). Virginia Wesleyan University. https://www.vwu.edu/about/news-and-events/features/story/exploring-sustainability-in-alaska

Exploring urban ecology in Japan. (2024, September 10). Virginia Wesleyan University. https://www.vwu.edu/about/news-and-events/features/story/exploring-urban-ecology-in-japan

Fernandez, M. (2025). *Generation Alpha Characteristics: Innovators or Trading Humanity for Technology?* Softonix.com, January 30, 2025; https://en.softonic.com/articles/generation-alpha-characteristics-stats

Fincher, M., & Katsinas, S. (2017). Testing the limits of the price elasticity of potential students at colleges and universities: Has the increased direct cost to the student begun to drive own higher education enrollment? *Journal of Higher Education Policy and Management, 39*(1), 31–39. DOI: 10.1080/1360080X.2016.1211975

Finney, C. (2023). Memory divine. In E. Sharkey (Ed.), *A darker wilderness: Black nature writing from soil to stars* (pp. [insert page range]). Milkweed Editions.

Firedogs Jumk Removal. (2019). *What Does Firedogs Do with Junk.* Indianapolis: IN. https://firedawgsjunkremoval.com/what-does-fire-dawgs-do-with-junk/

Fitzgerald, H. E., & Primavera, J. (2013). *Community Engagement and Higher Education: International and Local Perspectives.* MSU Press.

Flaherty, C. (2024). *How Students Feel About Campus Speech, in 5 Graphics,* Inside Higher Education, December 13, 2024; https://www.insidehighered.com/news/student-success/academic-life/2024/12/13/survey-students-thoughts-campus-speech-five-graphics

Fleck, A. (2023). *Cars Cause Biggest Share of Transportation CO2 emissions,* September 22, 2023, in Statista.com.

Florence, N. (1998). *bell hooks' engaged pedagogy: A transgressive education for critical consciousness.* Bergin & Garvey.

Foote, E. (1856). Circumstances Affecting the Heat of the Sun's Rays. *The American Journal of Science and Arts*, *22*, 382–383.

Frank, A. (2022, May 3). *Bard & Virginia Wesleyan win top honors in national collegiate investment competition.* IEN. https://www.intentionalendowments.org/bard_and_virginia_wesleyan_win_top_honors_in_national_collegiate_investment_competition

Frazier, C. M. (2020, October 1). Black feminist ecological thought: A manifesto. *Atmos.* https://atmos.earth/black-feminist-ecological-thought-essay

Freeborn, J. (2024). Microplastics in testicles may play a role in male infertility, study suggests. *Medical News Today*. May 24, 2024 https://www.medicalnewstoday.com/articles/microplastics-in-testicles-may-play-a-role- in-male-infertility-study-suggests

Freire, P. (1970). *Pedagogy of the oppressed* (Ramos, M. B., Trans.). Herder and Herder.

Fullan, M. (2005). *Leadership & sustainability: System thinkers in action*. Corwin Press.

Ganz, M. (2009). *Why David sometimes wins: Leadership, organization, and strategy in the California farm worker movement*. Oxford University Press. DOI: 10.1093/acprof:oso/9780195162011.001.0001

Garcia, C. (2021). *Week Junior survey finds Generation Alpha wants adults to listen more,* The Week Junior (With YouGov), November 10, 2021

Gatti, L. V., Basso, L. S., Miller, J. B., Gloor, M., Gatti Domingues, L., Cassol, H. L. G., Tejada, G., Aragão, L. E. O. C., Nobre, C., Peters, W., Marani, L., Arai, E., Sanches, A. H., Corrêa, S. M., Anderson, L., Von Randow, C., Correia, C. S. C., Crispim, S. P., & Neves, R. A. L. (2021). Amazonia as a carbon source linked to deforestation and climate change. *Nature*, *595*(7867), 388–393. DOI: 10.1038/s41586-021-03629-6 PMID: 34262208

Gavazzi, S. M., & Fox, M. (2015). *For Town and Gown: Building a Sustainable Relationship between a College and Its Host Community*. Myers Education Press.

George Washington University. (2008, June 1). *GW Task Force on Sustainability: Final report and recommendations* (p. 6). https://sustainability.gwu.edu/sites/g/files/zaxdzs5666/files/downloads/taskforcereportandrecommendations%20%281%29.pdf

Georgia Institute of Technology (2021). Georgia Institute of Technology QEP Impact Report, March 2021.

Georgia Institute of Technology. (2024a). Sustainability Next Plan. https://sustain.gatech.edu/sustainability-next-plan/

Georgia Institute of Technology. (2024b). Climate Action Plan. https://sustain.gatech.edu/georgia-tech-climate-action-plan/

Georgia Institute of Technology. (2024c). Center for teaching and learning. https://ctl.gatech.edu/

Georgiou, A. (2019). How much oxygen does the Amazon rainforest provide? *Newsweek,* August 27. https://www.newsweek.com/how-much-oxygen-amazon-rain-forest-1456274

Gingerich, J. (2019). *Fairness, Sustainability Defines Alpha Generation*, O'Dwyers PR, December 9, 2019, https://www.odwyerpr.com/story/public/13499/2019-12-09/fairness-sustainability-defines-alpha-generation.html

Goodnough, T. (2024, October 25). Expanding sustainability perspectives with poster art and Dakota language. University of Minnesota Morris. https://morris.umn.edu/news-events/news/expanding-sustainability-perspectives-poster-art-and-dakota-language

Gore, A. (2006). *An inconvenient truth.* Rodale.

Graham, E. (2014). Mandated trigger warnings threaten academic freedom. *Reclaiming the Narrative*, (November-December), 2014.

Green Building Council. (2021). LEED in higher education: A case study on the growth of green buildings on campuses. Retrieved from https://www.usgbc.org/resources/higher-education-leed-case-studies

Green colleges. (2024). The Princeton Review. https://www.princetonreview.com/college-rankings?rankings=green-colleges

Greer Environmental Sciences Center dedicated on Founders Day. (2017, September 7). Virginia Wesleyan University. https://www.vwu.edu/about/news-and-events/features/story/greer-environmental-sciences-center-dedicated-on-founders-day

Greer Environmental Sciences Center. (n.d.). Virginia Wesleyan University. https://www.vwu.edu/academics/school-of-mathematics-and-science/greer-environmental-sciences-center.php

Hackney, A. (2018). *Emerging Issues in Analytical Chemistry, Doping, Performance Enhancing Drugs, and Hormones in Sport.* Elsevier., DOI: 10.1016/B978-0-12-813442-9.00002-X

Hafner, K. (2024, July 16). *How hot does it get in Portsmouth? New mapping project aims to find out.* WHRO. https://www.whro.org/environment/2024-07-16/how-hot-does-it-get-in-portsmouth-new-mapping-project-aims-to-find-out

Hartocollis, A. (2023, November 3). Can humanities survive the budget cuts? *The New York Times*. https://www.nytimes.com/2023/11/03/us/liberal-arts-college-degree-humanities.html

Harvard University. (2024, May). *Report on institutional voice in the university* (p. 1). https://provost.harvard.edu/sites/hwpi.harvard.edu/files/provost/files/institutional_voice_may_2024.pdf

Haskins, C. H. (1957). The Rise of Universities.

Haywood, A. (2024). Amazon Rainforest deforestation and climate change. *National Geographic,* March 11. https://education.nationalgeographic.org/resource/amazon - deforestation-and-climate-change/

Hazzard, A. (2022, November 3). Native-run solar firm aims to lower heating emissions and costs. *Sahan Journal*.https://sahanjournal.com/climate-environment/solar-heating-8th-fire/

Henderson, C., Beach, A., & Finkelstein, N. (2011). Facilitating change in undergraduate STEM instructional practices: An analytic review of the literature. *Journal of Research in Science Teaching*, *48*(8), 952–984. DOI: 10.1002/tea.20439

Hirsch, J., Yow, R., & Wu, Y. C. S. (2023). Teaching students to collaborate with communities: Expanding engineering education to create a sustainable future. *Engineering Studies*, *15*(1), 30–49. DOI: 10.1080/19378629.2023.2176767

Holland, J. (2021). Nature's pharmacy: The remarkable plants of the Amazon Rainforest and what they cure. *The Telegraph*, February 19. https://www.telegraph.co.uk/travel/cruises/articles/how-to-be-a-botanical-buff/

Hooks, B. (1994). *Teaching to transgress: Education as the practice of freedom.* Routledge.

Hou, S. I., & Wilder, S. (2015). Changing pedagogy: Faculty adoption of service-learning: Motivations, barriers, and strategies among service-learning faculty at a public research institution. *SAGE Open*, *5*(1), 2158244015572283. DOI: 10.1177/2158244015572283

Howe, R. (2023, April 21). No more Silent Springs [Excerpt]. *richardhowe.com.* https://richardhowe.com/2023/04/21/no-more-silent-springs (Accessed September 1, 2024)

Humes, E. (2024, July 3). A conservative farm town went green without really trying. *The Wall Street Journal.*

Humes, E. (n.d.). Edward Humes: Pulitzer Prize winning author. Retrieved December 2, 2024, from https://www.edwardhumes.com/

Hurney, C. A., Nash, C., Hartman, C. J. B., & Brantmeier, E. J. (2016). Incorporating sustainability content and pedagogy through faculty development. *International Journal of Sustainability in Higher Education, 17*(5), 582–600. DOI: 10.1108/IJSHE-12-2014-0180

IEA. (2018). *The Future of Cooling: Opportunities for energy-efficient air conditioning*. IEA., DOI: 10.1787/9789264301993-

IEA. (2023). *Prospects for EV Development,* October 2023; www.iea.org)

Intergovernmental Panel on Climate Change (IPCC). (2023). Summary for policymakers. In Lee, H., & Romero, J. (Eds.), *Climate change 2023: Synthesis report. Contribution of Working Groups I, II and III to the Sixth Assessment Report of the Intergovernmental Panel on Climate Change* (pp. 1–34). IPCC., DOI: 10.59327/IPCC/AR6-9789291691647

Intergovernmental Panel on Climate Change. (2007). The Physical Science Basis: Contribution of Working Group I to the Fourth Assessment Report of the Intergovernmental Panel on Climate Change. Cambridge University Press. https://www.ipcc.ch/site/assets/uploads/2018/03/ar4-wg1-chapter1.pdf

Israel, B. A., Schulz, A. J., Parker, E. A., & Becker, A. B. (2005). Community-based participatory research: Policy recommendations for promoting a partnership approach in health research. *Education for Health, 14*(2), 182–197. PMID: 14742017

Jacobellis v. Ohio, 378 U.S. 184, 197 (Stewart, J., concurring) (1964).

January term 2024: Students dive into hands-on learning, intensive field experiences. (2024, January 18). Virginia Wesleyan University. https://www.vwu.edu/about/news-and-events/features/story/january-term-2024-students-dive-into-hands-on-learning-intensive-field-experiences

Jean-Charles, R. M. (2023). *Looking for other worlds: Black feminism and Haitian fiction*. University of Virginia Press.

Jenkins, K., Sovacool, B. K., & McCauley, D. (2020). Humanizing sociotechnical transitions through energy justice: An analytical framework. *Energy Policy*, •••, 117.

Johnson, M. (2024, September 23). UMN Morris mixes cows and solar on Midwest's largest agrivoltaic pasture. *Agweek* (ND).

Katerberg, W. (2016, February 12). The liberal arts—Present, past, and future. *Historical Horizons.*https://historicalhorizons.org/2016/02/12/the-liberal-arts-present-past-and-future/

Keeling, C. (1958). The concentration and isotopic abundances of atmospheric carbon dioxide in rural areas. *Geochimica et Cosmochimica Acta*, *13*(4), 322–334. DOI: 10.1016/0016-7037(58)90033-4

Kezar, A. (2006). Redesigning for collaboration in learning initiatives: An examination of four highly collaborative campuses. *The Journal of Higher Education*, *77*(5), 804–838. DOI: 10.1080/00221546.2006.11778945

Kezar, A. (2012). Bottom-Up/Top-Down Leadership: Contradiction or Hidden Phenomenon. *The Journal of Higher Education*, *83*(5), 725–760. DOI: 10.1080/00221546.2012.11777264

Kisker, C., Cohen, A., & Brawer, F. (2023). *The American Community College* (7th ed.). Jossey- Bass. (Original work published 2023), Retrieved from https://www.perlego.com/book/4213866

Kosnick, J., and MacInnis, B. (2024). *American Understanding of Climate Change*, published by Resources for the Future and Stanford University, July 15, 2024.

Kostka, K. (2020, October 2). University of Minnesota Morris carbon neutrality report. Ever-Green Energy. https://www.ever-greenenergy.com/posts/university-of-minnesota-morris-carbon-neutrality-report/

Kotter, J. P. (2007). Leading change: Why transformation efforts fail. In *Museum management and marketing* (pp. 20–29). Routledge.

Kuh, G. D. (2008). High-impact educational practices. *Peer Review : Emerging Trends and Key Debates in Undergraduate Education*, *10*(4), 30–31.

Kurland, Z., Hafner, K., Feder, E., ... (2023, November 9). The woman who demonstrated the greenhouse effect. *Scientific American.*https://www.scientificamerican.com/article/the-woman-who-demonstrated-the-greenhouse-effect/

Le, N. (2020). *The Impact of Fast Fashion on the Environment*. Princeton Student Climate Initiative. Princeton University. Princeton: NJ. https://psci.princeton.edu/tips/2020/7/20/the-impact-of-fast-fashion-on-the-environment

Leal Filho, W., Shiel, C., Paço, A., Mifsud, M., Ávila, L. V., Brandli, L. L., Molthan-Hill, P., Pace, P., Azeiteiro, U. M., Vargas, V. R., & Caeiro, S. (2019). Sustainable Development Goals and sustainability teaching at universities: Falling behind or getting ahead of the pack? *Journal of Cleaner Production*, *232*, 285–294. DOI: 10.1016/j.jclepro.2019.05.309

Lederman, D. (2023). Showcasing solutions for better campus discourse. *Insider Higher Education,* October 23rd. https://www.insidehighered.com/news/students/free- speech/2023/10/10/showcasing-solutions-promote-civil-discourse-campuses

Leiserowitz, A., Maibach, E., Rosenthal, S., Kotcher, J., Goddard, E., Carman, J., Myers, T., Verner, M., Marlon, J., Goldberg, M., Ettinger, J., Fine, J., & Thier, K. (2024). Climate Change in the American Mind: Politics & Policy, Fall 2024. Yale University and George Mason University. New Haven, CT: Yale Program on Climate Change Communication.

Lent, J. (2024, September 5). Revised transcription of a comment recorded (with permission).

Lerner, S. (2010). *Sacrifice Zones: The Front Lines of Toxic Chemical Exposure in the United States*. MIT Press. DOI: 10.7551/mitpress/8157.001.0001

Lindsey, R. (2024, April 9). Climate change: Atmospheric carbon dioxide. *NOAA Climate.gov*.https://www.climate.gov/news-features/understanding-climate/climate-change-atmospheric-carbon-dioxide

Longman, P. (2004). *The Empty Cradle : How Falling Birthrates Threaten World Prosperity and What to Do About It*. Basic Books.

Louv, R. (2005). *Last Child in the Woods: Saving Our Children from Nature-Deficit Disorder*. Algonquin Books.

Louv, R. (2011). *The Nature Principle: Human Restoration and the End of Nature-Deficit Disorder*. Algonquin Books.

Louv, R. (2016). *Vitamin N: The Essential Guide to a Nature-Rich Life*. Algonquin Books.

Louv, R. (2019). *Our Wild Calling: How Connecting with Animals Can Transform Our Lives - and Save Theirs*. Workman.

Lozano, R., Merrill, M. Y., Sammalisto, K., Ceulemans, K., & Lozano, F. J. (2017). Connecting competences and pedagogical approaches for sustainable development in higher education: A literature review and framework proposal. *Sustainability (Basel)*, *9*(10), 1889. DOI: 10.3390/su9101889

Lutto, A. (2022). *Crabby About Invasive Crabs*. U.S. Fish & Wildlife Service. https://www.fws.gov/story/2022-10/crabby-about-invasive-green-crabs

Madia, M. T. Carter, S Thomas. 2024. Higher education complementarity for climate action. Second Nature: Cambridge, MA. www.secondnature.org

Maduko, J. (2023). *CT State Transitional Strategic Plan,* May 1, 2023, p. 13; "www.ctstate.edu"

Marquart, C. (2021, July). Wind over Morris. West Central Research and Outreach Center, University of Minnesota. https://wcroc.cfans.umn.edu/research/renewable-energy/wind-over-morris

Matador Move Out. (2024). *Impact report.*https://www.canva.com/design/DAGF_QjU7bE/STfXaHA5UcqU29ivA53p1Q/edit

Maurrasse, D. J. (2001). *Beyond the Campus: How Colleges and Universities Form Partnerships with Their Communities*. Routledge.

Mayfield, D. (2019, August 9). *Virginia Wesleyan wins top conservation award from Chesapeake Bay Foundation.* The Virginian-Pilot. https://www.pilotonline.com/2017/12/06/virginia-wesleyan-wins-top-conservation-award-from-chesapeake-bay-foundation-2/

McAdam, D., & Scott, W. R. (2005). Organizations and movements. Social movements and organization theory, 4.

Mcmillin, J., & Dyball, R. (2009). Developing a whole-of-university approach to educating for sustainability: Linking curriculum, research and sustainable campus operations. *Journal of Education for Sustainable Development*, *3*(1), 55–64. DOI: 10.1177/097340820900300113

Measuring urban heat across Virginia. (2022, March 29). Virginia Wesleyan University. https://www.vwu.edu/about/news-and-events/features/story/measuring-urban-heat-across-virginia

Medlin, E. (2008). Sustainability and the APUCC. *Facilities Manager*, (March/April), 10–11.

Menon, M., & Paretti, M. (2024). Faculty Perspectives on Integrating Sustainable Development Into Engineering Education. *IEEE Transactions on Technology and Society*, *5*(3), 316–324. DOI: 10.1109/TTS.2024.3403482

Merzdorf, J. (2018, July 18). *Study of the world's beaches shows threats to protected areas-NASA Science.* NASA Science. https://science.nasa.gov/science-research/earth-science/study-of-the-worlds-beaches-shows-threats-to-protected-areas/

Meyerson, D. E., & Scully, M. A. (1995). Tempered Radicalism and the Politics of Ambivalence and Change. *Organization Science*, *6*(5), 585–600. DOI: 10.1287/orsc.6.5.585

Mikati, I., Benson, A. F., Luben, T. J., Sacks, J. D., & Richmond-Bryant, J. (2018). Disparities in distribution of particulate matter emission sources by race and poverty status. *American Journal of Public Health*, *108*(4), 480–485. DOI: 10.2105/AJPH.2017.304297 PMID: 29470121

Mission and vision. (n.d.) Virginia Wesleyan University. https://www.vwu.edu/academics/batten-honors-college/mission-vision.php

MIT Free Speech Alliance. (2024). Debate: Is Decarbonization Worth the Cost? Massachusetts Institute of Technology. November 14, 2024.

Mokski, E., Leal Filho, W., Sehnem, S., & Andrade Guerra, J. B. S. O. D. (2023). Education for sustainable development in higher education institutions: An approach for effective interdisciplinarity. *International Journal of Sustainability in Higher Education*, *24*(1), 96–117. DOI: 10.1108/IJSHE-07-2021-0306

Mori, R.Junior, Fien, J., & Horne, R. (2019). Implementing the UN SDGs in universities: Challenges, opportunities, and lessons learned. *Sustainability (New Rochelle, N.Y.)*, *12*(2), 129–133. DOI: 10.1089/sus.2019.0004

Morris Model. (n.d.). Planning. Retrieved December 2, 2024, from https://morrismodel.org/planning/

Mulà, I., Tilbury, D., Ryan, A., Mader, M., Dlouhá, J., Mader, C., Benayas, J., Dlouhý, J., & Alba, D. (2017). Catalysing change in higher education for sustainable development: A review of professional development initiatives for university educators. *International Journal of Sustainability in Higher Education*, *18*(5), 798–820. DOI: 10.1108/IJSHE-03-2017-0043

N.A. (2024). *Boston Museum of Science Survey Finds 'Generation Alpha' Taking Up the Challenge to Adapt and Act on Climate*, PR Newswire, April 22, 2024; https://www.prnewswire.com/news-releases/boston-museum-of-science-survey-finds-generation-alpha-taking-up-the-challenge-to-adapt-and-act-on-climate-302123223.html

N.A. (2024). *Overwhelming majority of young Americans worry about climate crisis*, The Guardian, October 17, 2024; referring a study by Avaaz published in The Lancet Planetary Health. https://www.theguardian.com/environment/2024/oct/17/young-americans-climate-change

NACE. (2014). *Career Readiness Competencies: Employer Survey Results*. National Association of Colleges and Employers. https://www.naceweb.org/career- readiness/competencies/career-readiness-competencies-employer-survey-results

Nash, J. C. (2024). *How we write now: Living with Black feminist theory*. Duke University Press.

National Geographic Society. (n.d.). *Mauna Loa Observatory*.https://education.nationalgeographic.org/resource/mauna-loa-observatory

National Oceanic and Atmospheric Administration. (n.d.). *Trends in atmospheric carbon dioxide*.https://gml.noaa.gov/ccgg/trends/data.html

National Research Council (NRC). (2012). Discipline-Based Education Research: Understanding and improving undergraduate learning in science and engineering. Washington, D.C., the National Academies Press. https://nap.nationalacademies.org/catalog/13362/discipline-based-education-research-understanding-and-improving-learning-in-undergraduate

National Wildlife Federation. (2015). *Greenprint: A Plan to Prepare Community College Students for Careers in the Clean Economy,* December, 2015; Cross Mwabe, G., Lamback, S., and Wright, M. editors.

Natkin, L. W., & Hill, L. M. (2020). Faculty integration of sustainability learning outcomes into curriculum: A case study of a faculty learning community and teaching practices. *New Directions for Teaching and Learning*, *161*(161), 123–138. DOI: 10.1002/tl.20377

Natkin, L. W., & Kolbe, T. (2016). Enhancing sustainability curricula through faculty learning communities. *International Journal of Sustainability in Higher Education*, *17*(4), 540–558. DOI: 10.1108/IJSHE-02-2015-0024

Natural Resources Defense Council. (n.d.). The environmental justice movement. https://www.nrdc.org/stories/environmental-justice-movement

Nature, S. (2025). *Climate Leadership Network*; https://secondnature.org/climate-leadership-network/

New England Recycling Council. (2019, November 19). *[Website]*. https://nerc.org

Newspaper exec who founded Weather Channel dies. (2009, October 30). 13NewsNow. https://www.13newsnow.com/article/news/local/newspaper-exec-who-founded-weather-channel-dies/291-375839746

Nguyen, T., Nguyen, T., Grote, U., Neubacher, F., Rahut, D., Do, M., & Paudel, G. (2023). Security risks from climate change and environmental degradation: Implications for sustainable land use transformation in the Global South. *Current Opinion in Environmental Sustainability*, *63*, 101322. Advance online publication. DOI: 10.1016/j.cosust.2023.101322

No need to fear these wasps, unless you're an emerald ash borer. (2024, August 1). Virginia Wesleyan University. https://www.vwu.edu/about/news-and-events/features/story/no-need-to-fear-these-wasps-unless-yoursquore-an-emerald-ash-borer

NOAA. (2023). *Gulf of Mexico 'dead zone' larger than average, scientists find.* National Oceanic and Atmospheric Association. Washington: DC. https://www.noaa.gov/news- release/gulf-of-mexico-dead-zone-larger-than-average-scientists-find

Novy, J. W., Banerjee, B., & Matson, P. (2021). A core curriculum for sustainability leadership. *Sustainability (Basel)*, *13*(19), 10557. DOI: 10.3390/su131910557

O'Callaghan, E. (2015). Eco-criticism and Elma Napier's sense of place. In Brown, J. D., & Rosenberg, L. R. (Eds.), *Beyond Windrush: Rethinking postwar Anglophone Caribbean literature* (pp. 113–126). University of Mississippi Press. DOI: 10.14325/mississippi/9781628464757.003.0007

Our Common Future. (1987). World Commission on Environment and Development.

Our history. (n.d.). Virginia Wesleyan University. https://www.vwu.edu/about/our-history.php

Our mission. (n.d.). Virginia Wesleyan University. https://www.vwu.edu/about/our-mission.php

Palmer, K. (2024, November 11). Recent Storms Stir Climate Anxiety Among College Students. *Inside Higher Ed.*http://www.insidehighered.com

Paravisini-Gebert, L. (2011). Deforestation and the yearning for lost landscapes in Caribbean Literatures. In DeLoughrey, E., Gosson, R. K., & Handley, G. B. (Eds.), *Post-colonial ecologies: Literatures of the environment* (pp. 99–116). DOI: 10.1093/acprof:osobl/9780195394429.003.0005

Patterson, R. (2015). *Greening Africana studies: Linking environmental studies with transforming Black experiences*. Temple University Press. DOI: 10.2307/j.ctt14bsx8d

Pearce, F. (2024). Mind the gaps: How the U.N. climate plan fails to follow the science. *Yale Environment 360*, 1/9/2024.

Peck, G. (2024, October 11). Personal communication.

Pelfrey, P. A. (2004). *A brief history of the University of California.* Univ of California Press. https://books.google.com/books?hl=en&lr=&id=qbEwDwAAQBAJ&oi=fnd&pg=PT7&dq=a+brief+history+of+the+university+of+california&ots=lR0pGj7Avw&sig=ownjKS6d9rw75OfiARCcohJtJqM

Pellow, D. N. (2016). Toward a critical environmental justice studies: Black Lives Matter as an environmental justice challenge. *Du Bois Review, 13*(2), 221–236. DOI: 10.1017/S1742058X1600014X

Pelton, T. (2014). *Polluted Runoff: How Investing in Runoff Pollution Control Systems Improves the Chesapeake Bay Region's Ecology, Economy, and Health.* Chesapeake Bay Foundation. Washington: D.C.https://www.cbf.org/document-library/cbf-reports/2014-Polluted-Runoff- Report-compressed20b2.pdf

Pessarrodona, A., Franco-Santos, R. M., Wright, L. S., Vanderklift, M. A., Howard, J., Pidgeon, E., Wernberg, T., & Filbee-Dexter, K. (2023). Carbon sequestration and climate change mitigation using macroalgae: A state of knowledge review. *Biological Reviews of the Cambridge Philosophical Society, 98*(6), 1945–1971. DOI: 10.1111/brv.12990 PMID: 37437379

Peterson, R. (2014). Permission to Speak: College Presidents, the Israeli Boycott, and Climate Change, *National Association of Scholars*, January 22, 2014.

Pew Research Center (2023). *Economy Remains the Public's Top Policy Priority; COVID-19 Concerns Decline Again*, February 2023.

Phillips, C. A., Caldas, A., Cleetus, R., Dahl, K. A., Declet-Barreto, J., Licker, R., & Merner, L. D. (2020). Compound climate risks in the COVID-19 pandemic: Intersectional perspectives on environmental justice and climate change. *Environmental Justice, 13*(3), 97–108.

Pope-Ruark, R. (2022). *Unraveling faculty burnout: Pathways to reckoning and renewal.* JHU Press. DOI: 10.56021/9781421445137

Presidents of Wesleyan. (2015). Virginia Wesleyan University. https://www.vwu.edu/about/president-and-leadership/presidents-of-wesleyan.php

Priestly, J. (1774). *Experiments and Observations on Different Kinds of Air.* Printed for J. Johnson.

Raintree (2024). *About The Rainforest.* https://rain-tree.com/facts.htm#:~:text=Unbelievably%2C%20more%20than%20200%2C000%20acr es,threatened%20as%20the%20destruction%20continues.

Ramísio, P. J., Pinto, L. M. C., Gouveia, N., Costa, H., & Arezes, D. (2019). Sustainability Strategy in Higher Education Institutions: Lessons learned from a nine-year case study. *Journal of Cleaner Production*, *222*, 300–309. DOI: 10.1016/j.jclepro.2019.02.257

Rasmussen, L. (2010). Lowell Rasmussen interview. University of Minnesota Morris Stories, 32. https://digitalcommons.morris.umn.edu/stories/32

Research.com. (2024). *U.S. College Statistics: 2024.*

Review, P. (2023). *College Hopes & Worries Survey*, March 15, 2023; https://www.princetonreview.com/press/college-hopes-worries-press-release-2023

Rhodes, F. H. T. (2006). Sustainability: The ultimate liberal art. *The Chronicle of Higher Education*, *53*(9), 71.

Richard, F. D., Berkey, B., & Burk, H. M. (2022). Motivation and Orientation: Faculty Perspectives on Development and Persistence in Service Learning and Community Engagement. *Journal of Community Engagement and Higher Education at Indiana State University*, *14*(1), 12–23.

Richardson, K., Steffen, W., Lucht, W., Bendtsen, J., Cornell, S. E., Donges, J. F., Drüke, M., Fetzer, I., Bala, G., von Bloh, W., Feulner, G., Fiedler, S., Gerten, D., Gleeson, T., Hofmann, M., Huiskamp, W., Kummu, M., Mohan, C., Nogués-Bravo, D., & Rockström, J. (2023). Earth beyond six of nine planetary boundaries. *Science Advances*, *9*(37), eadh2458. https://aug.edu/news-&-events/news/the-alba-%E2%80%93-csr-hellas-esg-barometer. DOI: 10.1126/sciadv.adh2458 PMID: 37703365

Richlin, L., & Cox, M. D. (2004). Developing scholarly teaching and the scholarship of teaching and learning through faculty learning communities. *New Directions for Teaching and Learning*, *2004*(97), 127–135. DOI: 10.1002/tl.139

Rittel, H., & Webber, M. (1973). Dilemmas in general theory of planning. *Policy Sciences*, *4*(2), 155–169. DOI: 10.1007/BF01405730

Roane, J. T. (2023). *Dark agoras: Insurgent Black life*. New York University Press.

Roque, B., Venegas, M., Kinley, R. de Nys, R., Duarte, T., Yang, X. & Kebreab, E. (2021). Red seaweed (Asparagopsis Taxiformis) supplementation reduces enteric methane by over 80 percent in beef steers. *PLoS One*, Mar 17;16(3).

Roque, B., Brooke, C., Ladau, J., Polley, T., Marsh, L., Najafi, N., Pandey, P., Singh, L., Kinley, R., Salwen, J. K., Eloe-Fadrosh, E., Kebreab, E., & Hess, M. (2019). Effect of the macroalgae Asparagopsis Taxiformis on methane production and rumen microbiome assemblage. *Animal Microbiome*, *1*(1), 3. DOI: 10.1186/s42523-019-0004-4 PMID: 33499933

Rosemark, S. (2024, October 11). Personal communication.

Rubens, A., Spigarelli, F., Cavicchi, A., & Rinaldi, C. (2017). Universities' third mission and the entrepreneurial university and the challenges they bring to higher education institutions. *Journal of Enterprising Communities: People and Places in the Global Economy*, *11*(3), 354–372. DOI: 10.1108/JEC-01-2017-0006

Rudolph, F. (1962). *The American College and University: A History*. Alfred A. Knopf.

Ruffin, K. (2010). *Black on earth: African American ecoliterary traditions*. University of Georgia Press. DOI: 10.1353/book11452

Ryan, A., & Cotton, D. (2013). Times of change: shifting pedagogy and curricula for future sustainability. In *The sustainable university* (pp. 151–167). Routledge.

Schueman, L. J. (2023, December 13). *How philanthropy can support climate justice*. One Earth. https://www.oneearth.org/how-philanthropy-can-support-climate-justice/

Schwartz, L. & Ritter, D. (2019). Civil discourse in the classroom: Simple approaches to tough conversations. *New Culture Wars*, November 2019.

SDSN, F. (2020). Accelerating education for the SDGs in universities: a guide for universities, colleges, and tertiary and higher education institutions.

Second Nature. (n.d.). 15-year celebration. Second Nature. Retrieved 10/15/24 from https://secondnature.org/15-year-celebration/

Serafini, P. G., de Moura, J. M., de Almeida, M. R., & de Rezende, J. F. D. (2022). Sustainable development goals in higher education institutions: A systematic literature review. *Journal of Cleaner Production*, *370*, 133473. DOI: 10.1016/j.jclepro.2022.133473

Shabecoff, P. (1988, June 24). Global warming has begun, expert tells Senate. *The New York Times*.https://www.nytimes.com/1988/06/24/us/global-warming-has-begun-expert-tells-senate.html

Sharkey, E. (2023). *A darker wilderness: Black nature writing from soil to stars*. Milkweed Editions.

Shearin, M. Z. (2010, June 24). *Tulip poplars populate campus.* Virginia Wesleyan University. https://www.vwu.edu/about/sustainability/tulip-poplars-populate-campus.php

Shearin, M. Z. (2010, September 22). *Being seen being green.* Virginia Wesleyan University. https://www.vwu.edu/about/sustainability/being-seen-being-green2.php

Skerry, B. (2024). The Gulf of Maine is warming fast: What does this mean for lobsters-and everything else? *National Geographic.* June 2024. https://www.nationalgeographic.com/environment/article/gulf-of-maine-climate-change- photographer

Smith, D., Finkelstein, N., Horii, C. V., Miller, E., & York, T. T. (2024). *Levers for change: Enacting a national agenda for undergraduate STEM education* (Scherer, L., & Veague, T., Eds.). American Association for the Advancement of Science.

Snow, D. A., Rochford, J., Worden, S. K., & Benford, R. D. (1986). Frame Alignment Processes, Micromobilization, and Movement Participation. *American Sociological Review, 51*(4), 464–481. DOI: 10.2307/2095581

Spencer, A., Ross, S., & Tyson, S. (2023). *How Americans view electric vehicles.* Pew Research Center.

Spring 2023 Marlin Business Conference: A Greener Tomorrow. (2023, April 13). Virginia Wesleyan University. https://www.vwu.edu/about/news-and-events/features/story/spring-2023-marlin-business-conference-a-greener-tomorrow

Spring, J. (2023). Replanting the Amazon could help save the world's climate. Here's why it's so hard to do. *Reuters.* June3rd. https://www.reuters.com/investigates/special-report/brazil- environment-reforestation/

State of the university: Record fundraising and endowment growth. (2021, August 20). Virginia Wesleyan University. https://www.vwu.edu/about/news-and-events/features/story/state-of-the-university-record-fundraising-and-endowment-growth

State, C. Colleges and Universities. (2022). *"GCC Garage Rooftop Solar Panels Provide Renewable Energy, Cost Savings,* November 28, 2022, "www.ctstate.edu/newsroom"

State, C. Colleges and Universities. (2022). *Board Agenda,* June 23, 2022, pp. 264-267.

State, C. Colleges and Universities. (2023). *CT State's NSF Grant to Help Increase Underrepresented Students in STEM Fields, October 11, 2023,* "www.ctstate.edu/newsroom"

State, C. Colleges and Universities. (2024). *CT State Asnuntuck First in Connecticut to be Selected for Metallica Scholars Initiative's Workforce Education Program," July 22, 2024,* "www.ctstate.edu/newsroom"

State, C. Colleges and Universities. (2024). *CT State Gateway Awarded $350,000 NSF Grant to Train Technicians for Evolving Needs of Automotive Industry*, July 18, 2024; "www.ctstate.edu/newsroom"

State, C. Colleges and Universities. (2024). *CT State Naugatuck Valley Awarded $1.99 million NSF grant to Support STEM Students,* September 11, 2024, "www .ctstate.edu/newsroom"

State, C. Colleges and Universities. (2025). *Get to Know Us*, www.cscu.edu

Statista. (n.d.). Global carbon dioxide emissions. https://www.statista.com/statistics/ 276629/global-co2-emissions/

Steinmetz, K. (2020, February 20). She coined the term intersectionality over 30 years ago. Here's what it means to her today. *Time Magazine.* https://time.com/ 5786710/kimberle-crenshaw-intersectionality/

Stephens, J. C., Hernandez, M. E., Román, M., Graham, A. C., & Scholz, R. W. (2008). Higher education as a change agent for sustainability in different cultures and contexts. *International Journal of Sustainability in Higher Education*, *9*(3), 317–338. DOI: 10.1108/14676370810885916

Sterling, S. (2011). Transformative learning and sustainability: Sketching the conceptual ground. *Learning and teaching in higher education, 5*(11), 17-33.

Stokes, D. E. (1997). *Pasteur's quadrant: Basic science and technological innovation.* The Brookings Institution.

Strang, D., & Soule, S. E. (1998). Diffusion in Organizations and Social Movements: From Hybrid Corn to Poison Pills. *Annual Review of Sociology*, *24*(1), 265–290. DOI: 10.1146/annurev.soc.24.1.265

Stubbs, W., Ho, S. S., Abbonizio, J. K., Paxinos, S., & Bos, J. J. A. (2021). Addressing the SDGs through an integrated model of collaborative education. In *Handbook on teaching and learning for sustainable development* (pp. 252–271). Edward Elgar Publishing. DOI: 10.4337/9781839104657.00025

Sugimura, L. (2021, November 23). UMN Morris improves composting with grinder in dining hall. University of Minnesota Morris. https://morris.umn.edu/news-events/ news/umn-morris-improves-composting-program-grinder-dining-hall

Summer scholars. (2018, August 7). Virginia Wesleyan University. https://www.vwu.edu/about/news-and-events/features/story/summer-scholars

Supply Chain Council of European Union. (2022, February 19). UMN Morris achieves a million pounds of compost. https://scceu.org/umn-morris-achieves-a-million-pounds-of-compost/

Sustainable CT. (2025). “sustainablect.org”

Tack, S. (2023). Teaching for liberation: The manifesto assignment as an example of bell hooks’ engaged pedagogy. *Australian Feminist Studies*, *38*(109), 141–156. DOI: 10.1080/08164649.2023.2255931

The American College of Greece. (2024) website: https://www.acg.edu/

The Batten legacy: Frank and Jane Batten. (n.d.). Virginia Wesleyan University. https://www.vwu.edu/academics/batten-honors-college/the-battens.php

The sustainability tracking, assessment, and rating system. (n.d.). STARS: A program of AASHE. https://stars.aashe.org/

The Sustainable Fashion Forum. (2023). *Is Polyester *Really* That Bad?* October 11, 2023. https://www.thesustainablefashionforum.com/pages/quick-question-what-is-polyester- and-is-it-really-that-bad-for-the-environment

Thomashow, M. (2020). Postface: Why Environmental and Sustainability Learning Matters (Now More Than Ever). *New Directions for Teaching and Learning*, *2020*(161), 189–193. DOI: 10.1002/tl.20381

Thompson, A. (2021, September 13). Here’s how much food contributes to climate change. *Scientific American.* https://www.scientificamerican.com/article/heres-how-much-food-contributes-to-climate-change/

Tight, Malcolm & Society for Research into Higher Education. (1988). *Academic freedom and responsibility / edited by Malcolm Tight.* Milton Keynes [Buckinghamshire] ; Philadelphia, Pa.: The Society for Research into Higher Education and Open University Press

Tilbury, D. (2011). Higher education for sustainability: a global overview of commitment and progress. *Higher education in the world*, *4*(1), 18 28.

Tomkinson, B., & Tomkinson, R. (2020). An Interdisciplinary Problem-Based Approach to Education for Sustainable Development. In *Teaching and Learning Strategies for Sustainable Development* (Vol. 19, pp. 29–40). Emerald Publishing Limited. DOI: 10.1108/S2055-364120200000019006

Tomlinson-Keasey, C. & Merced. (2001). *Cover Letter to the Draft Long Range Development Plan: Public draft*. The University.

Tree campus higher education. (n.d.). Arbor Day Foundation. https://www.arborday.org/programs/tree-campus-higher-education/

Trencher, G., Yarime, M., McCormick, K., Doll, C., & Kraines, S. (2014). Beyond the Third Mission: Exploring the Emerging University Function of Co-creation for Sustainability. *Science & Public Policy*, *41*(2), 151–179. DOI: 10.1093/scipol/sct044

Trzepacz, S., Lingås, D., Asscherickx, L., Peeters, K., van Duijn, H., & Akerboom, M. (2023). *LCA-based assessment of the management of European used textiles*. Norion Consult. https://euric.org/images/Position-papers/lca-based-assessment-of-the-management-of- european-used-textiles_corrected.pdf

U.S. Congress. (1970–1976). *Clean Air Amendments of1970, Safe Drinking Water Act of 1974, Toxic Substances Control Act of 1976, & Resource Conservation and Recovery Act of 1976*. Public Laws 91-604, 93-523, 94-469, and 94-580.

U.S. Green Building Council. (2024, September). Applying LEED to higher education projects. U.S. Green Building Council. https://support.usgbc.org/hc/en-us/articles/12091439417875-Applying-LEED-to-higher-education-projects

UN (2023) Global Sustainable Development Report 2023, Times of Crisis, Times of Change: Science for Accelerating Transformations to Sustainable Development

UNEP. (2015). *Plastics in Our Environment: Are We Polluting the Environment Through Our Personal Care?* United Nations Environment Programme. New York, NY. https://www.unep.org/resources/report/plastic-cosmetics-are-we-polluting-environment- through-our-personal-care

UNEP. (2024). *Our Planet Is Choking in Plastic*. United Nations Environment Programme. New York, NY. https://www.unep.org/interactives/beat-plastic-pollution/

United Nations (2015) Transforming Our World: The 2030 Agenda for Sustainable Development.

United Nations Department of Economic and Social Affairs. Population Division. (2008). *World population prospects: The 2008 revision* (only medium variants shown). https://web.archive.org

United Way of the National Capital Area. (2024). *The Gen Z Activism Survey,* March 5, 2024. https://unitedwaynca.org/blog/gen-z-activism-survey

University Global Coalition. (2022). About Us. https://universityglobalcoalition.org/about/

University Leaders for a Sustainable Future. (2015). *The Talloires Declaration.* https://ulsf.org

University of California, Institutional Research and Academic Planning. (2023, August). *Extra-curricular participation and student success.* https://www.ucop.edu/institutional-research-academic-planning/_files/extra-curricular-participation-and-student-success.pdf

University of Miami. (2025). *Rosenstiel School of Marine, Atmospheric, and Earth Science Coral Reef Futures Lab*https://coralreeffutures.earth.miami.edu/research/index.html

University of Minnesota Morris. (1958). Executive Committee Meeting Minutes, May 1958. UMN Morris Founding Documents collection, Briggs Library. https://digitalcommons.morris.umn.edu/founding_docs/21

University of Minnesota Morris. (1960). A Liberal Arts Education Pamphlet [Brochure].

University of Minnesota Morris. (n.d.). Sustainability Goals & Initiatives. Retrieved December 2, 2024, from https://morris.umn.edu/sustainability-umn-morris/sustainability-goals-initiatives

University of Minnesota Morris. (n.d.). Sustainable Academics & Coursework. Retrieved December 2, 2024, from https://morris.umn.edu/sustainability-umn-morris/sustainability-goals-initiatives/sustainable-academics-coursework

University Relations. (2014). Green Prairie Community Awarded Prestigious LEED Green Building Certification. Campus News Archive, 105. https://digitalcommons.morris.umn.edu/urel_news/105

USDA. (2024). *Sustainable Agriculture.* The United States Department of Agriculture. Washington: DC. https://www.nifa.usda.gov/topics/sustainable-agriculture

Values-aligned investments. (n.d.). Virginia Wesleyan University. https://www.vwu.edu/about/sustainability/values-aligned-investments.php#:~:text=Virginia%20Wesleyan%20University%20invests%20in,%2Dcarbon%2C%20and%20regenerative%20economy

Vandenbussche, J., Ritter, L., Callahan, K. M., & Westlund, E. E. (2021). Using strand committees to build faculty support for departmental change. *PRIMUS (Terre Haute, Ind.), 31*(3-5), 330–342. DOI: 10.1080/10511970.2020.1772916

VanDerziel, S., Gatta, M., Jahn, J., & Koncz, A. (2024). Job Outlook 2025. National Association of Colleges and Employers, November 2024.

Vasarhely, C. (2024). *The Impact of Plastic on Climate Change.* University of Colorado. Boulder: CO. https://www.colorado.edu/ecenter/2023/12/15/impact-plastic-climate- change

Vinson, D. (2024, February). The growing influence of LEED certification in higher education. Private University Products and News Magazine. Retrieved from https://pupnmag.com/article/the-growing-influence-of-leed-certification-in-higher-education/

Virginia Wesleyan University hosts IEN Fall Forum: a platform for sustainable investments and societal impact. (2023, October 23). Virginia Wesleyan University. https://www.vwu.edu/about/news-and-events/features/story/virginia-wesleyan-university-hosts-ien-fall-forum-a-platform-for-sustainable-investments-and-societal-impact

Virginia Wesleyan University STARS Report. (2023, January 19). https://reports.aashe.org/institutions/virginia-wesleyan-college-va/report/2023-01-19/

VWU develops custom sustainability course with Port of Virginia. (2022, October 5). Virginia Wesleyan University. https://www.vwu.edu/about/news-and-events/features/story/vwu-expands-partnership-with-port-of-virginia

VWU hosts German sustainability scholar. (2023, August 1). Virginia Wesleyan University. https://www.vwu.edu/about/news-and-events/features/story/vwu-hosts-german-sustainability-scholar

VWU receives 2020 Inspiring Programs in STEM Award. (2020, August 17). Virginia Wesleyan University. https://www.vwu.edu/about/news-and-events/features/story/vwu-receives-2020-inspiring-programs-in-stem-award

VWU recognized by Elizabeth River Project. (2022, December 6). Virginia Wesleyan University. https://www.vwu.edu/about/news-and-events/features/story/vwu-recognized-by-elizabeth-river-project

Wakefield, J. (2012). *UVM Celebrates End of Bottled Water Sales With Bottled Water 'Retirement Party,* December 5, 2012, https://www.uvm.edu/uvmnews/news/uvm-celebrates-end-bottled-water-sales-bottled-water-retirement-party

Walsh, D. C. (2019, December 3). It's time for higher education to "get real" about the climate crisis. *Medium.*https://medium.com/@dcdubs/its-time-for-higher-education-to-get-real-about-the-climate-crisis-d0247b99d969

Wampler, J. (2023, November 17). Surveying freshwater ecosystems. *Excelsis. (Fall 2023),* 6.

Wasson, A. (2022). UC Merced Through a Socio-Environmental Imagination Lens. *UC Merced Undergraduate Research Journal, 14*(1). https://escholarship.org/uc/item/0v48z732

Weiss, M., Barth, M., & von Wehrden, H. (2021a). The patterns of curriculum change processes that embed sustainability in higher education institutions. *Sustainability Science, 16*(5), 1579–1593. DOI: 10.1007/s11625-021-00984-1

Weiss, M., Barth, M., Wiek, A., & von Wehrden, H. (2021b). Drivers and Barriers of Implementing Sustainability Curricula in Higher Education—Assumptions and Evidence. *Higher Education Studies, 11*(2), 42–64. DOI: 10.5539/hes.v11n2p42

Wenger, E., McDermott, R. A., & Snyder, W. (2002). *Cultivating communities of practice: A guide to managing knowledge.* Harvard business press.

Wiggins, G. P., & McTighe, J. (2005). *Understanding by design.* Ascd.

Wilcken, L. (2023). Ecological wisdom in the marvelous realism of Vodou (by way of Jacques Alexis). *Journal of Haitian Studies, 29*(1), 90–96. DOI: 10.1353/jhs.2023.a922859

Williams, J. (2017). Collaboration, alliance, and merger among higher education institutions. OECD Education Working Papers, No. 160, OECD Publishing, Paris. DOI: 10.1787/19939019

Wilson, J., & Zimmerman, Z. (2023). *The End of Flat Power Demand Is Over.* Grid Strategies. https://gridstrategiesllc.com/wp-content/uploads/2023/12/National-Load-Growth-Report-2023.pdf

World Commission on Environment and Development. (1987). *Our common future.* Oxford University Press.

Wright, C., Ritter, L. J., & Wisse Gonzales, C. (2022). Cultivating a collaborative culture for ensuring sustainable development goals in higher education: An integrative case study. *Sustainability (Basel), 14*(3), 1273. DOI: 10.3390/su14031273

WWF. (2024). *Climate Change in the Amazon.* World Wildlife Federation. https://wwf.panda.org/discover/knowledge_hub/where_we_work/amazon/amazon_threats/climate_change_amazon/

Yniguez, S. (2024). *U.S. News Places UC Merced Among Top 30 Public Schools Following Top Social Mobility Ranking By Wall Street Journal | Newsroom.* https://news.ucmerced.edu/news/2024/us-news-places-uc-merced-among-top-30-public-schools-following-top-social-mobility-ranking

Yoakum, R. (2018, September 4). *Solar SmartFlower making power for Virginia Wesleyan University.* Williamsburg Yorktown Daily. https://wydaily.com/latest/local/2018/09/04/solar-smartflower-making-power-for-virginia-wesleyan-university/

Younger, S. (2023, August 9). *NASA study: Rising sea level could exceed estimates for U.S. coasts.* NASA Jet Propulsion Laboratory. https://www.jpl.nasa.gov/news/nasa-study-rising-sea-level-could-exceed-estimates-for-us-coasts/

Zahneis, M. (2022). A rare survey of faculty morale shows that the pandemic's effects continue to ripple. Chronicle of Higher Education. https://www.chronicle.com/article/a-rare-survey-of-faculty-morale-shows-that-the-pandemics-effects-continue-to-ripple

Zimmer, K. (2019). Why the Amazon doesn't really produce 20% of the world's oxygen. *National Geographic,* August, 28. https://www.nationalgeographic.com/environment/article/why-amazon-doesnt-produce- 20-percent-worlds-oxygen

Zork, C. (2018, February 13). *Three LEED projects contributing to Virginia's ranking on the Top 10 States.* U.S. Green Building Council. https://www.usgbc.org/articles/three-leed-projects-contributing-virginia-s-ranking-top-10-states

Zuleta, L. V. (2024, October). Personal communication.

About The Contributors

Elsa M. Núñez served as Eastern Connecticut State University's sixth president for 19 years. She became President of Eastern in 2006 following more than 20 years as a senior administrator at higher education institutions, including the City University of New York and the University of Maine System, where she served as Vice-Chancellor. Dr. Núñez also has been a tenured faculty member of English at Ramapo State College, the College of Staten Island (CUNY), Lehman College (CUNY), and Wheelock College in Massachusetts. She also held a faculty position as Professor of English at Eastern. Dr. Núñez is the author of two books and several book chapters and has published numerous articles on language acquisition, access, equity, academic attainment in higher education, and cultural differences in education and retention as it relates to the role of public universities in the 21st Century. Over the past 19 years, Dr. Núñez has firmly established Eastern's reputation as Connecticut's only public liberal arts university. Under her leadership, Eastern has climbed steadily in U.S. News and World Report's annual Best Colleges rankings and was recognized as one of the best colleges for Veterans in New England in 2022. Eastern has been recognized for many years as a Green Campus, and a number of academic programs have also achieved national recognition. In addition, the University is recognized as one of the best colleges in the Northeast by The Princeton Review. Under Dr. Núñez's stewardship, Eastern has a well-earned reputation for environmental sustainability. The University has been awarded "Green Campus" status by the Princeton Review 13 years in a row, its Science Building is Silver LEED certified by the U.S. Green Building Council, and Eastern recently received a Gold STARS (Sustainability Tracking, Assessment, and Rating System) rating from the Association for the Advancement of Sustainability in Higher Education (AASHE). Eastern has one of only three Gold ratings in Connecticut and is one of 138 colleges and universities nationwide to be Gold-rated. Present and past board appointments for regional and national organizations include Liberty Bank, The Nellie Mae Education Foundation, The President's Alliance on Higher Education and Immigration, The Community Foundation of Eastern Connecticut, and Second Nature. In her current appointment as President-in-Residence at The Harvard Graduate School of Education, she teaches public policy in higher education.

Patricia Szczys is Distinguished Professor and Director, Center for Sustainability Studies. Her scholarly and professional life has been committed to the conservation of endangered and threatened seabirds and to mentoring the next generation of environmental scientists. Dr. Szczys is the author of more than 20 peer-reviewed research papers and has given dozens of scholarly presentations. She served as the University Honors Program Director, the John C. Hicks Endowed Chair for Undergraduate Research, and the President of The Waterbird Society. Dr. Szczys currently leads sustainability initiatives at Eastern Connecticut State University by collaborating with Facilities and Operations, Student Affairs, and Academic Affairs to achieve measurable sustainability goals, provide high quality curricular integration across the disciplines, and engage the regional community in these efforts. Dr. Szczys received her B.S. from the University of North Dakota and her Ph.D. in Environmental Biology from the University of Massachusetts, Boston.

***

Stella Apostolaki, is Assistant Professor of Environmental Science at the Department of Science and Mathematics - Deree and Executive Director of the Centre of Excellence in Sustainability of The American College of Greece (ACG). She holds a BSc/(Hons) in Environmental Science, an MSc in Urban Water and Environmental Management and a PhD on sustainable urban drainage systems and river management options, from the University of Abertay Dundee, Scotland – U.K, under funding received by the Environment Agency for England & Wales, in collaboration with the Scottish Environment Protection Agency (SEPA), and HR Wallingford Consultants. She has a wide research experience, she is expert evaluator of Call for Proposals for European Commission, has worked as a researcher in collaboration with Universities and Authorities in Greece and in the U.K., and has participated in numerous EU funded projects. Main research interests: integrated water management and innovation, sustainable development, climate change, conservation of biodiversity, ecosystem restoration, public awareness - stakeholder engagement methods and public education, the SDGs and green urban planning.

Leah Bamberger joined Northeastern as Executive Director of Sustainability in 2022. In this role, she is leading the University's efforts to develop and implement a sustainability strategy that begins to align campus operations and development, and engage academics, research, and student resources with community priorities and concerns around sustainability. Leah comes to Northeastern from the City of Providence, Rhode Island, where she served as the Director of Sustainability for the past seven years. Under her leadership and in partnership with frontline community members, she co-created Providence's 2019 Climate Justice Plan that was recognized nationally and continues to serve as a model for communities all over the United States. Prior to her role in Providence, Leah managed the City of Boston's citywide sustainability initiative, Greenovate Boston, and led the development of Boston's 2014 Climate Action Plan. Leah holds a Masters in Regional Planning from the University of Massachusetts and a Bachelor of Science from the College of Charleston.

Tim Carter is president of Second Nature, a U.S.-based non-profit with a mission to accelerate climate action in, and through, higher education. As president, Dr. Carter provides strategic leadership for the organization, including a focus on leveraging the diverse strengths of the 400+ schools in Second Nature's Climate Leadership Network and University Climate Change Coalition. Prior to Second Nature, Dr. Carter was the founding director of Butler University's Center for Urban Ecology (CUE). Dr. Carter received his Ph.D. in Ecology with distinction from the Odum School of Ecology at the University of Georgia and completed his B.S. in Ecology and Evolutionary Biology from the University of Tennessee, Knoxville. He lives in Cambridge, Massachusetts and is an avid coffee home-roaster.

Kelly R. Cordova is the Chief of Staff/ Vice President of Strategic Initiatives at Virginia Wesleyan University where she advances university's goals across key areas, including Advancement, Marketing & Communications, Institutional Research, and Government Relations. Her rolefocuses on aligning programs, partnerships, and initiatives with the university's mission, ensuring they strategically position the institution for future growth and success. She earned her B.S. in mass communications from Virginia Commonwealth University and her M.B.A. from VWU. She serves as assistant editor for "President to President," an international thought series for college and university presidents.

Mark Fincher, Ph.D. is a faculty member and researcher on higher education finance, access, and the professoriate. He is specifically interested in how cost impacts attendance and how students are engaged impacts the quality and long-term impact of their education.

James R. Gomes is the Founding Director, *Emeritus*, of Clark University's Mosakowski Institute for Public Enterprise, where he forged connections between university research and public policy practice in fields such as health and human services, education, urban development, and the environment. During his time at

Clark, he also taught undergraduate courses as a professor of Practice in Political Science. A retired member of the Massachusetts and Federal Bars, he practiced law at the Boston firm of Hale and Dorr (now WilmerHale) and was a Massachusetts Assistant Attorney General. He played senior leadership roles for John Kerry during his term as Lieutenant Governor of Massachusetts and his first term in the United States Senate. In the field of environmental policy, he served as Massachusetts Undersecretary of Environmental Affairs under Governor Michael Dukakis, President and CEO of the Environmental League of Massachusetts, Chairman of the Massachusetts Environmental Trust, and Senior Advisor on climate issues to MIT's Vice President for Research. He holds a B.A. in Political Science from Trinity College (CT) and M.P.P. and J.D. degrees from Harvard. Harvard Law School named him a Wasserstein Fellow for his "outstanding contributions to public interest law."

David Horner's initiated a world-class study abroad program serving 250 universities and 1,200 students annually; increased five-fold the College's distinctive "Parallel Study Program" enrolling 1,000 Greek public university students annually; acquired and developed 27-acre campus to launch a Pre-K – 6 school; initiated a Sports Academy and ACG Fitness serving over 2,000 children and local community members annually; attained STARS Gold sustainability rating; achieved six-fold increase in annual fundraising and grew endowment to $270M.

Régine Michelle Jean-Charles is the Dean's Professor of Culture and Social Justice as well as Director of Africana Studies at Northeastern University. She is also professor of Women, Gender, and Sexuality Studies. A Black feminist literary scholar who works at the intersections of race, gender, and justice from a global perspective, her scholarship and teaching include work on Black France, African diasporic literatures, Caribbean Studies, Black women's writing, Haiti, and the Haitian diaspora. Dr. Jean-Charles holds a BA from the University of Pennsylvania as well as an AM and PhD from Harvard University. She is the author of three books—*Conflict Bodies: The Politics of Rape Representation in the Francophone Imaginary* (Ohio State University Press, 2014), *The Trumpet of Conscious Today* (Orbis Press, 2021) and *Looking for Other Worlds: Black Feminism and Haitian Fiction* (University of Virginia Press, 2022). She has also authored numerous publications that have appeared in books, edited volumes, peer-reviewed journals, and public media outlets like *Ms. Magazine* and *The Boston Globe*. Her most recent book, *Looking for Other Worlds,* received an Honorable Mention for the Modern Language Association's the James Russell Lowell Prize in 2023. Dr. Jean-Charles is currently at work on two book projects—*The Rape Culture Syllabus* is under contract with Columbia University Press, and *Tifi at the Center: Holding Haitian Girlhood.* She lives in Milton, Massachusetts with her family.

Steven Knapp is President and CEO of Carnegie Museums of Pittsburgh and President Emeritus of the George Washington University. A specialist in English Romanticism, literary theory, and the relation of literature to philosophy and religion, Dr. Knapp taught at the University of California, Berkeley before serving as Dean of Arts and Sciences and then Provost of the Johns Hopkins University. He has served on numerous boards and councils and is a fellow of the American Academy of Arts and Sciences, a member of the Council on Foreign Relations, and a member of the Modern Language Association. He currently chairs the Board of Directors of Second Nature. The author of three books and numerous articles, he earned his bachelor's degree at Yale University and his master's and doctoral degrees at Cornell University.

Scott D. Miller has served as President of Virginia Wesleyan University since 2015. Respected as one of the most entrepreneurial higher education executives in America, Dr. Miller is in his 34th year as a college president. Previously, he was President of Bethany College (2007-2015), Wesley College (1997-2007), and Lincoln Memorial University (1991-1997). In 2006, he was one of the 50 inaugural presidents in the American College and University Presidents' Climate Commitment (ACUPCC), later serving as Chair of the Climate Leadership Network and Board Member of Second Nature. He has authored or edited 21 books, 10 monographs, over 200 professional articles, and frequently writes higher education opinion pieces for newspapers and journals nationwide. For 20 years, he has served as executive editor of "President to President," an international thought series for college and university presidents. Dr. Miller earned his B.A. from West Virginia Wesleyan College, M.A. from the University of Dayton, Ed.S. from Vanderbilt University, and Ph.D. from The Union Institute & University.

Daniel T. Okoli is the vice chancellor and chief operating officer of the university of California, Merced, overseeing physical operations, planning and development; environmental health and safety; facilities management; physical and environmental planning and space planning; and sustainability. Dr. Okoli has served as vice president for capital planning and development at Northern Arizona University 2017-2022, championing sustainability efforts and building relationships with faculty to create experiential education opportunities for students in campus development projects. Dr. Okoli also held planning and development leadership roles at the University of Wisconsin Pace University and The Ohio State University. He has taught mathematics, physics, architecture, computer-aided drafting and design, and African American studies at the undergraduate level. He holds a Ph.D. from Colorado State University, an M.B.A. from the University of Arizona, a master's in architecture and an M.A. in humanities from The Ohio State University, and a B.S. in structural engineering and architecture from the University of Aston.

Janet Schrunk Ericksen serves as chancellor of the University of Minnesota Morris. A professor of English, Ericksen previously held the position of vice chancellor of academic affairs and dean. She holds a PhD from the University of Illinois at Urbana-Champaign, with specialization in early medieval English and Old Norse literature. Ericksen is a member of the University of Minnesota Academy of Distinguished Teachers, a recipient of the University of Minnesota Morris Alumni Association Teaching Award, and the author of Reading Old English Biblical Poetry (U Toronto Press 2020).

Rebecca L Sherrick is a summa cum laude graduate of Illinois Wesleyan University. She earned a PhD in history from Northwestern University and subsequently served as a tenured faculty member and senior administrator at Carroll University in Waukesha, Wisconsin. In 2000, she became president of Aurora University with campuses in Lake Geneva, Wisconsin and Aurora, Illinois. She led the institution through a transformative period of growth. Sherrick then served as interim president of Chicago's Saint Xavier University. Currently she is a member of The Registry and a senior advisor for WittKieffer.

Pavlos A. Vlachos (Ph.D) is the Theodore Papalexopoulos Chair in Sustainability and Professor of Marketing at Alba Graduate Business School (AGBS), The American College of Greece [with Tenure]. He serves as an Associate Editor at the Journal of Business Research (Sustainability/Ethics section). His work has appeared or is in press in Harvard Business Review, Organizational Behavior & Human Decision Processes, Journal of Management Studies, Human Relations, Human Resource Management (US), Journal of Organizational Behavior, Journal of the Academy of Marketing Science, Tourism Management, European Journal of Information Systems, Journal of Business Ethics, Industrial Marketing Management, International Journal of Human Resource Management, European Journal of Marketing, and Journal of Business Research among others.

Rebecca Watts Hull is assistant director, faculty development for sustainability education initiatives for the Center for Teaching and Learning (CTL) at the Georgia Institute of Technology, with an adjunct appointment in the School of History and Sociology. Rebecca works with faculty to incorporate Education for Sustainable Development into their course design and teaching practices. She collaborates with other units to lead strategic initiatives related to sustainability education, including Georgia Tech's Sustainability Next and Climate Action plans. Rebecca also partners with other colleges and universities to advance faculty development and peer-to-peer learning that expands and enriches the integration of sustainability across the curriculum among colleges and universities in Georgia and beyond. Before joining

Georgia Tech, Rebecca held diverse sustainability and educational leadership roles in the public, private and nonprofit sectors with responsibilities that included science and environmental curriculum design, environmental project management, and community-based environmental advocacy. Rebecca earned an M.S. and Ph.D. from Georgia Tech in History and Sociology of Technology and Science, focusing her research on social movements and organizational change, and she also holds an M.S. in Natural Resources and Environment from the University of Michigan. Her college teaching has included Sustainability Leadership at Emory University, Environment and Sustainability Studies at Agnes Scott College, and American Environmental History, Social Movements, Community Organizing, and Organizing for Social Change at Georgia Tech.

Antony Wormack is a tenured educational professional with over 15 years of experience in higher education administration and public policy development within both public and private institutions. He holds a Doctor of Law and Policy from Northeastern University, a Master's in Organizational Management from Eastern Connecticut State University, and a Bachelor of Science in Education from Lesley University. Dr. Wormack has held various leadership roles at Naugatuck Valley Community College, including Director of Career Services and Coordinator of Academic Advising Services. Prior to that, he served as a Program Coordinator at the University of Connecticut and an Assistant Director at the Hartford Consortium for Higher Education. Dr. Wormack has also been actively involved in committee work, including co-chairing the CARE Team/Title IX Committee and the DEI Campus Committee, and has presented at several conferences.

Index

A

C

E

F

G

H

I

L

M

N

P

Q

R

S

T

U

W

Z

www.ingramcontent.com/pod-product-compliance
Ingram Content Group UK Ltd.
Pitfield, Milton Keynes, MK11 3LW, UK
UKHW050418051026
15470UKWH00022B/1701

9 798369 354544